PROGRAMMING FOR
CHEMICAL ENGINEERS
USING C, C++, AND MATLAB®

PROGRAMMING FOR CHEMICAL ENGINEERS USING C, C++, AND MATLAB®

Raul Raymond A. Kapuno, Jr.

INFINITY SCIENCE PRESS LLC
Hingham, Massachusetts
New Delhi

Publisher: David Pallai

INFINITY SCIENCE PRESS LLC
11 Leavitt Street
Hingham, MA 02043
Tel. 877-266-5796 (toll free)
Fax 781-740-1677
info@infinitysciencepress.com
www.infinitysciencepress.com

This book is printed on acid-free paper.

R.R. Kapuno, Jr. *Programming for Chemical Engineers Using C, C++, and MATLAB®*.
ISBN: 978-1-934015-09-4

The publisher recognizes and respects all marks used by companies, manufacturers, and developers as a means to distinguish their products. All brand names and product names mentioned in this book are trademarks or service marks of their respective companies. Any omission or misuse (of any kind) of service marks or trademarks, etc. is not an attempt to infringe on the property of others.

Library of Congress Cataloging-in-Publication Data

Kapuno, Raul Raymond.
  Programming for chemical engineers using C, C++, and MATLAB. / Raul
Raymond Kapuno, Jr.
    p. cm.
  Includes index.
  ISBN 978-1-934015-09-4
  1. Chemical engineering--Computer programs. 2. C (Computer program
language) 3. C++ (Computer program language) 4. MATLAB. I. Title.
  TP184.K37 2008
  660.0285'536--dc22
                        20080006056 7 8 9 5 4 3 2 1
089101234

Our titles are available for adoption, license or bulk purchase by institutions, corporations, etc. For additional information, please contact the Customer Service Dept. at 877-266-5796 (toll free).

Requests for replacement of a defective CD-ROM must be accompanied by the original disc, your mailing address, telephone number, date of purchase and purchase price. Please state the nature of the problem, and send the information to INFINITY SCIENCE PRESS LLC, 11 Leavitt Street, Hingham, MA 02043. The sole obligation of INFINITY SCIENCE PRESS LLC to the purchaser is to replace the disc, based on defective materials or faulty workmanship, but not based on the operation or functionality of the product.

To my wife and daughter

ABOUT THE AUTHOR

Raul Raymond Kapuno, Jr. is an instructor at the Cebu Institute of Technology (Philippines) where he teaches courses in chemical engineering and computer programming. He holds degrees in chemical and industrial engineering and has published several journal articles. His professional experience includes Coca-Cola and Daewoo Motors.

ACKNOWLEDGMENTS

Aside from my interest in chemical engineering, I have always been equally fascinated with computer programming. It is through this interest that I decided to write a book, merging these two fields of knowledge. The idea was further reinforced by the need to have a reference book for my Computer Application courses.

Just as with any endeavor, this idea would not have materialized without the help of significant people. Hence, I would like to extend my gratitude to Engr. Amis Pacamalan, head of the ChE department; to Engr. Camila Yagonia, and the rest of the ChE faculty of the Cebu Institute of Technology. I would like to give special credit to Engr. Coleta Esplanada, Engr. Concordia Bacalso, Engr. Jofranz Gallego, Engr. Jeveeh Azucena, Mr. Carlo Pestano, and my ChE 422 students for taking the time to review my manuscript. Also, to Mr. Jefty Negapatan who helped me with some of the computer programs. And most importantly, I would like to thank my parents, my wife Alpha, and my daughter Raili for supporting me all throughout—from the book's conception to its final completion.

PREFACE

Designed for chemical engineering students and industry professionals, this book details the development of reusable computer programs by guiding the reader through the process of

- establishing the general theoretical concept;
- determining the applicable numerical methods;
- testing the algorithm through manual calculation;
- writing and debugging the computer program based on the algorithm; and
- validating the result, using statistical analysis.

While each program developed is explained in detail, there are also tips and warning messages provided to assist the readers in the learning process.

Features

- Designed to help merge knowledge of computer programming and chemical engineering principles
- Uses the three most popular programming languages (C, C++, and MATLAB) currently being used in the chemical engineering curriculum and throughout the industry
- Includes complete program listings that can be run immediately by copying into a text editor and compiling them
- Accompanied by a CD-ROM featuring source code and executables
- Explains each program in detail to illustrate the flow and functions of the different commands
- Features a supplemental Web site and Instructor's Resource Disc for use as a textbook

All programs in the book are written in the three most popular languages (C, C++, and MATLAB) currently used in chemical engineering curriculum and in the industry. Because the book is written by a chemical engineer, practitioners and students will learn to write programs for appropriate subject matter.

TABLE OF CONTENTS

INTRODUCTION

PURPOSE OF THIS BOOK

Today, people use computers to help them solve problems that are too complex. When solved manually, these problems sometimes take too much time. But with the advent of high-speed computers, this computing can be done in seconds.

Through the years, more and more chemical engineering students, instructors, and professionals are exposed to different computer programming languages. It was with this thought that this book was written. This book will help merge a user's knowledge of computer programming with chemical engineering principles, making the user more equipped to write computer programs, which can be used and reused.

This book also aims to provide step-by-step assistance in using computers and developing software for solving problems that cover the following areas of analyses:

- Material Balance
- Energy Balance
- Fluid Flow
- Mass and Heat Transfer
- Optimization
- Reaction Kinetics

The book begins by establishing the general theoretical concepts, then determining the appropriate numerical methods applicable, testing the algorithm through manual calculations, writing and debugging the computer program based on the algorithm, and finally, validating the result using statistical analysis.

Examples are also included as references, but please note that examples are intentionally provided *not* to include all aspects of the chemical engineering study in order to help develop one's computer programming skills for other areas not covered by the book.

COMPUTER PROGRAMMING LANGUAGES USED

The three computer programming tools employed in the book: C, C++, and MATLAB, are the most commonly used programs in the chemical engineering curriculum today.

The first two, C and C++, represent two styles of programming—structured (C) and object oriented (C++). These approaches allow programmers to develop a better habit of programming than the spaghetti code in which BASIC programs are designed. It is also easy to transfer programs written in C or C++, from one machine to another, with different operating environments.

Note that all C and C++ programs in the book were tested and compiled using the Open Watcom C/C++ compiler. The Open Watcom C/C++ is an Open Source multi-platform C and C++ compiler. It packs with it a comprehensive tool for developing and debugging 16-bit and 32-bit applications for DOS, extended DOS, Novell NLMs, 16-bit OS/2, 32-bit Windows 3.x, Windows 95/98/Me, Win32w, and Windows NT/2000/XP. The compiler can be downloaded, as well as all its accompanying documentation from this site: http://www.openwatcom. org/.

There are other C/C++ compilers available for free on the Internet. This list can be viewed at the Free Byte Web site: *http://www.freebyte.com/programming/ cpp/#cppcompilers*

Appendix A provides details on the Open Watcom C/C++, as well as instructions on using other C/C++ compilers.

The third programming tool used is MATLAB. MATLAB stands for MATrix LABoratory, which is a high-level language for mathematical computing. It provides an environment that integrates programming, visualization, and computation of problems. MATLAB was developed by MathWorks and their site can be visited at: *http://www.mathworks.com/*.

HOW TO USE THIS BOOK

This book was organized with the assumption that the reader has some background in computer programming, is familiar with the logical sequence of developing computer programs, and knows computers and their limitations. Although the book will provide a brief overview of the programming languages used, it will not go into detail covering all the functions available, but only those that are going to be used within the book.

It is also assumed that readers have a good background in chemical engineering study, so the book focuses more on how to merge readers' knowledge of making application programs that can be used by chemical engineers.

All programs presented in the book are already complete and can be run immediately by just copying the code into a text editor and compiling it using the available C/C++ compilers or running it in the MATLAB Command Window. After each program listing, a detailed explanation is given to help readers understand the program flow and the functions of the different commands. Tips and warning messages are also provided to assist readers. At the end of each chapter, laboratory exercises are given to check readers' understanding of what was just discussed and to practice their programming skills based on their learning.

Each example problem, program listing, figure, table, and equation has a corresponding reference code for easy identification. These codes are composed of numbers representing the chapter where it is located, a dot symbol, and then the sequence number as it appears in the chapter.

Also, during the detailed explanation of the program listing, line numbers are used to refer to a specific line in the program, which serves as reference during its discussion. These numbers, however, should not be included in actual encoding and should not be considered as part of the computer source code.

SPECIAL NOTE FROM THE AUTHOR

The data and information within the book have been obtained from different literature sources. While reasonable effort has been made in the collection of data (such as physical properties of various materials) and testing of the computer programs, the author disclaims any warranty, express or implied, as to the accuracy or reliability of the data or calculations.

All results of calculations obtained from the computer programs yield approximate results, which might not be suitable for a design that requires accurate information. It is the sole responsibility of the reader to validate the data obtained from the book or from the computer programs included in this book, and to determine whether the results of these programs are accurate and suitable for any specific purpose. No guarantee of accuracy or fitness for any purpose is express or implied by the author.

PART I

USING
C AND C++

1

REVIEW ON C PROGRAMMING

L et's begin by reviewing what we know so far about the C computer language. Although C is a powerful programming language, it is quite easy to learn.

This chapter will review C programming that involves:

Constant and Variable Name Declaration
Constant Name Declaration
Variable Name Declaration

Basic Output and Input Statements
Output Statements
Input Statements

Operators
Arithmetic Operators
Assignment Operators
Relational Operators
Logical Operators

Conditional Statements
`if` - `else` Statement
Nested `if` - `else if` Statement
`switch/case` Statement

Looping Statements
   for Loop Statement
   while Loop Statement
   do - while Loop Statement
Arrays
Creating Functions

 All program listings discussed in this chapter can be found on the CD-ROM.

```
\Program Listings\Chapter 1\PL1_1.c        Program Listing 1.1
\Program Listings\Chapter 1\PL1_2.c        Program Listing 1.2
\Program Listings\Chapter 1\PL1_3.c        Program Listing 1.3
\Program Listings\Chapter 1\PL1_4.c        Program Listing 1.4
\Program Listings\Chapter 1\PL1_5.c        Program Listing 1.5
\Program Listings\Chapter 1\PL1_6.c        Program Listing 1.6
\Program Listings\Chapter 1\PL1_7.c        Program Listing 1.7
\Program Listings\Chapter 1\PL1_8.c        Program Listing 1.8
\Program Listings\Chapter 1\PL1_9.c        Program Listing 1.9
\Program Listings\Chapter 1\PL1_10.c       Program Listing 1.10
\Program Listings\Chapter 1\multiply.h     Multiply Header File
```

CONSTANT AND VARIABLE NAME DECLARATION

In order for any constant or variable name to be used in a C program, it must first be declared. However, there are some limitations when naming constants and variables. These include:

- Names must be composed of letters (a to x, A to Z), numbers (0 to 9), and underscores (_).
- Names cannot have more than two words with space(s) in between.
- Existing commands or reserved words such as *switch, do, for, while, break, case,* etc., cannot be used as constant or variable names.
- C is case sensitive, so uppercase and lowercase are treated differently.
- Only the first 31 characters of a name are significant.

Such limitations should be be kept in mind to effectively declare constant and variable names.

Name Declaration

To declare a constant name, we use the directive `#define`. For example, in declaring R (the universal gas constant in cal/gmole **K**), we declare it as

```
#define R 1.9872
```

This defines a symbolic constant R whose value is `1.9872` throughout the entire program. This symbolic constant will be replaced by the value specified.

Variable Name Declaration

Variable names must correspond to data types. Table 1.1 provides the list of data types for which variable names can be defined.

Table 1.1: Data types.

Type	Specifier	Meaning	Range/Example
int	%d	whole number	-32768 to +32767
float	%f	number w/ decimal	$\pm$3.4E-38 to $\pm$3.4E38
char	%c	single letter/ character	'r', '◦'
double	%lf	larger numbers w/ decimal	$\pm$1.7E-308 to $\pm$1.7E308
char (string)	%s	string of characters	"hello world"

Variable and constant names should denote what the variable or constant represents.

For example, to declare a variable with an integer data type, a floating point, or a character, you would use the following formats:

```
int num;
float temp;
char response;
```

Grouping similar data type variables in a list is also possible:

```
int num1, num2, sum;
float temp, pressure, weight;
char response, letter;
```

It is good practice to initialize the variables with values to prevent receiving warning messages from the compiler. Un-initialized variables contain garbage

information, and if not replaced will cause erroneous output, which is difficult to debug or trace.

BASIC OUTPUT AND INPUT STATEMENTS

Output Statements

The most common output statement used is the `printf()` function with the format:

```
printf(<format string>,<variable>..);
```

This function prints any variable, equation result, word string, or other information on the screen. The format string is any string of words that are enclosed by a double quotation ("`similar to this`"). Printing the value of a variable can be accomplished by placing the format symbol within the quoted string, followed by the variable name.

```
printf("The temperature of the room is %5.3f.", temp);
```

Based on the example, the format specifier, in this case `%5.3f`, can be expressed in terms of

the marker (%): This is required ,and should always be included when using a format specifier.
the field width (5 or any number): An optional part, this specifies the total length of the value to display.
the precision (3 or any number): An optional part used to specify the number of decimal places in which the value will be presented or displayed.
the data type (d, f, g, c or s): A required part where letters indicate the data type of the variable. (Also refer to Table 1.1.)

To have more control over the output, additional characters can be used and inserted within the quoted string. Table 1.2 gives some of the control characters available in the C language.

Table 1.2: Control characters.

Characters	Meaning
\a	alert character
\b	backspace

\n	new line/next line
\r	carriage return/enter key
\t	horizontal tab
\v	vertical tab
\\	back slash

To clear the screen in preparation for the screen printout, the function `system("cls")` is used. This function is useful to eliminate clutter on the screen before printing.

Input Statements

Before we can print out anything, or process any data, we need to request the values needed. To do this, we use input statements such as `scanf()` and `getch()`.

The `scanf()` is the input equivalent of the output statement `printf()`, with the format:

```
scanf(<format string>, <variable_address>);
```

Remember that items after the format string must be a variable address not a variable name, with an ampersand (`&`) symbol:

```
scanf("Enter pressure :", &pressure);
```

Another important input function is the `getch()` function. This function allows input of a single character without displaying the typed character on screen. Its main purpose is to obtain the character value of the selected key and assign it to the corresponding variable:

```
char response;
response = getch();
```

Most of the time we use the system function command `system("pause")` to halt the program until a key is pressed. This allows time to view the result before the program is terminated and the DOS window closes.

Other functions included in the standard and console input/output header files (`stdio.h` and `conio.h`) are listed in Appendix C.

OPERATORS

Once we obtain values from applying the input statements, we can process the information using various operators.

Arithmetic Operators

C supports common arithmetic operations by providing operators for addition, multiplication, and others (see Table 1.3 for a complete list). The % operator is also in the list. This is also known as the modulus operator. It simply provides the remainder in a division operation. We will encounter this and other arithmetic operators as we execute mathematical computations within the program.

Note, however, that an arithmetic operation, specifically the addition (+) operator, can also be applied to strings. For example:

```
string firstname, surname, fullname;
firstname = "John";
surname = " Smith";
fullname = firstname+surname;
```

By appending the `surname` to the `firstname`, the `fullname` variable will now contain the string "John Smith".

Also, precedence between arithmetic operators follows the acronym MDAS or if the % operator is to be considered, MDRAS (* / % + -).

Table 1.3: List of C operators.

Operator	Symbol	Meaning
Arithmetic	+	addition
	–	subtraction
	*	multiplication
	/	division
	%	remainder/modulus
Assignment	=	equal
	+=	plus equal
	-=	minus equal
	*=	multiply equal
	/=	divide equal
	%=	modulus equal
	--	decrement by 1
	++	increment by 1
Relational	>	greater than
	>=	greater than or equal to
	<	less than
	<=	less than or equal to
	==	equal to
	!=	not equal to

Logical	&&	and
	||	or
	!	not
	!=	not equal to

Assignment Operators

The assignment operator may be the most important operator in C. It is represented by the equal sign (=). We have already seen this used in the previous sections.

In assignment operation, the value on the right of the equal sign is going to be the value of the variable found on the left side. For example:

```
pressure = 14.7
```

In cases where there are two variables, such as

```
pressure = psia = 14.7;
```

the order of assigning is from right to left, so that 14.7 will be equated to variable psia and, in turn, will be equated to pressure, making both variables have the same value of 14.7.

Other expressions of assignment operation combined with arithmetic operation can be seen in Table 1.3. So, another way of expressing sum=sum+number is sum+=number, or product=product*number will be product*=number. If the variable is to be incremented by one, these can be expressed either by number=number+1 or number+=1 or number++.

Relational Operators

Relational operators allow us to compare two values (operand) and provide a result whether the comparison is true or false. If true, the function will return a binary value of 1; if false, the resulting binary value is 0. These are mostly used in conditional statements. See Table 1.3 for a list of other relational operators.

A statement with a relational operator in between two operands is known as a relational expression.

Logical Operators

Logical operators work with logical values to determine whether the given relational expressions are true or false. Refer to Table 1.3 for a summary.

The && operator or logical AND operator takes two relational expressions and returns true if both expressions are true. It will return false if only one or none of the expressions are true.

Unlike the `&&` operator, the `||` operator (also known as the logical OR operator) returns true even if only one expression is true. However, it will be false if both expressions are already false. The `!` operator or the logical NOT operator, on the other hand, reverses the return value, returning false if the expression is true.

⚠ **WARNING** *The use of the assignment operator (=) should not be mistaken as the "equal to" (==) relational operator. In a condition statement, the equal to (==) operator is used not the equal (=) assignment operator. On the other hand, in a mathematical operation it is the equal operator (=) that must be used.*

Let's start creating computer programs and apply what we have reviewed so far.

Example Problem 1.1. Make a C program that can calculate the product of two numbers.

Program Listing 1.1: Product of two numbers

```
1   #include <stdio.h>
2   #include <stdlib.h>
3   int main()
4   {
5   int product,num1,num2;
6   printf(" Enter two whole numbers. \n");
7   printf("\n  First Number: ");
8   scanf("%d",&num1);
9   printf("\n  Second Number: ");
10  scanf("%d",&num2);
11  product=num1*num2;
12  printf("\n The product is %d.\n",product);
13  system("pause");
14  return(0);
15  }
```

Line 1: The `#include` statement tells the C compiler to use the `<stdio.h>` header file, which contains the standard input/output functions `printf()` and `scanf()`.

Line 2: Another header file included is the `<stdlib.h>`, which contains the `system()` function.

Line 3: The `main()` function is required in every C program. This is where the main operation of the program occurs.

Line 5:   Declaration of variables `product`, `num1`, and `num2` as integers.

Lines 6, 7, 9: The `printf()` function prints the statement on the screen. The control character `\n` ensures that the statement occupies a new line before being printed.

Lines 8, 10: The input function `scanf()` request two values from the user and stores these values in the `num1` and `num2` variables.

Line 11: The two variables are processed using the multiplication (`*`) operation to give a resulting value, which in turn is assigned to a variable `product`.

Line 12: The `product` is printed out on the screen using `printf()`.

Line 13: With `system("pause")`, the program pauses while waiting for the user to hit any key before proceeding.

Line 14: The main program returns a `null` value before terminating.

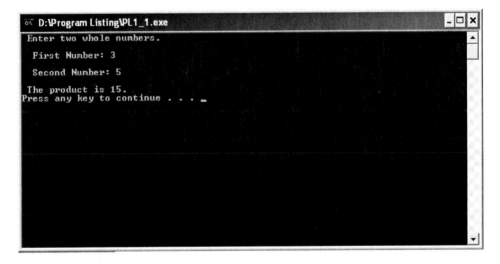

FIGURE 1.1:   Output of Program Listing 1.1.

Example Problem 1.2. Make a C program that can compute the average of three input numbers.

Program Listing 1.2: Average of three numbers

```
1   #include <stdio.h>
2   #include <stdlib.h>
3   int main()
4   {
5   double num1,num2,num3;
```

```
6   double ave;
7   printf(" Enter three numbers\n");
8   printf("\n First number: ");
9   scanf("%lf",&num1);
10  printf("\n Second number: ");
11  scanf("%lf",&num2);
12  printf("\n Third number: ");
13  scanf("%lf",&num3);
14  ave=(num1+num2+num3)/3;
15  printf("\n The average is %lf.\n",ave);
16  system("pause");
17  return(0);
18  }
```

Lines 1, 2: Again, the #include statement tells the C compiler to use the <stdio.h> and <stdlib.h> header files, which contain the standard input/output functions printf(), scanf() and the system() function.

Line 3: The main() function is where the main operation of the program occurs.

Lines 5, 6: Declaration of variables num1, num2, num3, and ave as double data type.

Line 7: The printf() function prints out the statement on the screen. The control character \n ensures that the statement occupies a new line before being printed.

Lines 9, 11, 13: The input function scanf() requests three values from the user and stores these values in the num1, num2, and num3 variables.

Line 14: The three variables are processed using addition (+) and division (/)

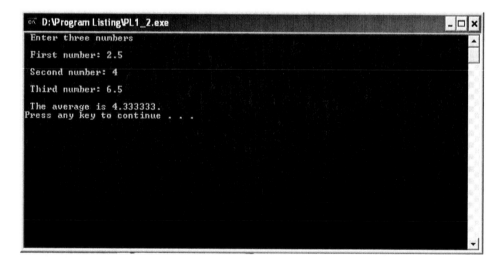

FIGURE 1.2: Output of Program Listing 1.2.

operations to give a resulting value, which in turn assigns the result to variable `ave`.

Line 15: The `ave` is printed out on the screen using `printf()`.

Line 16: With `system("pause")`, the program pauses while waiting for the user to hit any key before proceeding.

Line 17: Then the main program returns a `null` value before terminating.

⚠ **WARNING** *Always remember to end each statement with a semicolon (;). Lots of time can be spent tracing errors only to find out a missing semicolon is the cause.*

CONDITIONAL STATEMENTS

Conditional statements in the program instruct the computer to make decisions based on a given set of choices.

There are three types of conditional statements we can use: the `if - else`, the nested `if - else if`, and the `switch/case` statement.

`if - else` Statement

This conditional statement follows the format:

```
if (condition1)
    statement 1;
else (condition2)
    statement 2;
```

Example Problem 1.3. Write a C program that determines if the inputted Reynolds number indicates laminar flow (less than 2100) or not (greater than or equal to 2100).

Program Listing 1.3: Using the if - else Statement

```
1   #include <stdio.h>
2   #include <stdlib.h>
3   #define LIMIT 2100
4   int main()
5   {
6   int nre;
7   printf("\n Enter Reynolds number: ");
8   scanf("%d" ,&nre);
9   if(nre<LIMIT)
10  printf("\n The flow is laminar.");
```

```
11 else
12 printf("\n The flow is not laminar.");
13 system("pause");
14 return(0);
15 }
```

Line 3: The #define directive is used to declare the value of the constant name LIMIT.

Line 6: Declaration of variable nre as an integer.

Line 8: The input function scanf() requests the Reynolds number from the user and stores this in the nre variable.

Lines 9 to 12: These lines cover the if - else statements. When conditions are met, the corresponding printf() function is initiated to print the desired output message.

FIGURE 1.3: Output of Program Listing 1.3.

Nested if - else if Statement

This conditional statement follows the format:

```
if (condition1)
    statement 1;
else if (condition2)
    statement 2;
else if (condition 3)
```

```
      statement3;
else statement4;
```

Example Problem 1.4. Write a C program that accepts a temperature in °C and then determines the phase of water (solid, liquid, or gas).

Program Listing 1.4: Using the Nested `if - else if` Statement

```
1   #include <stdio.h>
2   #include <stdlib.h>
3   int main()
4   {
5   int temp;
6   printf("\n Enter the temperature in deg C: ");
7   scanf("%d" ,&temp);
8   if(temp>100)
9   printf("\n The water is in gas phase.\n");
10  else if(temp <=100 && temp>0)
11  printf("\n The water is in liquid phase.\n");
12  else
13  printf("\n Water is in solid phase.\n");
14   system("pause");
15   return(0);
16  }
```

Line 5: Declaration of variable `temp` as an integer.

Line 7: The input function `scanf()` requests a value from the user and stores this in the `temp` variable.

Line 8 to 13: These lines cover the nested `if - else if` statements. When conditions are met, the corresponding `printf()` function is initiated to print the desired output statement.

Line 10: The line uses the logical operator `(&&)`. This instructs the computer that the condition will only be true when `temp` is less than or equal to 100 and greater than 0.

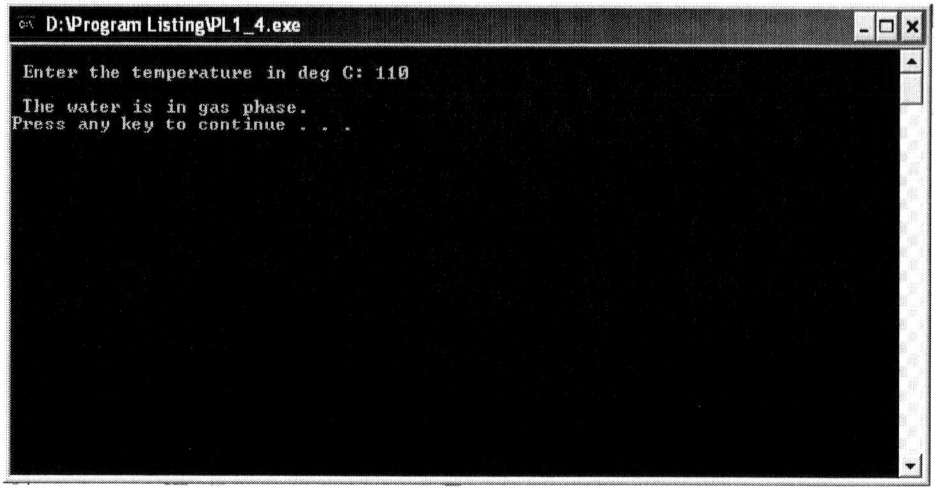

FIGURE 1.4: Output of Program Listing 1.4.

switch/case Conditional Statement

This conditional statement follows the format:

```
switch (var-expression) {
case const-value: statement 1; break;
case const-value: statement 2; break;
case const-value: statement 3; break; ...
case const-value: statement n; break;
default: statement;
}
```

Example Problem 1.5. Make a C program to perform conversion of temperature from one unit to another. Allow the user to choose from a selection of units of temperature.

Program Listing 1.5: Using the switch/case Statement

```
1   #include <stdio.h>
2   #include <stdlib.h>
3   #include <conio.h>
4   int main()
5   {
6   double init_temp, final_temp;
7   char unit;
8   printf("\n Enter value to be converted: ");
```

```
 9  scanf("%lf",&init_temp);
10  printf("\n Select initial and final units:");
11  printf("\n A) deg C to F");
12  printf("\n B) deg C to K");
13  printf("\n C) deg F to R");
14  printf("\n D) deg F to C");
15  printf("\n E) K to deg C");
16  printf("\n F) R to deg F \n");
17  unit=getche();
18  switch(unit){
19  case 'A':case 'a': final_temp=(init_temp*1.8)+32;
    break;
20  case 'B': case 'b': final_temp=(init_temp+273);
    break;
21  case 'C': case 'c': final_temp=(init_temp+460);
    break;
22  case 'D': case 'd': final_temp=(init_temp-32)/1.8;
    break;
23  case 'E': case 'e': final_temp=(init_temp-273);break;
24  case 'F': case 'f': final_temp=(init_temp-460);break;
25  default: printf("\n Choice unavailable");break;
26  }
27  printf("\n\nThe converted value is: %lf.\n", final_
    temp);
28  system("pause");
29  return 0;
30  }
```

Line 3:   The #include statement tells the C compiler to use the <conio.h> header file, which contains the console input/output function getch().

Lines 6, 7: Declaration of variable init_temp and final_temp as a double and unit as a character.

Line 9:   The input function scanf() requests a value from the user and stores this in the init_temp variable.

Lines 10 to 16: Print out the selection menu.

Line 17: The getche() function gets the selected option and stores it in the unit character variable, reflecting the selected option on the screen.

Lines 18 to 26: These lines cover the switch/case statements. When conditions are met, the corresponding printf() function is initiated to print the desired output message. The break command is used to skip the other conditions. Note that both uppercase and lowercase letters are set as selections because of C's case sensitivity.

Line 27: The result is then printed on screen using the `printf()` function.

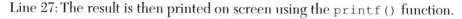

FIGURE 1.5:   Output of Program Listing 1.5.

LOOPING STATEMENTS

Looping statements in a program instruct the computer to repeat certain statements or a series of statements a specified number of times or until a certain condition is met.

for Loop Statement

This loop statement follows the format:

```
for (init; condition; increment/decrement)
{
    statements;
}
```

Example Problem 1.6. Write a C program that will compute for the concentration of A in a zero order A → B reaction from the start (0 sec) up to one minute at 10 seconds interval given Ca_0 15 gmol/L and k is 0.0567 gmol/L-sec.

Program Listing 1.6: Using the `for` Loop Statement
```
1   #include <stdio.h>
```

```
2   #include <stdlib.h>
3   int main()
4   {
5   double conc, init_con,k;
6   int time;
7   init_con = 15.00;
8   k = 0.0567;
9   printf("\nTIME(sec)  CONC.(g mol/L)\n");
10  for (time=0; time<=60; time = time+10)
11  {
12   conc = init_con - (k*time);
13   printf("  %d\t\t%lf\n",time,conc);
14  }
15  system("pause");
16  return 0;
17  }
```

Lines 5 to 6: Declares variables k, init_con, and conc as doubles, and time as an integer.

Lines 7 to 8: Initialize init_con with a value of 15.00 and k with 0.0567.

Lines 10 to 14: These lines cover the for loop statement. In the loop, printf() is used to print the resulting conc. The loop stops when the value of time reaches more than 60.

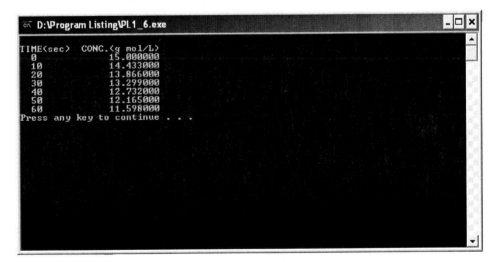

FIGURE 1.6: Output of Program Listing 1.6.

`while` Loop Statement

This looping statement follows the format:

```
initialization;
while (condition) {
   statements:
   increment/decrement;
}
```

Address the same problem in Example Problem 1.6.

Program Listing 1.7: Using the `while` Loop Statement

```
1   #include <stdio.h>
2   #include <stdlib.h>
3   int main()
4   {
5    double conc, init_con,k;
6    int time;
7    init_con = 15.00;
8    k = 0.0567;
9    time =0;
10   printf("\nTIME(sec)  CONC.(g mol/L)\n");
11   while (time<=60)
12   {
13    conc = init_con - (k*time);
14    printf(«  %d\t\t%lf\n»,time,conc);
15    time=time+10;
16    }
17   system("pause");
18   return 0;
19  }
```

Lines 11 to 16: Similarly, these lines cover the while `loop` statement. In the loop, `printf()` is utilized to print the resulting `conc`. The loop stops when value of `time` reaches more than 60.

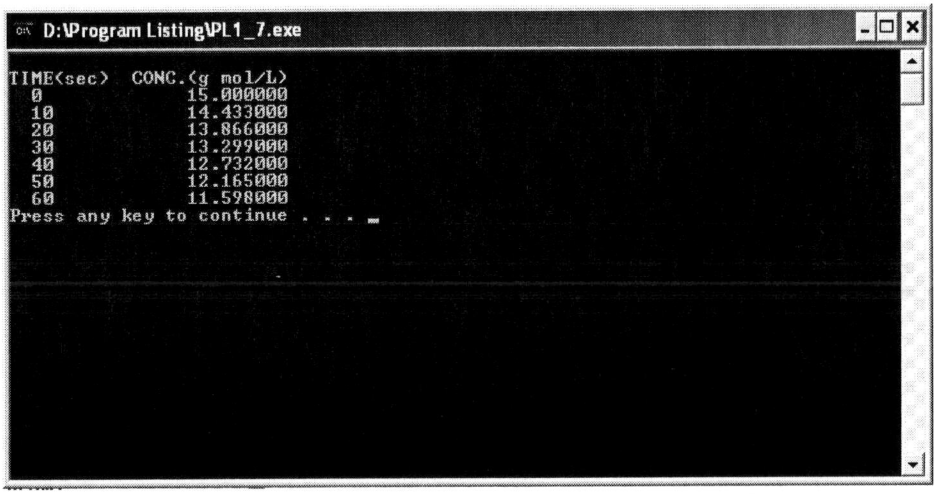

FIGURE 1.7:   Output of Program Listing 1.7.

do-while Loop Statement

This looping statement follows the format:

```
initialization;
do {
    statements;
    increment/decrement;
} while (condition);
```

Note that in do - while loop statement, the body (enclosed in {}) is executed at least once regardless of the initial condition.

Address the same problem in Example Problem 1.6.

Program Listing 1.8: Using the do - while Loop Statement

```
1   #include <stdio.h>
2   #include <stdlib.h>
3   int main()
4   {
5     double conc, init_con,k;
6     int time;
7     init_con = 15.00;
8     k = 0.0567;
9     time =0;
10    printf("\nTIME(sec)  CONC.(g mol/L)\n");
```

```
11  do
12  {
13    conc = init_con - (k*time);
14    printf("  %d\t\t%lf\n",time,conc);
15    time=time+10;
16  }while (time<70);
17  system("pause");
18  return 0;
19 }
```

Lines 11 to 16: These lines cover the do - while statement. Note that looping
        is done as long as time is less than 70.

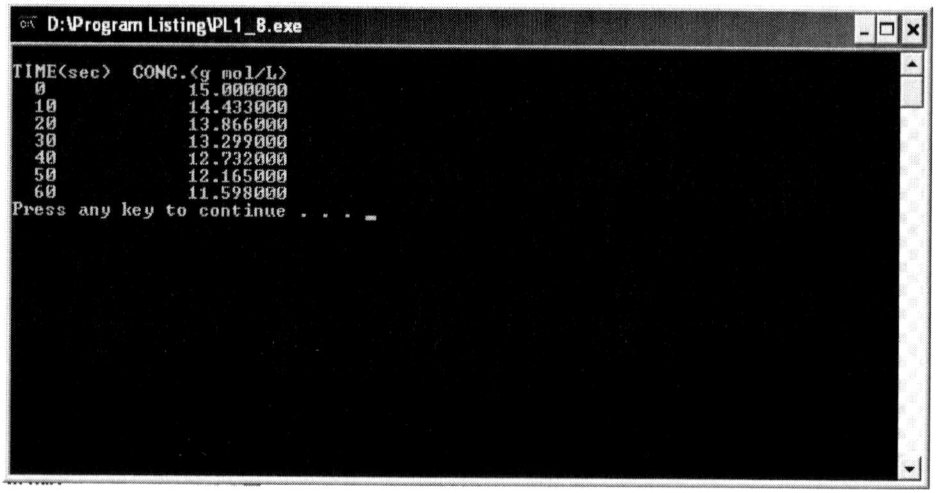

FIGURE 1.8:   Output of Program Listing 1.8.

ARRAYS

An array is a type of variable that can store one or more values having similar data
types with reference to only one variable name. This can be identified through
a pair of square brackets ([ ]).

Syntax for one-dimensional type of arrays:

```
Datatype array_name[index];
array[0]=10; array[1]=20; array[2]=30;
```

This can also be initialized as:

```
int array[3]={10,20,30};
```

Syntax for two-dimensional type of arrays:

```
Datatype array_name [row] [col];

array[0][0] = 10; array[0][1] = 20; array[0][2] = 30;
array[1][0] = 40; array[1][1] = 50; array[1][2] = 60;
array[2][0] = 70; array[2][1] = 80; array[2][2] = 90;
```

Example Problem 1.7. Write a C program that will print out on screen a temperature conversion table. This table should present the given values of temperature in degrees Centigrade (°C) with corresponding equivalents in degrees Fahrenheit (°F), Kelvin, and Rankine.

Program Listing 1.9: Using Arrays

```
1    #include <stdio.h>
2    #include <stdlib.h>
3    int main()
4    {
5     int temp[4][4],i;
6     for(i=0;i<4;i++)
7     {
8      temp[i][0] = (i*50)+100;
9      temp[i][1] = (temp[i][0]*1.8)+32;
10     temp[i][2] = temp[i][0]+273;
11     temp[i][3] = temp[i][1]+460;
12    }
13    printf("\nTable of Temperature Conversion\n");
14    printf(" deg C    deg F    K        R\n");
15    for(i=0;i<4;i++)
16    printf("\n %d  %d  %d  %d\n", temp[i][0], temp[i][1],
      temp[i][2], temp[i][3]);
17    system("pause");
18    return 0;
19   }
```

Line 5: Declaration of variable `i` as an integer and `temp[][]` as an array integer.

Lines 6 to 12: The `for` loop statement is used to initialize the `temp[][]` array with different integer values based on the mathematical operation performed.

Lines 15 to 16: Prints out the values of the array `temp[][]` again using the for loop statement.

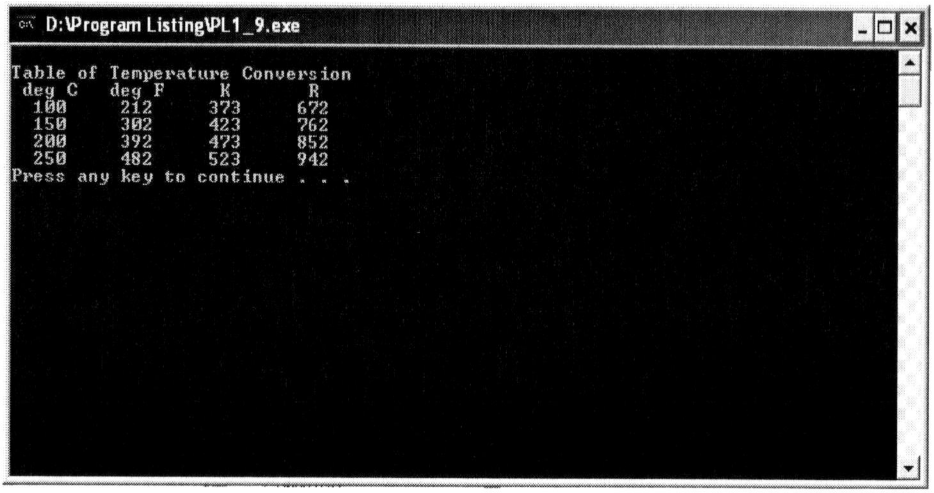

FIGURE 1.9: Output of Program Listing 1.9.

CREATING FUNCTIONS

Functions are the building blocks of C programming. They are a group of statements or instructions performing a specific task. Functions follow a general form:

```
return_type function_name (parameters)
{
 body of the function;
 return(value);
}
```

The `return()` function can be used to:

- cause an abrupt exit from a function it is in,
- or return a value.

All functions, except those declared as type `void`, return a value. The `return()` function explicitly specifies this value. In fact, we can save some of the functions that are quite useful in a header file (a file with an `.h` extension). In this way, we can just `#include` it whenever we make a program and call the function when needed.

Example Problem 1.8. Revise Program Listing 1.1 to make use of a new function `multiply()`, which is included in a separate `multiply.h` header file.

Program Listing 1.10: The program listing for the new header file `multiply.h`.

```
#include <stdio.h>
int multiply(int n1, int n2)
{
  int ans;
  ans = n1*n2;
  return(ans);
}
```

The listing of the main program:

```
1  #include <stdio.h>
2  #include <stdlib.h>
3  #include "multiply.h"
4  int main()
5  {
6   int product,num1,num2;
7   printf("\n Enter two whole numbers. \n");
8   printf("\n First Number: ");
9   scanf("%d", &num1);
10  printf("\n Second Number: ");
11  scanf("%d", &num2);
12  product=multiply(num1,num2);
13  printf("\n The product is %d.",product);
14  system("pause");
15  return 0;
16 }
```

Line 3: The `multiply.h` header is linked in the program so that the `multiply()` function can be used.

Line 12: In this line, the `multiply()` function is utilized, processing the values of `num1` and `num2`. In turn, it receives a value after the mathematical operation and stores it in the `product` variable.

Line 13: Prints out the resulting `product`.

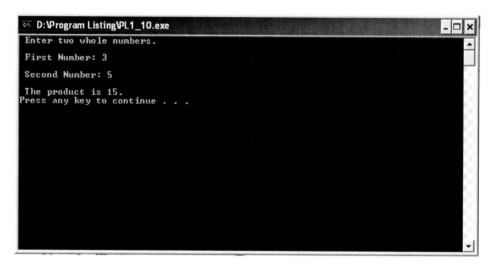

FIGURE 1.10:   Output of Program Listing 1.10.

LABORATORY EXERCISES

1) Create a computer program that will convert density of g/ml to lb/ft³ for the range of 0.5 to 1.5 g/ml at each 0.1-g increment. Print the result of each conversion on screen.

2) Eight kilograms (8000 g) of carbon dioxide gas is filled up in a cylinder to a pressure of 60 atm with a temperature of -80°C. Apply the van der Waals equation to determine the volume of the gas in the tank.

$$V^3 - \left(nb + \frac{nRT}{P}\right)V^2 + \frac{n^2 a}{P}V - \frac{n^3 ab}{P} = 0$$

The critical constant for carbon dioxide:

$$a = 3.60 \times 10^6 \text{ atm} \left(\frac{cm^3}{g\text{-}mol}\right)^2 \qquad b = 42.8 \left(\frac{cm^3}{g\text{-}mol}\right)$$

Write a computer program to solve for V through iteration. Hint: Use the initial value of V derived from $PV = nRT$.

3) Write a computer program to convert pressure in lb/in² into mm Hg and atm at a 5 lb/in² increment intervals where starting and ending values are requested from the user.

4) Improve Program Listing 1.3 by requesting the user to input all needed

information to solve for the Reynolds number and print out the result including whether it indicates laminar flow or not. Use equation:

$$N_{Re} = \frac{Dv\rho}{\mu}$$

where D is the diameter of the tube, v is the velocity of the fluid, ρ is the density of the fluid, and μ is the viscosity of the fluid.

5) Modify Program Listing 1.6 to be more flexible by allowing the user to provide values for Ca_0 and k.

Chapter 2 NUMERICAL COMPUTATION

O ne of the many applications of computers is arithmetic operations. In fact, the computer is sometimes considered an overgrown, powerful calculator that can handle large numbers of tedious calculations. However, computers cannot do all these calculations without human intervention. Computers need instructions in the form of computer codes to perform complex numerical computations.

In this chapter we develop computer programs that involve:

Solving Simultaneous Linear Equations
    Matrix Algebra
    Cramer's Rule and Determinants
    Gauss-Jordan Method
    Numerical Methods
    Jacobi Method/Gauss-Seidel Method
Regression Analysis
    Linear Regression
    Linear Regression for Nonlinear Relationship
    Polynomial Regression
Validation through Statistical Analysis
    Correlation Coefficient and Coefficient of Determination

Standard Error of Estimate
Significance Test
Interpolation Analysis
Linear Interpolation
Lagrange Polynomial Interpolation
Linear Programming
Simplex Method

All program listings discussed in this chapter can be found on the CD-ROM.

\Program Listings\Chapter 2\PL2_1.c	Program Listing 2.1
\Program Listings\Chapter 2\PL2_2.c	Program Listing 2.2
\Program Listings\Chapter 2\PL2_3.c	Program Listing 2.3
\Program Listings\Chapter 2\PL2_4.c	Program Listing 2.4
\Program Listings\Chapter 2\PL2_5.c	Program Listing 2.5
\Program Listings\Chapter 2\PL2_6.c	Program Listing 2.6
\Program Listings\Chapter 2\PL2_7.c	Program Listing 2.7
\Program Listings\Chapter 2\PL2_8.c	Program Listing 2.8

SOLVING SIMULTANEOUS LINEAR EQUATIONS

In chemical engineering, solving for simultaneous linear equations is frequently required especially in material and energy balances. The most common procedure for finding the solution uses matrix algebra and numerical methods.

Matrix Algebra

A set of linear algebraic equations can be represented as follows:

$$a_{11} x_1 + a_{12} x_2 + \ldots a_{1n} x_n = b_1$$
$$a_{21} x_1 + a_{22} x_2 + \ldots a_{2n} x_n = b_2$$
$$\vdots$$
$$a_{n1} x_1 + a_{n2} x_2 + \ldots a_{nn} x_n = b_n$$

where:

$x_1, x_2, \ldots x_n$ are unknown variables

$a_{11}, a_{12} \ldots a_{nn}$ are known constant coefficients

$b_1, b_2 \ldots b_n$ are known constants

Cramer's Rule and Determinants

Cramer's rule is a solution technique that is preferred when dealing with a small number of equations. The rule states that each unknown in a system of linear algebraic equations may be expressed as a fraction of two determinants with denominator d and with the numerator obtained from d by replacing the column of coefficients of the unknown in the equation by constants b_1, b_2, and b_3.

$$a_{11}\,x_1 + a_{12}\,x_2 + a_{13}\,x_3 = b_1$$
$$a_{21}\,x_1 + a_{22}\,x_2 + a_{23}\,x_3 = b_2$$
$$a_{31}\,x_1 + a_{32}\,x_2 + a_{33}\,x_3 = b_3$$

We can solve for the unknowns by

$x_1 = (x_{1a}/d)$; $x_2 = (x_{2a}/d)$; $x_3 = (x_{3a}/d)$ where:

$$
d = \begin{array}{|ccc|cc}
 & (-) & (-) & (-) & \\
a_{11} & a_{12} & a_{13} & a_{11} & a_{12} \\
a_{21} & a_{22} & a_{23} & a_{21} & a_{22} \\
a_{31} & a_{32} & a_{33} & a_{31} & a_{32} \\
 & (+) & (+) & (+) &
\end{array}
$$

$$
\begin{aligned}
d = (a_{11}{\times}a_{22}{\times}a_{33}) + (a_{12}{\times}a_{23}{\times}a_{31}) + (a_{13}{\times}a_{21}{\times}a_{32}) - (a_{13}{\times}a_{22}{\times}a_{31}) \\
- (a_{11}{\times}a_{23}{\times}a_{32}) - (a_{12}{\times}a_{21}{\times}a_{33})
\end{aligned}
\tag{2.1}
$$

$$
x_{1a} = \begin{array}{|ccc|cc}
 & (-) & (-) & (-) & \\
b_1 & a_{12} & a_{13} & b_1 & a_{12} \\
b_2 & a_{22} & a_{23} & b_2 & a_{22} \\
b_3 & a_{32} & a_{33} & b_3 & a_{32} \\
 & (+) & (+) & (+) &
\end{array}
$$

$$
\begin{aligned}
x_{1a} = (b_1{\times}a_{22}{\times}a_{33}) + (a_{12}{\times}a_{23}{\times}b_3) + (a_{13}{\times}b_2{\times}a_{32}) \\
- (a_{13}{\times}a_{22}{\times}b_3) - (b_1{\times}a_{23}{\times}a_{32}) - (a_{12}{\times}b_2{\times}a_{33})
\end{aligned}
\tag{2.2}
$$

$$
x_{2a} = \begin{array}{|ccc|cc}
 & (-) & (-) & (-) & \\
a_{11} & b_1 & a_{13} & a_{11} & b_1 \\
a_{21} & b_2 & a_{23} & a_{21} & b_2 \\
a_{31} & b_3 & a_{33} & a_{31} & b_3 \\
 & (+) & (+) & (+) &
\end{array}
$$

$$
\begin{aligned}
x_{2a} = (a_{11}{\times}b_2{\times}a_{33}) + (b_1{\times}a_{23}{\times}b_3) + (a_{13}{\times}b_2{\times}a_{32}) \\
- (a_{13}{\times}b_2{\times}a_{31}) - (b_1{\times}a_{23}{\times}b_3) - (a_{12}{\times}b_2{\times}a_{33})
\end{aligned}
\tag{2.3}
$$

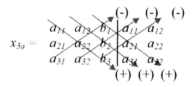

$$x_{3a} = (a_{11}{\times}a_{22}{\times}b_3)+(a_{12}{\times}b_2{\times}a_{31})+(b_1{\times}a_{21}{\times}a_{32})-(b_1{\times}a_{22}{\times}a_{31})$$
$$-(a_{11}{\times}b_2{\times}a_{32})-(a_{12}{\times}a_{21}{\times}b_3) \tag{2.4}$$

Now let us apply what we have just learned with an example problem.

Example Problem 2.1. Given the following linear equations, determine the values of x_1, x_2, and x_3 using Cramer's rule.

$$3x_1 + x_2 + x_3 = 25$$
$$x_1 - 3x_2 + 2x_3 = 10$$
$$2x_1 + x_2 - x_3 = 6$$

Using Eqs 2.1, 2.2, 2.3, and 2.4:

$$d = [(3)(-3)(-1)+(1)(2)(2)+(1)(1)(1)] - [(2)(-3)(1)+(1)(1)(-1)+(3)(1)(2)]$$
$$d = 15$$

$$x_{1a} = [(25)(-3)(-1)+(1)(2)(6)+(1)(10)(1)] - [(6)(-3)(1)+(10)(1)(-1)+(25)(1)(2)]$$
$$x_{1a} = 75$$

$$x_{2a} = [(3)(10)(-1)+(25)(2)(2)+(1)(1)(6)] - [(2)(10)(1)+(1)(25)(-1)+(3)(6)(2)]$$
$$x_{2a} = 45$$

$$x_{3a} = [(3)(-3)(6)+(1)(10)(2)+(25)(1)(1)] - [(2)(-3)(25)+(1)(1)(6)+(3)(1)(10)]$$
$$x_{3a} = 105$$

Solving for the unknowns: $x_1 = 75/15 = 5$ $x_2 = 45/15 = 3$ $x_3 = 105/15 = 7$

When creating the program, we need to identify all the required information to complete Eqs 2.1 to 2.4. We are going to apply the output/input statements and assign all inputted values to different variables. We will then convert the equations into programming codes to allow the computer to do all the mathematical operations.

Program Listing 2.1: Cramer's Rule Program.

```
1   #include <stdio.h>
2   #include <stdlib.h>
3   int main()
4   {
5    int r,c;
6    double a[3][3],b[3];
7    double x1a,x2a,x3a,x1,x2,x3,d;
8    printf("\n Solving for 3 Equations - 3 Unknowns\n");
9    for (r=0; r<3; r++)
10   {
11    for(c=0; c<3; c++)
12    {
13     printf(" Enter a[%d][%d]: ",r,c);
14     scanf("%lf",&a[r][c]);
15    }
16    printf(" Enter b[%d]: ",r);
17    scanf("%lf",&b[r]);
18   }
19  d=(a[0][0]*a[1][1]*a[2][2])+(a[0][1]*a[1][2]*a[2][0])
    +(a[0][2]*a[1][0]*a[2][1])-(a[0][2]*a[1][1]*a[2][0])-
    (a[0][0]*a[1][2]*a[2][1])-(a[0][1]*a[1][0]*a[2][2]);
20  x1a=(b[0]*a[1][1]*a[2][2])+(a[0][1]*a[1][2]*b[2])
    +(a[0][2]*b[1]*a[2][1])-(a[0][2]*a[1][1]*b[2])-
    (b[0]*a[1][2]*a[2][1])-(a[0][1]*b[1]*a[2][2]);
21  x2a=(a[0][0]*b[1]*a[2][2])+(b[0]*a[1][2]*a[2][0])
    +(a[0][2]*a[1][0]*b[2])-(a[0][2]*b[1]*a[2][0])-
    (a[0][0]*a[1][2]*b[2])-(b[0]*a[1][0]*a[2][2]);
22  x3a=(a[0][0]*a[1][1]*b[2])+(a[0][1]*b[1]*a[2][0])
    +(b[0]*a[1][0]*a[2][1])-(b[0]*a[1][1]*a[2][0])-
    (a[0][0]*b[1]*a[2][1])-(a[0][1]*a[1][0]*b[2]);
23   x1=x1a/d; x2=x2a/d; x3=x3a/d;
24   printf("\n X1 = %lf, X2 = %lf, X3 = %lf\n",x1,x2,x3);
25   system("pause");
26  return 0;
27  }
```

Lines 1 and 2: The #include statements tell the C compiler to use the
<stdio.h> and <stdlib.h> header files, which contain the standard
input/output functions printf() and scanf() and the function
system().

Line 3: The `main()` function is where the main operation of the program occurs.

Line 5: Declaration of variables `r` and `c` as integers.

Lines 6, 7: Declaration of variables `a[][]`, `b[]`, `x1a`, `x2a`, `x3a`, `x1`, `x2`, `x3`, and `d` as doubles.

Line 8: The `printf()` function prints out the statement on the screen. The control character `\n` ensures that the statement occupies a new line before being printed.

Lines 9 to 18: The `for` loop statements are used to get the values of the constant coefficients `a[][]` and the constant `b[]`.

Lines 19 to 22: Process the input data using Eq 2.1, 2.2, 2.3, and 2.4 to get the values of `d`, `x1a`, `x2a`, and `x3a`, respectively.

Line 23: Computes for unknown variables `x1`, `x2`, and `x3`.

Line 24: The resulting values are printed on the screen.

Line 25: With `system("pause")`, the program pauses while waiting for the user to hit any key before proceeding.

Line 26: The main program returns a null value before terminating.

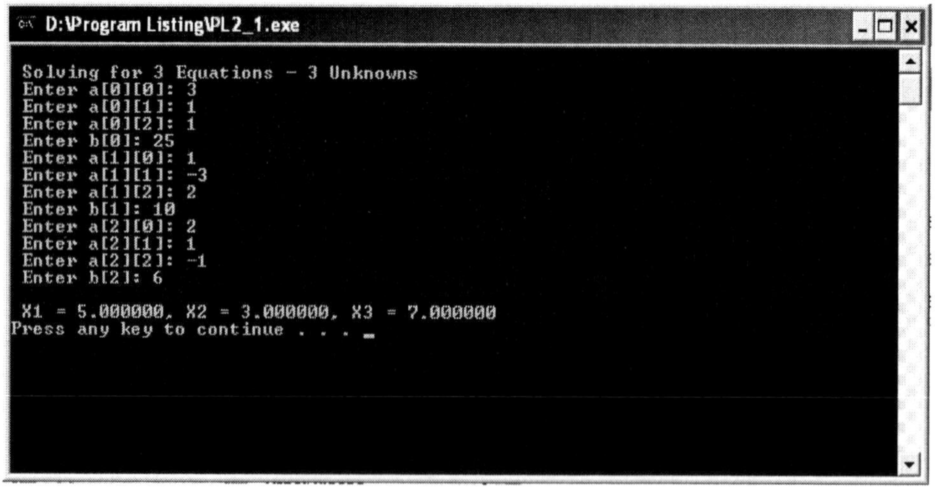

FIGURE 2.1: Output of Program Listing 2.1.

Gauss-Jordan Method

The Gauss-Jordan method transforms the coefficient matrix into another matrix that is much easier to solve. The new system represented by the augmented matrix basically has the same solution set as the original system of linear equations. Inverting a matrix makes Gauss-Jordan elimination an efficient and effective method.

Given a square matrix $[A]$:

$$[A] = \begin{bmatrix} a_{11} & a_{12} & a_{13} \\ a_{21} & a_{22} & a_{23} \\ a_{31} & a_{32} & a_{33} \end{bmatrix}$$

The inverse can be written as

$$[A]^{-1} = \begin{bmatrix} a_{11}^{-1} & a_{12}^{-1} & a_{13}^{-1} \\ a_{21}^{-1} & a_{22}^{-1} & a_{23}^{-1} \\ a_{31}^{-1} & a_{32}^{-1} & a_{33}^{-1} \end{bmatrix}$$

To obtain the inverse we apply the Gauss-Jordan method:

$$\left| \begin{array}{ccc|ccc} a_{11} & a_{12} & a_{13} & 1 & 0 & 0 \\ a_{21} & a_{22} & a_{23} & 0 & 1 & 0 \\ a_{31} & a_{32} & a_{33} & 0 & 0 & 1 \end{array} \right| \rightarrow \left| \begin{array}{ccc|ccc} 1 & 0 & 0 & a_{11}^{-1} & a_{12}^{-1} & a_{13}^{-1} \\ 0 & 1 & 0 & a_{21}^{-1} & a_{22}^{-1} & a_{23}^{-1} \\ 0 & 0 & 1 & a_{31}^{-1} & a_{32}^{-1} & a_{33}^{-1} \end{array} \right|.$$

The Gauss-Jordan method is used in order to reduce the coefficient matrix to an identity matrix $[I]$:

$$[I] = \begin{bmatrix} 1 & 0 & 0 \\ 0 & 1 & 0 \\ 0 & 0 & 1 \end{bmatrix}.$$

When this is accomplished, the right-hand side of the augmented matrix will contain the inverse. Once $[A]^{-1}$ is obtained, we multiply it by the corresponding right-hand-side vector:

$$[b] = \begin{bmatrix} b_1 \\ b_2 \\ b_3 \end{bmatrix}.$$

And because performing matrix multiplication is much quicker and easier than inversion, we do the inversion once, then derive the additional solution using multiplication.

$$x_1 = a_{11}^{-1} b_1 + a_{12}^{-1} b_2 + a_{13}^{-1} b_3$$
$$x_2 = a_{21}^{-1} b_1 + a_{22}^{-1} b_2 + a_{23}^{-1} b_3$$
$$x_3 = a_{31}^{-1} b_1 + a_{32}^{-1} b_2 + a_{33}^{-1} b_3$$

Example Problem 2.2. Given the same linear equations, determine the values of x_1, x_2, and x_3 using Gauss-Jordan elimination (the inversion and multiplication routine).

$$3x_1 + x_2 + x_3 = 25$$
$$x_1 - 3x_2 + 2x_3 = 10$$
$$2x_1 + x_2 - x_3 = 6$$

We get the inverse matrix using the Gauss-Jordan method:

$$
\begin{vmatrix}
3 & 1 & 1 & 1 & 0 & 0 \\
1 & -3 & 2 & 0 & 1 & 0 \\
2 & 1 & -1 & 0 & 0 & 1
\end{vmatrix}
\begin{matrix}
\rightarrow \text{row3} \\
\rightarrow \text{row1} \\
\rightarrow \text{row2}
\end{matrix}
$$

$$
\begin{vmatrix}
1 & -3 & 2 & 0 & 1 & 0 \\
2 & 1 & -1 & 0 & 0 & 1 \\
3 & 1 & 1 & 1 & 0 & 0
\end{vmatrix}
\begin{matrix}
\\
\rightarrow \text{row1} \times (-2) + \text{row2} \\
\rightarrow \text{row1} \times (-3) + \text{row3}
\end{matrix}
$$

$$
\begin{vmatrix}
1 & -3 & 2 & 0 & 1 & 0 \\
0 & 7 & -5 & 0 & -2 & 1 \\
0 & 10 & -5 & 1 & -3 & 0
\end{vmatrix}
\begin{matrix}
\\
\rightarrow \text{row2} \times (1/7) \\
\\
\end{matrix}
$$

$$
\begin{vmatrix}
1 & -3 & 2 & 0 & 1 & 0 \\
0 & 1 & -5/7 & 0 & -2/7 & 1/7 \\
0 & 10 & -5 & 1 & -3 & 0
\end{vmatrix}
\begin{matrix}
\rightarrow \text{row2} \times 3 + \text{row1} \\
\\
\rightarrow \text{row2} \times (-10) + \text{row3}
\end{matrix}
$$

$$
\begin{vmatrix}
1 & 0 & -1/7 & 0 & 1/7 & 3/7 \\
0 & 1 & -5/7 & 0 & -2/7 & 1/7 \\
0 & 0 & 15/7 & 1 & -1/7 & -10/7
\end{vmatrix}
\begin{matrix}
\\
\\
\rightarrow \text{row3} \times (7/15)
\end{matrix}
$$

$$
\begin{vmatrix}
1 & 0 & -1/7 & 0 & 1/7 & 3/7 \\
0 & 1 & -5/7 & 0 & -2/7 & 1/7 \\
0 & 0 & 1 & 7/15 & -1/15 & -10/15
\end{vmatrix}
\begin{matrix}
\rightarrow \text{row3} \times (1/7) + \text{row1} \\
\rightarrow \text{row3} \times (5/7) + \text{row2} \\
\\
\end{matrix}
$$

$$\begin{vmatrix} 1 & 0 & 0 & 7/105 & 14/105 & 35/105 \\ 0 & 1 & 0 & 35/105 & -35/105 & -35/105 \\ 0 & 0 & 1 & 7/15 & -1/15 & -10/15 \end{vmatrix} \qquad [b] = \begin{bmatrix} 25 \\ 10 \\ 6 \end{bmatrix}$$

By doing multiplication we get the answers:

$x_1 = (7 \times 25)/105 + (14 \times 10)/105 + (35 \times 6)/105 = 5$
$x_2 = (35 \times 25)/105 - (35 \times 10)/105 - (35 \times 6)/105 = 3$
$x_3 = (7 \times 25)/15 - (1 \times 10)/15 - (10 \times 6)/15 = 7$

The two major routines to be incorporated into the program are matrix inversion and multiplication. By doing these routines, we made use of `for` loop statements to perform all the iterative mathematical operations.

Program Listing 2.2: Gausss-Jordan Program.

```
1   #include <stdio.h>
2   #include <stdlib.h>
3   int main()
4   {
5    int r,c,i;
6    double a[6][6],b[6];
7    double d,x[6],e1;
8    printf("\n Solving for 3 Equations - 3 Unknowns\n");
9    for (r=0; r<3; r++)
10     {
11     for (c=0; c<3; c++)
12      {
13      printf(" Enter a[%d][%d]: ",r,c);
14      scanf("%lf",&a[r][c]);
15      }
16     printf(" Enter b[%d]: ",r);
17     scanf("%lf",&b[r]);
18     x[r]=0;
19     }
20    for(r=0;r<3;r++)
21    {
22     d=-1/a[r][r];
23     for(c=0;c<3;c++)
24      {
```

```
25      if (c==r) (c++);
26      a[r][c]=a[r][c]*d;
27      }
28      d=-d;
29      for(i=0;i<3;i++)
30      {
31        if(i==r)(i++);
32        e1=a[i][r];
33        for(c=0;c<3;c++)
34        {
35          if(c==r)(a[i][r]=a[i][r]*d);
36          else(a[i][c]=a[i][c]+a[r][c]*e1);
37        }
38      }
39      a[r][r]=d;
40      }
41      for(i=0;i<3;i++)
42      {
43        for(c=0;c<3;c++)
44        x[i]=x[i]+b[c]*a[i][c];
45      }
46      for(i=0;i<3;i++)
47      printf("variables x[%d]= %lf\n",i,x[i]);
48      system("pause");
49      return 0;
50    }
```

Line 5: Declaration of variables r, c, i as integers.

Lines 6, 7: Declaration of variables a[][],b[],x[], d, e1 as doubles.

Line 8: The printf() function prints out the statement on the screen. The control character \n ensures that the statement occupies a new line before being printed.

Lines 9 to 19: The for loop statements are used to get the values of the constant coefficients a[][] and the constant b[].

Line 18: Initializes all xs with 0 values.

Lines 20 to 40: Perform matrix inversion.

Lines 41 to 45: Perform matrix multiplication.

Lines 46 to 47: The resulting values are printed on the screen.

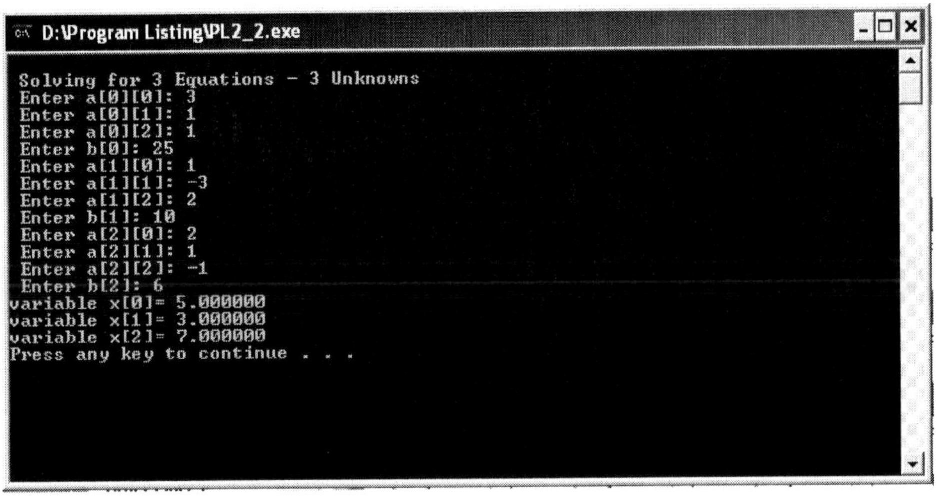

FIGURE 2.2:   Output of Program Listing 2.2.

Numerical Method

The numerical method uses iteration to solve for unknowns. By incorporating this into the computer program we take advantage of the computer's power to do basic repetitive computation. In this section, we are going to cover two examples of this method—the Jacobi and Gauss-Seidel methods.

The Jacobi Method and Gauss-Seidel Methods

Using the same *n* linear equations presented earlier, we rearrange the equations to express each unknown in terms of the others:

$$x_1 = \frac{b_1 - (a_{12}x_2 + ...a_{1n}x_n)}{a_{11}}$$

$$x_2 = \frac{b_2 - (a_{21}x_1 + a_{23}x_3 + ...a_{2n}x_n)}{a_{22}} \tag{2.5}$$

$$x_n = \frac{b_n - (a_{n1}x_1 + ...a_{nn-1}x_{n-1})}{a_{nn}}$$

We start the process by assuming each value of the unknowns on the right-hand side of the above equation. Since the value zero is most convenient, we substitute it to Eq 2.5. The first substitution yields the values of the unknowns as

$$x_1 = b_1/a_{11}; \ x_2 = b_2/a_{22}; \ ... x_n = b_n/a_{nn.} \tag{2.6}$$

These values are then substituted back on the right-hand side of the equation, giving another set of values of x. We repeat the process until the values obtained are almost equal to the previous values.

Both the Jacobi and Gauss-Seidel methods have similar processes except that in the Gauss-Seidel method, the new values of the unknowns are inserted into the next equation as they are generated. In this way, the method requires less iteration than the Jacobi method.

Let's proceed with an example.

Example Problem 2.3. Given the same linear equations, determine the values of x_1, x_2, and x_3 using the Gauss-Seidel iterative method.

$$3x_1 + x_2 + x_3 = 25$$
$$x_1 - 3x_2 + 2x_3 = 10$$
$$2x_1 + x_2 - x_3 = 6$$

Iteration 1:

$$x_1 = 25/3 = 8.333333$$
$$x_2 = [10 - (1)(8.333333)]/-3 = -0.555556$$
$$x_3 = [6 - (2)(8.333333) - (1)(-0.555556)]/-1 = 10.111111$$

Iteration 2:

$$x_1 = \{25 - [(1)(-0.555556) + (1)(10.111111)]\}/3 = 5.148148$$
$$x_2 = \{10 - [(1)(5.148148) + (2)(10.111111)]\}/-3 = 5.123456$$
$$x_3 = \{6 - [(2)(5.148148) + (1)(5.123456)]\}/-1 = 9.419752$$

Table 2.1: Results of the Gauss-Seidel Iterative Method.

Iteration	x_1	x_2	x_3
2	5.148148	5.123456	9.419752
3	3.485597	4.108367	5.079561
4	5.270690	1.809938	6.351318
5	5.612915	2.771851	7.997680
6	4.743490	3.579617	7.066596
7	4.784596	2.972596	6.541788
:			
67	5.000000	3.000000	7.000000

In developing the Gauss-Seidel program we use loop statements to process a sequence of instructions. We also provide a condition that will trigger the end of the iteration if satisfied.

Program Listing 2.3: Gauss-Seidel Program.

```
1   #include <stdio.h>
2   #include <stdlib.h>
```

```
3   #include <math.h>
4   int main()
5   {
6    int r,c;
7    double a[3][3],b[3];
8    double x[3],x_temp[3],x_prev[3],flag;
9    printf("\n Solving for 3 Equations - 3 Unknowns\n");
10   for (r=0; r<3; r++)
11   {
12    for (c=0; c<3; c++)
13    {
14     printf(" Enter a[%d][%d]: ",r,c);
15     scanf("%lf",&a[r][c]);
16    }
17    printf(" Enter b[%d]: ",r);
18    scanf("%lf",&b[r]);
19   }
20   for (r=0; r<3; r++)
21   {
22    x[r]=0;
23    x_temp[r]=0;
24   }
25   for (r=0; r<3; r++)
26   {
27    for (c=0; c<3; c++)
28    {
29     x_temp[r]=x_temp[r]+(a[r][c]*x[c]);
30    }
31    x[r]=(b[r]-x_temp[r])/a[r][r];
32    x_temp[r]=0;
33    x_prev[r]=x[r];
34   }
35   do{
36    for (r=0; r<3; r++)
37    {
38     for (c=0; c<3;c++)
39     {
40      if(c==r) x_temp[r]=x_temp[r]+(a[r][c]*0);
41      else x_temp[r]=x_temp[r]+(a[r][c]*x[c]);
42     }
43     x[r]=(b[r]-x_temp[r])/a[r][r];
44     x_temp[r]=0;
```

```
45    flag=fabs(x_prev[r]-x[r]);
46    x_prev[r]=x[r];
47    }
48    }while(flag>0.00000000001);
49    printf("\n X1 = %lf,  X2 = %lf,  X3 = %lf\n",x[0],x[1],x[2]) ;
50    system("pause");
51    return 0;
52  }
```

Line 6: Declaration of variables r, c, and i as integers.

Lines 7 to 8: Declaration of variables a[][],b[], x[],x_temp[],x_prev[], and flag as long floating points or doubles.

Line 9: The printf() function prints out the statement on the screen. The control character \n ensures that the statement occupies a new line before being printed.

Lines 10 to 19: The for loop statements are used to get the values of the constant coefficients a[][] and the constant b[].

Lines 20 to 24: Initializes all xs and x_temps with 0 values.

Lines 25 to 34: Perform the first loop of the mathematical operation based on Eqs 2.5 and 2.6.

Lines 35 to 48: The do - while loop performs successive iterations until the condition is met.

Line 48: The condition is set so as not to stop the iteration when the difference between the current x value and the previous x value is greater than 0.00000000001.

Line 49: the resulting values are printed on the screen.

```
D:\Program Listing\PL2_3.exe

Solving for 3 Equations - 3 Unknowns
Enter a[0][0]: 3
Enter a[0][1]: 1
Enter a[0][2]: 1
Enter b[0]: 25
Enter a[1][0]: 1
Enter a[1][1]: -3
Enter a[1][2]: 2
Enter b[1]: 10
Enter a[2][0]: 2
Enter a[2][1]: 1
Enter a[2][2]: -3
Enter b[2]: 6

X1 = 7.000000, X2 = 1.000000, X3 = 3.000000
Press any key to continue . . .
```

FIGURE 2.3: Output of Program Listing 2.3.

REGRESSION ANALYSIS

Each data set consists of the independent variable x and the corresponding value of the dependent variable y. The purpose of regression analysis is to seek out mathematical relationship between the dependent variable y and the independent variable x. The relationship selected predicts the value of the dependent variable with the least error.

There are two types of regression analyses covered in this section; linear and polynomial.

Linear Regression

Sometimes the use of a straight-line relationship may be adequate to represent physical data over a limited range. The objective is to determine the values of the coefficients a and b in the equation

$$y = a + bx \tag{2.7}$$

Based on the series of x and y values given, we can compute for these coefficients using Eqs 2.8 and 2.9:

$$a = \frac{\sum y_i}{n} - \frac{b \sum x_i}{n} \tag{2.8}$$

$$b = \frac{\sum x_i y_i - \left(\dfrac{\sum x_i \sum y_i}{n} \right)}{\sum x_i^2 - \left[\dfrac{\left(\sum x_i\right)^2}{n} \right]} \tag{2.9}$$

The best way to determine if the data points follow a straight line is by plotting them on an x-y diagram. Statistical analysis can also be used to validate linearity.

To further illustrate linear regression analysis, let's consider this example problem.

Example Problem 2.4. Given the following x and y values, determine the equation of the line that best fits the data $(x, y) = (1, 4.3), (3, 4.64), (5, 4.98), (7, 5.32)$, and $(9, 5.7)$.

Solve for the sum of x, y, xy, and x^2.

Table 2.2 : Summary of x, y, xy, and x^2 values

x	y	xy	x^2
1	4.3	4.3	1
3	4.64	13.92	9
5	4.98	24.9	25
7	5.32	37.24	49
9	5.7	51.3	81
Sum 25	24.94	131.66	165

Likewise, the resulting a and b:

$$b = (131.66 - [(25)(24.94)/5])/ (165 - [(25)^2/5])$$
$$b = 0.174$$

$$a = (24.94/5) - (0.1740(25))/5$$
$$a = 4.118$$

Therefore, the line equation is $y = 4.118 + 0.174x$.

Program Listing 2.4: Linear Regression Program.

```
1    #include <stdio.h>
2    #include <stdlib.h>
3    int main()
4    {
5     int n,i;
6     double x,y,sumx,sum_sqx,sumy,sumxy,b,a;
7     printf("\nEnter number of x and y pairs: ");
8     scanf("%d",&n);
9     sumx=0;sum_sqx=0;sumy=0;sumxy=0;
10    for(i=0;i<n;i++)
11    {
12     printf("Enter x: ");
13     scanf("%lf",&x);
14     printf("Enter y: ");
15     scanf("%lf",&y);
16     sumx=sumx+x;
17     sum_sqx=sum_sqx+(x*x);
18     sumy=sumy+y;
19     sumxy=sumxy+(x*y);
20    }
21    b=(sumxy-(sumx*sumy)/n)/(sum_sqx-(sumx*sumx)/n);
```

```
22   a=(sumy/n)-(b*sumx/n);
23   printf("\nEquation of line is Y = %lf + %lfX\n",a,b);
24   system("pause");
25   return 0;
26 }
```

Line 5: Declaration of variables n, i as integers.

Line 6: Declaration of variables x, y, sumx, sum_sqx, xy, sumxy, b, and a as doubles.

Lines 7, 8: Request the number of x and y pairs to input and store this value to n.

Line 9: Initializes sumx, sum_sqx, sumy and sumxy to 0.

Lines 10 to 20: The for loop statement is used to request the values of x and y based on the given n, then perform initial computations.

Line 21: Performs mathematical operations based on Eq 2.9.

Line 22: Then performs another mathematical operation based on Eq 2.8.

Line 23: The resulting line equation is printed on the screen.

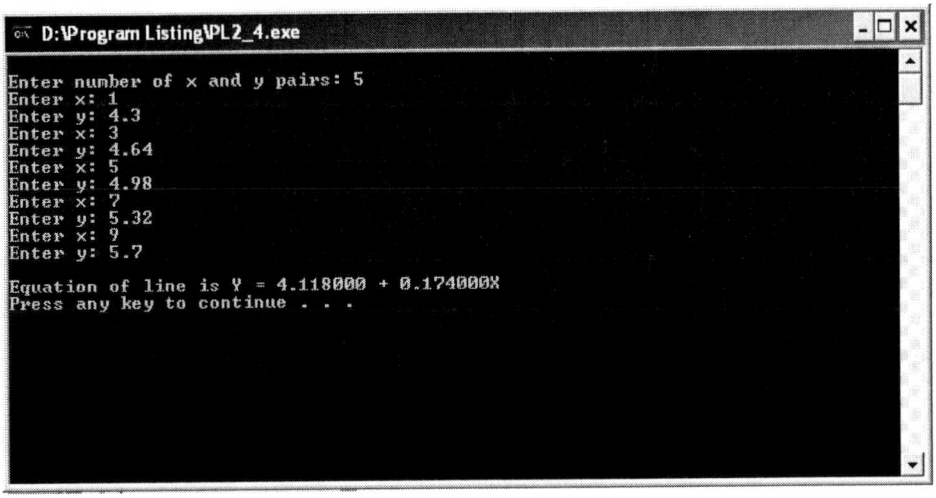

FIGURE 2.4:   Output of Program Listing 2.4.

Linear Regression for Nonlinear Relationships

Linear regression analysis is not only limited to linear relationships. There are also data that when plotted on an *x-y* diagram, are initially curvilinear, but with some adjustments on the *x* and *y* coordinates, can be expressed as linear. Four types of equations exhibit such characteristics. These are:+

Exponential Equations

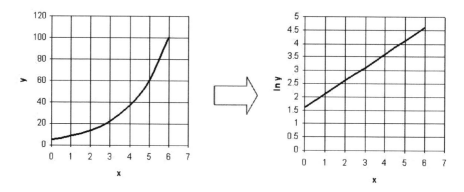

FIGURE 2.5: Exponential to Linear Relationship.

$\ln y = \ln a + bx \ln e$, since $\ln e = 1$
$\ln y = \ln a + bx$ (slope = b, y-intercept = $\ln a$)
Taking the anti-ln of both sides, we obtain

$$y = ae^{bx}.$$ **(2.10)**

Example Problem 2.5. Given the following $(x, y) = (1, 1), (2, 2), (3, 4), (4, 8.2),$ $(5, 16.6)$, and $(6, 33.3)$ determine the exponential equation that can best fit the x-y relationship.

Obtaining the natural logarithmic value of y, we get:

Table 2.3: Summary of x, y, and $\ln y$ values

x	y	$\ln y$
1	1	0.00000
2	2.03	0.70804
3	4.08	1.40610
4	8.22	2.10657
5	16.56	2.80699
6	33.34	3.50676

Using the Linear Regression Program (Program Listing 2.4), we established the line equation:

$\ln y = -0.697369 + 0.7x.$

Taking the anti-ln of both sides by getting the exponents of the values
$$y = 0.498 \ e^{0.7x} \ .$$

Logarithmic Equations

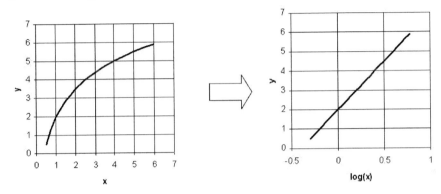

FIGURE 2.6: Logarithmic to Linear Relationship.

$$y = a + b \ log(x) \qquad\qquad (2.11)$$

Example Problem 2.6. Given the following $(x, y) = (1, 1)$, $(2, 1.9)$, $(3, 2.43)$, $(4, 2.81)$, $(5, 3.1)$, and $(6, 3.33)$ determine the logarithmic equation that can best fit the x-y relationship.

Obtaining the log value of x, we get:

Table 2.4: Summary of x, $log(x)$, and y values

x	$log\ (x)$	y
1	0	1
2	0.30103	1.9
3	0.477121	2.43
4	0.60206	2.81
5	0.69897	3.1
6	0.778151	3.33

Using the Linear Regression Program (Program Listing 2.4), we established the line equation:

$$y = 0.999248 + 3.0log(x)$$

Power Equations

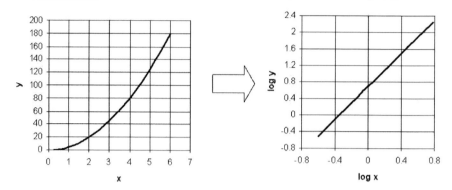

FIGURE 2.7: Power to Linear Relationship.

$\log y = \log a + b \log x$ (slope $= b$, y-intercept $= \log a$)
Taking the anti-log of both sides, we obtain

$$y = ax^{b.}$$ **(2.12)**

Example Problem 2.7. Given the following $(x, y) = (1, 1)$, $(2, 2.83)$, $(3, 5.20)$, $(4, 8)$, $(5, 11.18)$, and $(6, 14.7)$ determine the power equation that can best fit the x-y relationship.

Obtaining the log values of both x and y, we get:

Table 2.5 : Summary of x, $\log(x)$, y, and $\log(y)$ values

x	$\log(x)$	y	$\log(y)$
1	0	1	0
2	0.30103	2.83	0.451786
3	0.477121	5.20	0.716003
4	0.60206	8	0.90309
5	0.69897	11.18	1.048442
6	0.778151	14.7	1.167317

Using the Linear Regression Program (Program Listing 2.4), we established the line equation:

$$\log(y) = 0.000127 + 1.499958 \log(x)$$

Taking the anti-log of both sides, we get:

$$y = 1.000127\, x^{1.499958}$$

Saturation Rate Equations

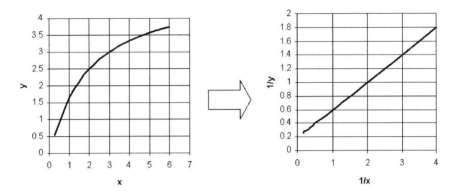

FIGURE 2.8: Saturation Rate to Linear Relationship.

$$\frac{1}{y} = \frac{1}{a} + \frac{b}{ax} \ (\text{slope} = b/a, \text{y-intercept} = 1/a)$$

Taking the reciprocal of both sides, we obtain

$$y = a\,\frac{x}{(b+x)}.$$ (2.13)

Example Problem 2.8. Given the following $(x, y) = (1, 0.38)$, $(2, 0.6)$, $(3, 0.75)$, $(4, 0.86)$, $(5, 0.94)$, and $(6,1)$ determine the saturation rate equation that can best fit the x-y relationship.

Obtaining the reciprocal values of both x and y, we get:

Table 2.6: Summary of x, $1/x$, y, and $1/y$ values

x	$1/x$	y	$1/y$
1	1	0.38	2.666667
2	0.5	0.6	1.666667
3	0.333333	0.75	1.333333
4	0.25	0.86	1.166667
5	0.2	0.94	1.066667
6	0.166667	1	1

Using the Linear Regression Program (Program Listing 2.4), we established the line equation:

$$1/y = 0.666667 + 2.00/x.$$

Taking again the reciprocal of both sides, we get:

$$y = 1.5x/(3.0 + x).$$

Polynomial Regression

When linear regression is not give a good fit, higher order polynomial regression may be needed. Finding the best polynomial curve that is best suited to fit the data can be done through polynomial regression.

The least-square procedure can be readily applied to fit the data to the nth order polynomials.

$$y = a_0 + a_1 x + a_2 x^2 + \ldots + a_n x^n$$

For each data point there exists a value of the independent variable x and the dependent variable y predicted by the relationship; $a_0, a_1, a_2, \ldots a_n$ are parameters whose values have to be determined through analysis of the data. Thus, if there are m data points, we can write the polynomial relationship for each one as

$$y_0 = a_0 + a_1 x_0 + a_2 x_0^2 + \ldots a_n x_0^n$$
$$y_1 = a_0 + a_1 x_1 + a_2 x_1^2 + \ldots a_n x_1^n$$
$$y_2 = a_0 + a_1 x_2 + a_2 x_2^2 + \ldots a_n x_2^n$$
$$\vdots$$
$$y_{m-1} = a_0 + a_1 x_{m-1} + a_2 x_{m-1}^2 + \ldots a_n x_{m-1}^n$$

We can get the derivative with respect to each of the coefficients of the polynomial and set these equations to zero. Rearranging further, we can then derive the following set of normal equations.

$$a_0 m + a_1 \Sigma x_i + a_2 \Sigma x_i^2 + \ldots a_n \Sigma x_i^n = \Sigma y_i$$
$$a_0 \Sigma x_i + a_1 \Sigma x_i^2 + a_2 \Sigma x_i^3 + \ldots a_n \Sigma x_i^{n+1} = \Sigma x_i y_i$$
$$a_0 \Sigma x_i^2 + a_1 \Sigma x_i^3 + a_2 \Sigma x_i^4 + \ldots a_n \Sigma x_i^{n+2} = \Sigma x_i^2 y_i \qquad \textbf{(2.14)}$$
$$\vdots$$
$$a_0 v x_i^n + a_1 \Sigma x_i^{n+1} + a_2 \Sigma x_i^{n+2} + \ldots a_n \Sigma x_i^{2n} = \Sigma x_i^n y_i$$

Variables a_0 to a_n can then be determined by applying the method for solving simultaneous linear equations. Let us consider this example problem to further illustrate how to derive polynomial equations through regression.

Example Problem 2.9. Given the following (x, y) = (1, 1), (2, 7), (3, 11), (4, 13), (5, 15), (6,17), (7, 19), (8, 20), (9, 21), (10, 22), and (11, 23) determine the 2nd order polynomial equation that best fits the data.

The sum of x, y, x^2, x^3, x^4, xy, and x^2y are presented in Table 2.7.

Table 2.7: Tabulation of x, y, x^2, x^3, x^4, xy, and x^2y.

x	y	x^2	x^3	x^4	xy	x^2y
1	1	1	1	1	1	1
2	7	4	8	16	14	28
3	11	9	27	81	33	99
4	13	16	64	256	52	208
5	15	25	125	625	75	375
6	17	36	216	1296	102	612
7	19	49	343	2401	133	931
8	20	64	512	4096	160	1280
9	21	81	729	6561	189	1701
10	22	100	1000	10000	220	2200
11	23	121	1331	14641	253	2783
Sum 66	169	506	4356	39974	1232	10218

$$\sum x_i = 66$$

$$\sum y_i = 169$$

$$\sum x_i y_i = 1232$$

$$\sum x^2_i y_i = 10218$$

$$\sum x_i^2 = 506$$

$$\sum x_i^3 = 4356$$

$$\sum x_i^4 = 39974$$

$m = 11$

$$11a_0 + 66a_1 + 506a_2 = 169$$
$$66a_0 + 506a_1 + 4356a_2 = 1232$$
$$506a_0 + 4356a_1 + 39974a_2 = 10218$$

Solving the simultaneous linear equation, we can get the following values of the coefficients:

$$a_0 = -1.739394$$

$$a_1 = 4.387413$$

$$a_2 = -0.200466$$

Translating to a polynomial equation we have $y = -1.739394 + 4.387413x - 0.200466x^2$.

Program Listing 2.5: Polynomial Regression Program.

```
1    #include<stdio.h>
2    #include<stdlib.h>
3    int main()
4    {
5     int r,c,num,i,j,deg,n,k;
6     double a[40][40],b[40],yval[20],xval[20];
7     double d,x[40],e1,sumx[20],sumxy[20],tempx;
8     printf("\nEnter the order of polynomial: ");
9     scanf("%d",&deg);
10    printf("Enter the number of x and y pairs: ");
11    scanf("%d",&n);
12    for(i=0;i<n;i++)
13    {
14     printf("\nEnter x: ");
15     scanf("%lf",&xval[i]);
16     printf("Enter y: ");
17     scanf("%lf",&yval[i]);
18    }
19    for(j=0;j<((2*deg)+1);j++)
20    {
21     sumx[j]=0;
22     sumxy[j]=0;
23     tempx=1;
24     for(i=0;i<n;i++)
25     {
26      for(k=0;k<(j+1);k++)
27      {
28       if (k==0) tempx=1;
29       else tempx=tempx*xval[i];
30      }
```

```
31    sumx[j]=sumx[j]+tempx;
32    sumxy[j]=sumxy[j]+(tempx*yval[i]);
33   }
34  }
35  num=deg+1;
36  for (r=0; r<num; r++)
37  {
38   for (c=0; c<num; c++)
39    {
40     a[r][c]=sumx[r+c];
41    }
42   b[r]=sumxy[r];
43   x[r]=0;
44   }
45   for(r=0;r<num;r++)
46   {
47    d=-1/a[r][r];
48    for(c=0;c<num;c++)
49    {
50     if (c==r)(c++);
51     a[r][c]=a[r][c]*d;
52    }
53   d=-d;
54   for(i=0;i<num;i++)
55   {
56    if(i==r)(i++);
57    e1=a[i][r];
58    for(c=0;c<num;c++)
59    {
60     if(c==r)(a[i][r]=a[i][r]*d);
61     else(a[i][c]=a[i][c]+a[r][c]*e1);
62    }
63   }
64   a[r][r]=d;
65   }
66  for(i=0;i<num;i++)
67  {
68   for(c=0;c<num;c++)
69   x[i]=x[i]+b[c]*a[i][c];
70  }
71  printf("\nY = %lf + %lfX",x[0],x[1]);
```

```
72  for(i=2;i<num;i++) printf("+ %lfX^%d",x[i],i);
73  printf("\n");
74  system("pause");
75  return 0;
76  }
```

Line 5:   Declaration of variables r, c, num, i, j, deg, n, and k as integers.

Lines 6, 7: Declaration of variables xval[], yval[],d, x[], sumx[20], tempx, sumxy, b[], and a[][] as doubles.

Lines 8 to 11: Request the order of polynomials (stored in deg variable), and the number of x and y pairs (stored in n variable).

Lines 12 to 18: The for loop statement is used to request the values of x and y based on the given n.

Lines 19 to 44: Performs initial computations to get all coefficient constants and right-hand side constants of Eq 2.14.

Lines 45 to 70: Then performs matrix inversion and multiplication to solve for the simultaneous linear Eq 2.14.

Lines 71 to 73: The resulting polynomial equation is printed on the screen.

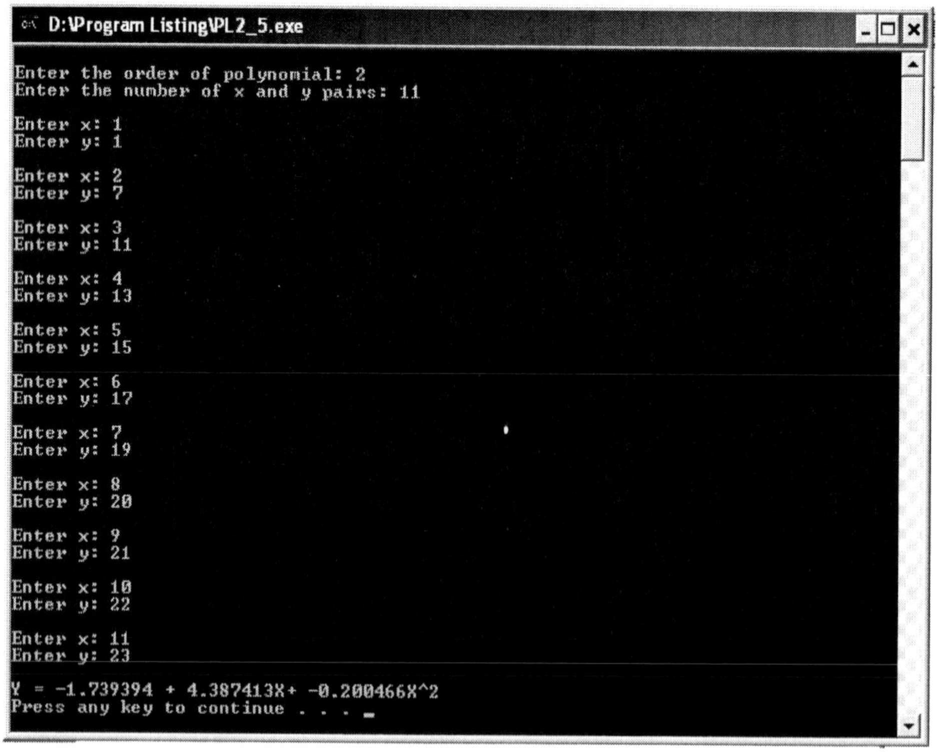

FIGURE 2.9: Output of Program Listing 2.5.

VALIDATION THROUGH STATISTICAL ANALYSIS

Correlation Coefficient and Coefficient of Determination

As mentioned earlier, one way of validating if the polynomial equation represents a good fit is through the coefficient of determination.

First, we determine the sum of squares around the mean for the dependent variable (y) and label the sum as S_t. This is the amount of spread in the dependent variable that exists prior to regression. After performing the regression, we compute for S_r, which is the sum of squares of the residual around the regression curve. This represents the spread that remains after regression.

$$S_t = \sum(y - \bar{y})^2 \tag{2.15}$$

$$S_r = \sum(y - y_e)^2 \tag{2.16}$$

Where $\bar{y}$ is the average and y_e is the y derived from the equation. The difference of S_t and S_r can be normalized to the total error to get

$$r^2 = (S_t - S_r)/S_t \tag{2.17}$$

Where r^2 is the coefficient of determination and r is the correlation coefficient. For a perfect fit, $S_r = 0$ and $r^2 = 1$, signifying that at this condition the curve equation explains entirely the variability. Let us try this example problem.

Example Problem 2.10. Find the coefficient of determination and the correlation coefficient of the equation derived in Example Problem 2.9: $y = -1.739394 + 4.387413x - 0.200466x^2$.

Let y be the actual y values given in Example Problem 2.9
       y_e be the y values obtained using the derived polynomial equation

Table 2.8: Tabulation to determine $(y - \bar{y})^2$ and $(y - y_e)^2$.

x	y	y_e	$(y - \bar{y})^2$	$(y - y_e)^2$
1	1	2.447553	206.314050	2.095410
2	7	6.233568	69.950413	0.587418
3	11	9.618651	19.041322	1.908125
4	13	12.602802	5.586777	0.157766
5	15	15.186021	0.132231	0.034604
6	17	17.368308	2.677686	0.135651

7	19	19.149663	13.223140	0.022399
8	20	20.530086	21.495868	0.280991
9	21	21.509577	31.768595	0.259669
10	22	22.088136	44.041322	0.007768
11	23	22.265763	58.314050	0.539104

$\bar{y}$ = 15.363636
S_t = 472.545455
S_r = 6.028904

Therefore, the coefficient of determination r^2 and the correlation coefficient r are as follows:

r^2 = 0.987242
r = 0.993600

Standard Error of Estimate

If we designate ye as the y estimate derived using the curve equation, the measure of the scatter about the regression curve can be presented as

$$S_{y,x} = \sqrt{\frac{\sum(y - y_e)^2}{\upsilon}} \qquad (2.18)$$

where $S_{y,x}$ is the standard error of estimate. Subscript y,x denotes that the error is for a predicted value of y corresponding to a particular value of x. The denominator υ is the number of degrees of freedom, which can be determined through n given data points. That is, for linear equation $\upsilon = n - 2$, for 2nd order $\upsilon = n - 3$, etc.

Since we introduced $\sum(y - y_e)^2$ in the previous section (Eq 2.16) as S_r, we can rewrite eq 2.18 as

$$S_{y,x} = \sqrt{\frac{S_r}{\upsilon}}. \qquad (2.19)$$

In selecting the curve equation with the best fit, we choose the one with the smallest standard error of estimate. In fact, $S_{y,x}$ equals zero, which signifies that the curve passes through all the data points and is therefore the best fit for the given points.

Example Problem 2.11 Using the given information in Example Problem 2.9, determine the standard error of estimate $S_{y,x}$ of equation y = -1.739394 + 4.387413x – 0.200466x^2.

Using Table 2.8 we can again determine S_r. Substituting other values to Eq 2.19 we get

$$S_{y,x} = \sqrt{\frac{S_r}{n\text{-}3}} = \sqrt{\frac{6.028904}{8}} = 0.868109 \ .$$

If conflicting conclusions are obtained from deriving the coefficient of determination (r^2) and the standard error of estimate ($S_{y,x}$), adding more data points will resolve this inconsistency. Never disregard common sense during evaluation.

Significance Tests

A significance test is conducted to establish the degree of fitness of the model from the data. Through this, we can evaluate the significance of the derived linear or polynomial equation. This test determines whether the population coefficient of correlation, estimated by the sample correlation coefficient r, is equal to zero. When the value of r approaches 1, there is a strong relationship between the data generated from the model and the actual data provided. However, when it approaches zero, the relationship becomes almost nonexistent. This will help us decide whether to use the linear/polynomial equation we established.

Generally, a statistic involving student's t distribution can be used:

$$t = \frac{r\sqrt{\upsilon}}{\sqrt{1-r^2}} \tag{2.20}$$

where:

υ = degree of freedom ($\upsilon = n - (i+1)$ for i^{th} order polynomial)
r^2 = coefficient of determination
r = correlation coefficient

The hypotheses being tested in regression by the t-test for significance are the following:

H_0:   Correlation coefficient is zero; no correlation is established between the values generated by the derived equation and the actual data provided.

H_1:   Correlation coefficient is not zero; there is correlation established between the values generated by the derived equation and the actual data provided.

When $t \leq t_{critical}$ we accept H_0, but when $t > t_{critical}$ we reject it. The $t_{critical}$ can be determined using student's t distribution table (Appendix J) using a given degrees of freedom and percent (%) confidence interval.

It is desirable to reject H_0 because by doing so we are stating that the regression model (the derived equation) explains the variation of the dependent variable based on the independent variable. Thus, we can conclude that the values derived from the equation are usable.

Let us consider again Example Problem 2.9. We determine, based on the previous computation, the following:

$\upsilon = 8$

$r^2 = 0.987242$

$r = 0.993600$

$$t = \frac{r\sqrt{\upsilon}}{\sqrt{1-r^2}} = \frac{0.9936\sqrt{8}}{\sqrt{1-0.987242}} = 24.881$$

From Appendix J, we determined $t_{critical}$ ($\upsilon = 8$, *level of significance (two tail areas)* = 0.001 or *confidence level* = 99.95%) as 5.04, and since $t > t_{critical}$, we reject the null hypothesis H_0.

Therefore, we can conclude that equation $y = -1.739394 + 4.387413x - 0.200466x^2$ can be used to represent the x-y values provided in the example problem.

INTERPOLATION ANALYSIS

Sometimes chemical engineers are required to estimate intermediate values between precise data points. The most common method used for this purpose is interpolation. There are two types of interpolation covered in this section—linear and Lagrange polynomial interpolation.

Linear Interpolation

The simplest form of interpolation is to connect two data points with a straight line. This technique is called linear interpolation and can be represented by the following equation:

$$\frac{y-y_0}{x-x_0} = \frac{y_i-y_0}{x_i-x_0}, \text{ rearranging we get,}$$

$$y = y_0 + \frac{y_i-y_0}{x_i-x_0}(x-x_0). \tag{2.21}$$

In general, the smaller the interval between the data points the better the approximation.

Example Problem 2.12. Given the saturation temperature of steam and the corresponding pressure, determine the pressure at which steam is saturated at 212°F.

Steam saturation temperature at 14.123 psia is 210°F.
Steam saturation temperature at 15.561 psia is 215°F.

Through linear interpolation, we can solve for the pressure when the steam saturation temperature is 212°F.

$$Pressure\ (psia) = 14.123 + \frac{15.561 - 14.123}{215 - 210}(212 - 210)$$

$$Pressure\ (psia) = 14.6982$$

Program Listing 2.6: Linear Interpolation Program.

```
1   #include <stdio.h>
2   #include <stdlib.h>
3   int main()
4   {
5    int n,i;
6    double x[2],y[2],ansy,valx;
7    printf(«\nEnter two (2) x and y pairs»);
8    n = 2;
9    ansy=0;
10   for(i=0;i<n;i++)
11   {
12    printf("\nEnter x: ");
13    scanf("%lf",&x[i]);
14    printf("Enter y: ");
15    scanf("%lf",&y[i]);
16   }
17   printf("\nEnter the value of x for interpolation: ");
18   scanf("%lf",&valx);
19   ansy = y[0] + (((y[1]-y[0])/(x[1]-x[0]))*(valx-x[0]));
20   printf(«\nThe interpolated value is: %lf\n»,ansy);
21   system("pause");
22   return 0;
23  }
```

Line 5: Declaration of variables i and n as integers.
Line 6: Declaration of variables x[], y[], ansy, and valx as doubles.
Lines 10 to 16: The for loop statement is used to request the two pair values of x and y.

Lines 17, 18: Request for the given value of x to which the independent variable y is unknown.

Line 19: Performs mathematical computation based on Eq 2.21.

Line 20: The resulting value of y is printed on the screen.

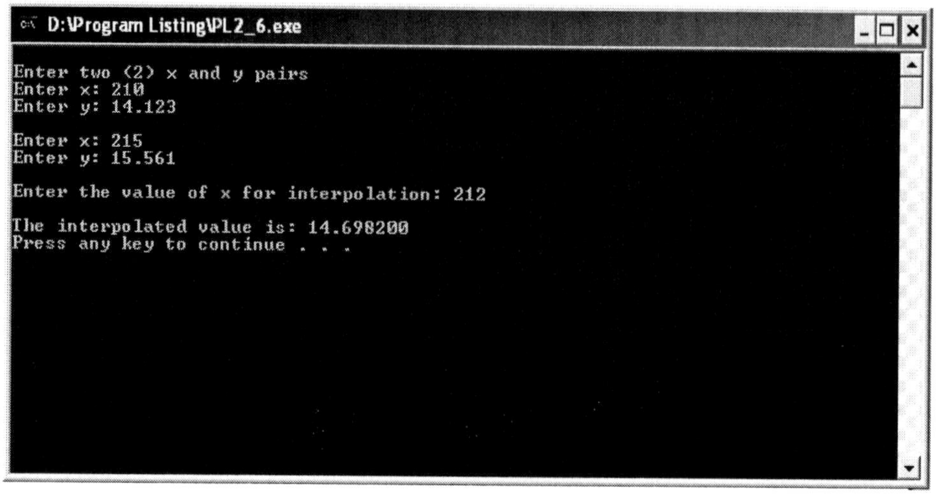

FIGURE 2.10: Output of Program Listing 2.6.

Lagrange Polynomial Interpolation

For data that cannot be represented linearly, Lagrange polynomial interpolation can be used. This method of interpolation can be both applied to even or unevenly spaced data. Unlike regression, we don't need to solve for any system of equations to evaluate the polynomial. Furthermore, by creating the computer program, a considerable amount of computation will be performed by the computer even if we are dealing with a higher degree of polynomials.

The Lagrange interpolating polynomial can be presented concisely as

$$y = \sum_{i=0}^{n} L_i(x) y_i \qquad (2.22)$$

where:

$$L_i(x) = \prod_{j=0, j \neq i}^{n} \frac{x - x_j}{x_i - x_j}. \qquad (2.23)$$

The Π designates the product of the equation. Let us try this example problem.

Example Problem 2.13. Given the values of four x and y pairs: $(x, y) = (0, 0)$, $(0.02, 0.134)$, $(0.08, 0.365)$, and $(0.15, 0.517)$, determine the value of y when x is 0.06.

$$L_0(0.06) = \frac{0.06 - 0.02}{0 - 0.02} \times \frac{0.06 - 0.08}{0 - 0.08} \times \frac{0.06 - 0.15}{0 - 0.15} = -0.3$$

$$L_1(0.06) = \frac{0.06 - 0}{0.02 - 0} \times \frac{0.06 - 0.08}{0.02 - 0.08} \times \frac{0.06 - 0.15}{0.02 - 0.15} = 0.6923$$

$$L_2(0.06) = \frac{0.06 - 0}{0.08 - 0} \times \frac{0.06 - 0.02}{0.08 - 0.02} \times \frac{0.06 - 0.15}{0.08 - 0.15} = 0.6428$$

$$L_3(0.06) = \frac{0.06 - 0}{0.15 - 0} \times \frac{0.06 - 0.02}{0.15 - 0.02} \times \frac{0.06 - 0.08}{0.15 - 0.08} = -0.0352$$

$$y = (-0.3)(0) + (0.6923)(0.134) + (0.6428)(0.365) + (-0.0352)(0.517)$$

$$y = 0.309232$$

This algorithm can be transformed to a computer program easily. The program listing for Lagrange interpolation is presented in Program Listing 2.7.

Program Listing 2.7: Lagrange Polynomial Interpolation Program.

```
1   #include <stdio.h>
2   #include <stdlib.h>
3   #include <conio.h>
4   int main()
5   {
6     int i,j,n;
7     double xval, yval, x[20], y[20],L[20];
8     char res;
9     res='n';
10    printf("\nEnter number of x and y pairs: ");
11    scanf("%d",&n);
12    for(i=0;i<n;i++)
13    {
```

```
14    printf("\nEnter x[%d]: ",i);
15    scanf("%lf",&x[i]);
16    printf("Enter y[%d]: ",i);
17    scanf("%lf",&y[i]);
18    }
19    do{
20      printf("\nEnter value of x for unknown y: ");
21      scanf("%lf",&xval);
22      yval=0;
23      for(i=0;i<n;i++)
24      {
25        L[i]=1;
26        for(j=0;j<n;j++)
27        {
28          if (j!=i) L[i]=L[i]*((xval-x[j])/(x[i]-x[j]));
29        }
30        yval=yval+(L[i]*y[i]);
31      }
32      printf("\nThe value of y is %lf",yval);
33      printf("\nTry again (y/n)?\n");
34      res=getch();
35    }while(res=='Y' || res=='y');
36  system("pause");
37  return 0;
38 }
```

Line 6: Declaration of variables i, j, and n as integers.

Line 7: Declaration of variables x[], y[], L[], xval, and yval as doubles.

Lines 8, 9: Declaration of res as a character and initialize it as 'n'.

Lines 10, 11: Request the number of x and y pairs and stores the values in n variable.

Lines 12 to 18: The for loop statement is used to request the values of x and y based on the given n.

Lines 20, 21: Request the given value of x to which the independent variable y is unknown.

Lines 23 to 31: Perform mathematical computation based on Eqs 2.22 and 2.23.

Line 32: The resulting value of y is printed on the screen.

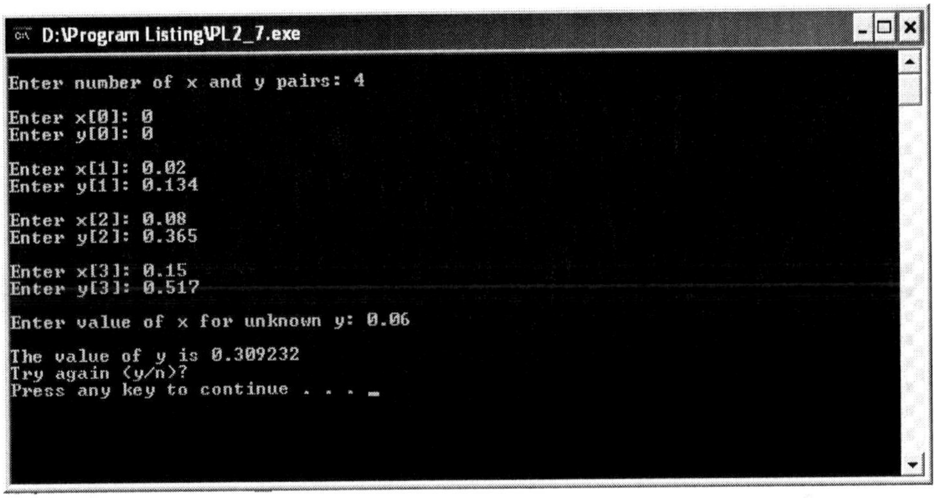

FIGURE 2.11: Output of Program Listing 2.7.

⚠WARNING *Do not use Lagrange polynomial interpolation if the data are too far apart. Always validate the result before applying this interpolation method in a program.*

LINEAR PROGRAMMING

Linear programming (or optimization) uses a mathematical model to describe problems that require allocating limited resources from numerous activities in the best possible way. An efficient solution to optimization problems following the linear programming model is the simplex method.

Simplex Method

The simplex method follows a systematic solution procedure of iteration until a desired result has been obtained. It uses the variables in a matrix form; and with the notion that the optimum solution occurs at one of the extreme points of the linear equations, it will look for that point by employing iteration starting from the extreme point of the region of a feasible solution.

The basic form can be presented as follows:

$$z = c_1 x_1 + c_2 x_2 + c_3 x_3 + \ldots c_n x_n$$

where:

z = objective function
c = coefficients of objective function

Any number of constraints (equality or inequality) can be provided, but the simplex method only considers cases where constraints are bounded by a maximum value and values of xs are positive.

$$a_{11}x_1 + a_{12}x_2 + a_{13}x_3 + \dots a_{1n}x_n \leq b_1$$

In cases where objective functions are to be minimized, b will be negative.

The simplex method starts by converting the functional inequality constraints to equivalent equality constraints. This conversion is done using slack variables resulting in an augmented form of the model.

$$a_{11}x_1 + a_{12}x_2 + a_{13}x_3 + \dots a_{1n}x_n + x_{n+m} = b_1$$

The new form is much more convenient for algebraic manipulation and for identification of an extreme point.

A matrix table called simplex tableau is then constructed:

Table 2.9: Simplex Tableau

z	x_1	x_2	x_3	x_4	x_5		
1	$-c_1$	$-c_2$	$-c_3$	0	0	0	row 1
0	a_{11}	a_{12}	a_{13}	1	0	b_1	row 2
0	a_{21}	a_{22}	a_{23}	0	1	b_2	row 3

Looking at the prepared tableau, we select the variable with the most negative coefficient c in row 1 and designate the column of coefficients of x as the pivotal column. We divide each number in the constant column by the corresponding number in the pivotal column, ignoring zero or negative numbers. We identify the pivotal row to have the smallest quotient and the pivot number as the number, which is at the intersection of the pivotal row and pivotal column.

We then perform pivoting operation by dividing the pivotal row by the pivot number. For each nonlimiting row multiply the newly transformed pivotal row by the negative of the coefficient in the pivotal column of the nonlimiting row. Add the result to the nonlimiting row. Repeat the process of identifying the most negative coefficient in row 1 until all the coefficients are nonnegative. An optimal solution then results.

Take a look at this example problem.

Example Problem 2.14. Maximize: $z = 80x_1 + 100x_2$ (Objective Function)

Constraints:

$0.5x_1 + 0.5x_2 \leq 25$
$0.2x_1 + 0.6x_2 \leq 10$
$0.8x_1 + 0.4x_2 \leq 14$

$x_1 \geq 0, x_2 \geq 0$

Converting constraints into equalities by introducing slack variables:

$0.5x_1 + 0.5x_2 + x_3 = 25$
$0.2x_1 + 0.6x_2 + x_4 = 10$
$0.8x_1 + 0.4x_2 + x_5 = 14$

Drawing up the simplex tableau:

Table 2.10: Initial Simplex Tableau

Z	x_1	x_2	x_3	x_4	x_5		
1	-80	-100	0	0	0	0	
0	0.5	0.5	1	0	0	25	→ 25/0.5 = 50
0	0.2	0.6	0	1	0	10	→ 10/0.6 = 16.67
0	0.8	0.4	0	0	1	14	→ 14/0.4 = 35

From the table, -100 is the most negative number and 16.67 is the least quotient. This will be where the pivotal column and row is (identified by shaded area). The pivot number therefore is 0.6.

The next step is to perform pivoting operations.
From the pivotal row (row 3):

Row 3: [0.2/0.6=0.33], [0.6/0.6=1], [0/0.6=0], [1/0.6=1.66], [0/0.6=0], [10/0.6=16.66]

Row 1: [-80+(100)(0.33) = -46.66], [-100+(100)(1)=0], [0+(100)(0)=0], [0+(100)(1.66)=166.6], [0+(100)(0)=0], [0+(100)(16.66)= 1666.66]

Row 2: [0.5-(0.5)(0.33)=0.66], [0.5-(0.5)(1)=0], [1-(0.5)(0)=1], [0-(0.5)(1.67)= -0.83], [0-(0.5)(0)=0],[25-(0.5)(16.66)=16.66]

Row 4: [0.8-(0.4)(0.33)=0.66], [0.4-(0.4)(1)=0], [0-(0.4)(0)=0], [0-(0.4)(1.66)= -0.66], [1-(0.4)(0)=1], [14-(0.4)(16.67)=7.33]

This will result in a new simplex tableau.

Table 2.11: Second Simplex Tableau

Z	x_1	x_2	x_3	x_4	x_5		
1	-46.66	0	0	166.6	0	1666.66	
0	0.66	0	1	-0.83	0	16.66	→ 16.66/0.66 = 25
0	0.33	1	0	1.66	0	16.66	→ 16.66/0.33 = 50
0	0.66	0	0	-0.66	1	7.33	→ 7.33/0.66 = 11

The most negative is -46.66; this is where the new pivotal column will be, while the pivotal row, having the least quotient, is row 4.

From the new pivotal row (row 4):

Row 4: [0.66/0.66=1], [0/0.66=0], [0/0.66=0], [-0.66/0.66=-1], [1/0.66=1.5], [7.33/0.66=11]

Row 1: [-46.66+(46.66)(1) = 0], [0+(46.66)(0)=0], [0+(46.66)(0)=0], [166.66+(46.66)(-1)=120], [0+(46.66)(1.5)=70], [1666.66+(46.66)(11)= 2180]

Row 2: [0.66-(0.66)(1)=0], [0-(0.66)(0)=0], [1-(0.66)(0)=1], [-0.83-(0.66)(-1)=-0.16], [0-(0.66)(1.5)=-1], [16.66-(0.66)(11)=9.33]

Row 3: [0.33-(0.33)(1)=0], [1-(0.33)(0)=1], [0-(033)(0)=0], [1.66-(0.33)(-1)=2], [0-(0.33)(1.5)=-0.5],[16.66-(0.33)(11)=13]

By doing the operation again, we obtain the final tableau:

Table 2.12: Final Simplex Tableau

Z	x_1	x_2	x_3	x_4	x_5	
1	0	0	0	120	70	2180
0	0	0	1	-0.16	-1	9.33
0	0	1	0	2	-0.5	13
0	1	0	0	-1	1.5	11

This will be the optimum solution, where: $x_1 = 11$, $x_2 = 13$ and $z = 2180$.

Program Listing 2.8: Simplex Method Program.

```
1   #include <stdio.h>
2   #include <stdlib.h>
3   #include <math.h>
4   int main()
5   {
6   int nc,ne,nopt,pivot1,pivot2,xerr=0,i,j;
7   double cf1,cf2,simtab[20][20],flag,aux,xmax;
8   char res;
9   printf("\n LINEAR PROGRAMMING - SIMPLEX METHOD\n\n");
10  printf("<A> Maximize or <B> Minimize: ");
11  scanf("%c",&res);
12  printf("\n Enter Number of Variables in Objective
    Function: ");
13  scanf("%d",&ne);
14  printf(" Enter Number of Constraints: ");
15  scanf("%d",&nc);
16  if (res == 'A' || res=='a')
17    cf1 = 1.0;
18  else
19    cf1 = -1.0;
20  printf("\n Enter Coefficients of Function:\n");
21  for (j = 1; j<=ne; j++)
22  {
23    printf("\tCoefficient #%d = ",j);
24    scanf("%lf",&cf2);
25    simtab[1][j+1] = cf2 * cf1;
26  }
27  cf2=0;
28  simtab[1][1] = cf2 * cf1;
29  for (i = 1; i<=nc; i++)
30  {
31    printf("\n Constraint #%d:\n",i);
32    for (j = 1; j<=ne; j++)
33    {
34      printf("\tCoefficient #%d = ",j);
35      scanf("%lf",&cf2);
36      simtab[i + 1][j + 1] = -cf2;
37    }
```

```
38    printf("\tValue at right side = ");
39    scanf("%lf",&simtab[i+1][1]);
40    }
41    printf("\n_____\n\n");
42    for(j=1; j<=ne; j++)  simtab[0][j+1] = j;
43    for(i=ne+1; i<=ne+nc; i++)  simtab[i-ne+1][0] = i;
44    do{
45     xmax = 0.0;
46     for(j=2; j<=ne+1; j++)
47     {
48      if (simtab[1][j] > 0.0 && simtab[1][j] > xmax)
49      {
50       xmax = simtab[1][j];
51       pivot2 = j;
52      }
53     }
54    flag = 1000000.0;
55    for (i=2; i<=nc+1; i++)
56    {
57     if (simtab[i][pivot2] < 0.0)
58     {
59      aux = fabs(simtab[i][1] / simtab[i][pivot2]);
60      if (aux < flag)
61      {
62       flag = aux;
63       pivot1 = i;
64      }
65     }
66    }
67    aux = simtab[0][pivot2];
68    simtab[0][pivot2] = simtab[pivot1][0];
69    simtab[pivot1][0] = aux;
70    for (i=1; i<=nc+1; i++)
71    {
72     if (i != pivot1)
73     {
74      for (j=1; j<=ne+1; j++)
75      {
76       if (j != pivot2) simtab[i][j] -= simtab[pivot1][j]
          * simtab[i][pivot2] / simtab[pivot1][pivot2];
77      }
```

```
78    }
79   }
80   simtab[pivot1][pivot2]  =  1.0  /  simtab[pivot1]
     [pivot2];
81   for (j=1; j<=ne+1; j++)
82   {
83    if (j != pivot2) simtab[pivot1][j] *= fabs(simtab[p
     ivot1][pivot2]);
84   }
85   for (i=1; i<=nc+1; i++)
86   {
87    if (i != pivot1) simtab[i][pivot2] *= simtab[pivot1]
     [pivot2];
88   }
89   for (i=2; i<=nc+1; i++)
90   if (simtab[i][1] < 0.0)  xerr = 1;
91   nopt = 0;
92   if (xerr != 1)
93   {
94    for (j=2; j<=ne+1; j++)
95    if (simtab[1][j] > 0.0)  nopt = 1;
96   }
97   }while(nopt==1);
98   if (xerr == 0)
99   {
100   for (i=1; i<=ne; i++)
101   for (j=2; j<=nc+1; j++)
102   {
103    if (simtab[j][0] == 1.0*i)
104    printf(" Variable #%d: %f\n",i,simtab[j][1]);
105   }
106   printf("\nValue of Objective: %f\n",simtab[1][1]);
107  }
108  else printf("No Feasible Solution.\n");
109  system("pause");
110  return 0;
111  }
```

Line 6: Declaration of variables `nc`, `ne`, `nopt`, `pivot1`, `pivot2`, `xerr`, `i`, and `j` as integers.

Line 7: Declaration of variables `cf1`, `cf2`, `simtab[][]`, flag, `aux`, and `xmax` as doubles.

Line 8: Declaration of `res` as a character.

Lines 9 to 40: Request all the needed values to fill up the simplex tableau using looping statements.

Lines 42 to 69: Perform matrix pivoting.

Lines 70 to 88: Perform mathematical routines based on the elimination method.

Lines 89 to 97: Check for the optimal solution.

Lines 98 to 108: Finally, the results are printed on the screen.

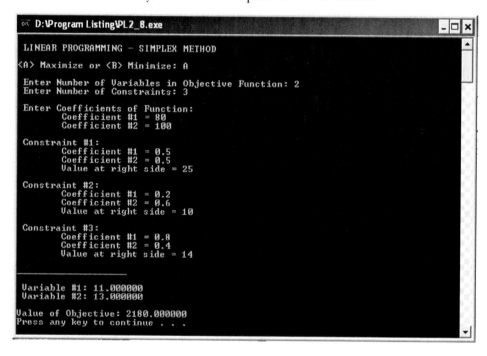

FIGURE 2.12: Output of Program Listing 2.8.

LABORATORY EXERCISES

1) Change the Gauss-Seidel program (Program Listing 2.3) to make use of the Jacobi method. Compare the number of iterations required to solve the following simultaneous linear equations:

a)
$$8x_1 + 3x_2 + 4x_3 = 231$$
$$2x_1 + 9x_2 + 3x_3 = 143$$
$$3x_1 + 8x_2 + 7x_3 = 264$$

b)
$$4x_1 + 1x_2 + 6x_3 = 195$$
$$5x_1 + 2x_2 + 7x_3 = 209$$
$$4x_1 + 7x_2 + 5x_3 = 199$$

2) A catalytic reaction A → 4R is studied in a plug flow using various amounts of catalyst and 15 liters/ hr of pure A feed. The result of analysis based on the rate equation $4 \ln (1/ 1\text{-}X_A) - 3X_A = k (Vr/15)$ is presented as follows:

Vr/ 15 (liters catalyst)(hr)/ liters fluid	$4 \ln (1/ 1\text{-}X_A) - 3X_A$
0.001	0.0726
0.002	0.1817
0.004	0.3720
0.006	0.5585
0.008	0.7479

Determine the slope k in (liters fluid)/(liter catalyst)(hr) using linear regression. Verify fitness by solving the r^2 and $S_{y,x}$.

3) Modify the Gauss-Jordan program (Program Listing 2.2) to accept up to 16 equations and unknowns. Test by solving the following simultaneous linear equations:

$6x_1 + 2x_2 + 7x_3 + 2x_4 + 4x_5 + 2x_6 + 5x_7 + 7x_8 + 8x_9 + 9x_{10} + 4x_{11} + 3x_{12} + 6x_{13} + 7x_{14} + 8x_{15} + 5x_{16} = 479$

$4x_1 + 9x_2 + 2x_3 + 5x_4 + 5x_5 + 8x_6 + 1x_7 + 7x_8 + 4x_9 + 2x_{10} + 8x_{11} + 1x_{12} + 1x_{13} + 6x_{14} + 3x_{15} + 8x_{16} = 393$

$9x_1 + 4x_2 + 5x_3 + 1x_4 + 6x_5 + 7x_6 + 1x_7 + 5x_8 + 4x_9 + 5x_{10} + 3x_{11} + 8x_{12} + 1x_{13} + 9x_{14} + 6x_{15} + 2x_{16} = 401$

$7x_1 + 2x_2 + 1x_3 + 9x_4 + 1x_5 + 8x_6 + 6x_7 + 4x_8 + 7x_9 + 9x_{10} + 8x_{11} + 6x_{12} + 8x_{13} + 7x_{14} + 8x_{15} + 5x_{16} = 479$

$2x_1 + 5x_2 + 8x_3 + 8x_4 + 5x_5 + 1x_6 + 4x_7 + 5x_8 + 4x_9 + 1x_{10} + 7x_{11} + 4x_{12} + 4x_{13} + 3x_{14} + 3x_{15} + 1x_{16} = 365$

$8x_1 + 3x_2 + 3x_3 + 1x_4 + 5x_5 + 6x_6 + 5x_7 + 9x_8 + 1x_9 + 8x_{10} + 6x_{11} + 2x_{12} + 7x_{13} + 8x_{14} + 8x_{15} + 4x_{16} = 461$

$9x_1 + 7x_2 + 9x_3 + 2x_4 + 8x_5 + 9x_6 + 9x_7 + 3x_8 + 4x_9 + 7x_{10} + 5x_{11} + 4x_{12} + 5x_{13} + 6x_{14} + 4x_{15} + 7x_{16} = 521$

$6x_1 + 9x_2 + 8x_3 + 4x_4 + 3x_5 + 1x_6 + 3x_7 + 5x_8 + 2x_9 + 6x_{10} + 9x_{11} + 8x_{12} + 3x_{13} + 3x_{14} + 7x_{15} + 4x_{16} = 418$

$5x_1 + 1x_2 + 3x_3 + 5x_4 + 2x_5 + 4x_6 + 9x_7 + 4x_8 + 7x_9 + 5x_{10} + 6x_{11} + 3x_{12} + 4x_{13} + 1x_{14} + 6x_{15} + 8x_{16} = 419$

$1x_1 + 5x_2 + 5x_3 + 8x_4 + 8x_5 + 5x_6 + 8x_7 + 9x_8 + 3x_9 + 2x_{10} + 4x_{11} + 6x_{12} + 7x_{13} + 8x_{14} + 3x_{15} + 7x_{16} = 502$

$8x_1 + 3x_2 + 6x_3 + 4x_4 + 7x_5 + 8x_6 + 2x_7 + 3x_8 + 2x_9 + 6x_{10} + 7x_{11} + 5x_{12} + 8x_{13} + 3x_{14} + 1x_{15} + 5x_{16} = 420$

$2x_1 + 4x_2 + 5x_3 + 3x_4 + 8x_5 + 9x_6 + 7x_7 + 2x_8 + 9x_9 + 4x_{10} + 5x_{11} + 7x_{12} + 1x_{13} + 2x_{14} + 6x_{15} + 6x_{16} = 469$

$5x_1 + 8x_2 + 6x_3 + 2x_4 + 1x_5 + 7x_6 + 8x_7 + 2x_8 + 3x_9 + 7x_{10} + 6x_{11} + 9x_{12} + 3x_{13} + 1x_{14} + 7x_{15} + 7x_{16} = 438$

$7x_1 + 7x_2 + 7x_3 + 1x_4 + 6x_5 + 6x_6 + 2x_7 + 7x_8 + 4x_9 + 2x_{10} + 3x_{11} + 8x_{12} + 8x_{13} + 9x_{14} + 6x_{15} + 5x_{16} = 450$

$1x_1 + 2x_2 + 1x_3 + 5x_4 + 5x_5 + 5x_6 + 7x_7 + 3x_8 + 5x_9 + 1x_{10} + 6x_{11} + 9x_{12} + 6x_{13} + 5x_{14} + 2x_{15} + 2x_{16} = 394$

$6x_1 + 1x_2 + 2x_3 + 6x_4 + 4x_5 + 2x_6 + 3x_7 + 1x_8 + 2x_9 + 5x_{10} + 1x_{11} + 6x_{12} + 3x_{13} + 6x_{14} + 5x_{15} + 4x_{16} = 287$

4) Develop a C program that can perform quadratic interpolation from three given points $\{(x_0, y_0), (x_1, y_1), (x_2, y_2)\}$ and a value of x to solve the unknown value of y from the equation:

$$y = a_0 + a_1 x + a_2 x^2$$

where:

$$a_0 = b_0 - b_1 x_0 + b_2 x_0 x_1$$
$$a_1 = b_1 - b_2 x_0 - b_2 x_1$$
$$a_2 = b_2$$

with the given points (x_0, y_0), (x_1, y_1), (x_2, y_2)
$$b_0 = y_0$$

$$b_1 = \frac{y_1 - y_0}{x_1 - x_0}$$

$$b_2 = \frac{\dfrac{y_2 - y_1}{x_2 - x_1} - \dfrac{y_1 - y_0}{x_1 - x_0}}{x_2 - x_0}$$

3

PHYSICAL PROPERTIES—PREDICTION AND APPROXIMATION

Graphical solutions are almost always used to characterize a behavior of a system. The physical properties of different materials are usually presented in textbooks in the form of graphs or tables. When we want to derive specific values from these graphs and tables and the information is not directly available, we use mathematical computations. The objective of this chapter is to apply what we learned in Chapter 2 to convert graphical solutions to mathematical equations that are easily incorporated into a computer program.

In this chapter, we will learn to create C programs involving:

Phase Equilibrium

Binary System
Antoine Equation
Benzene-Toluene Equilibrium System

Saturated Steam Data

Saturated Pressure
Density and Specific Volume
Enthalpy
Entropy

Humidity

Relative Humidity
Dew Point
Absolute Humidity

Duhring Line (BPE)

NaOH-Water System

Enthalpy-Concentration

NaOH-Water System

Friction Factor

Reynolds Number/Friction Factor Relationship

All program listings discussed in this chapter can be found on the CD-ROM.

\Program Listings\Chapter 3\PL3_1.c	Program Listing 3.1
\Program Listings\Chapter 3\PL3_2.c	Program Listing 3.2
\Program Listings\Chapter 3\PL3_3.c	Program Listing 3.3
\Program Listings\Chapter 3\PL3_4.c	Program Listing 3.4
\Program Listings\Chapter 3\PL3_5.c	Program Listing 3.5
\Program Listings\Chapter 3\PL3_6.c	Program Listing 3.6
\Program Listings\Chapter 3\benz_tol.h	Benzene-Toluene Header File
\Program Listings\Chapter 3\humidity.h	Humidity Header File
\Program Listings\Chapter 3\pipe.h	Pipe Header File
\Program Listings\Chapter 3\steam.h	Steam Header File

PHASE EQUILIBRIUM

Binary System

In a phase equilibrium system, data are presented in x (the liquid-phase composition) and y (the vapor-phase composition) coordinates. Taking the pressure constant, one variable (say x) can be changed independently and the other variable (the temperature and y) will follow. Plotting these data points will give us the equilibrium curve. For a binary mixture, the equilibrium relationship can be represented by a single curve.

Phase equilibrium information for a binary system is well represented through an *x-y* diagram as shown in Figure 3.1 for benzene-toluene. But to

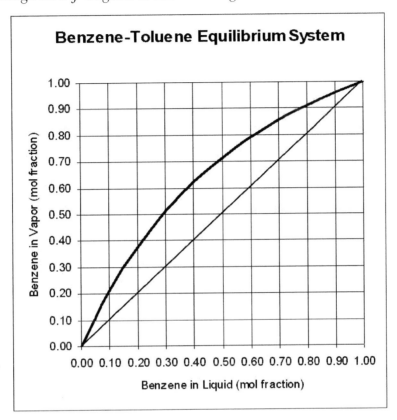

FIGURE 3.1: Benzene- toluene equilibrium curve.

construct such a diagram, we need to determine the different values of *y* for given values of *x*. One way is through the use of an Antoine equation.

Antoine Equation

The Antoine equation, a widely used equation for vapor pressure-temperature determination of a liquid-vapor phase equilibrium system, has the form:

$$\log p_a^\circ = A_a - \frac{B_a}{T + C_a}; \ \log p_b^\circ = A_b - \frac{B_b}{T + C_b}. \tag{3.1}$$

Considering we are dealing with a binary system, the subscripts a and b are the components. The partial pressure of a ($p_a^{\,\cdot}$) can be obtained given the mole fraction of component a in liquid (x).

$$p_a = x p_a^{\,\cdot}; \; p_b = (1\text{-}x)p_b^{\,\cdot}. \tag{3.2}$$

The mole fraction of component a in vapor (y) can then be computed through

$$y = \frac{p_a}{p_a + p_b}. \tag{3.3}$$

The algorithm for determining y with a given value of x is as follows:

- Establish a constant value for total pressure to which the x-y relation should be based.
- For a given value of x, assume a value of the temperature (T).
- Calculate $\log p^{\cdot}$ for each component $(a$ and $b)$ using Eq 3.1, then derive $p_a^{\,\cdot}$ and $p_b^{\,\cdot}$ by getting the anti-log of the two obtained values. Refer to Appendix I for the Antoine equation constants for the A, B, and C values of each substance.
- Then compute p_a and p_b using Eq 3.2.
- Get the sum of p_a and p_b and compare the result with the constant value of the total pressure. If it is not equal, assume another value of T. Repeat the process until the values converge.
- When the correct temperature (T) has finally been obtained, calculate y using Eq 3.3.
- The process can be repeated using other values of x to get the corresponding y values.

To further illustrate, let us consider the following problem.

Example Problem 3.1. Determine the value y in a benzene-toluene system with x taken at 0.1 intervals from 0.0 to 1.0. For each x value, find T (in K) to which $(p_a + p_b = 1.013 \text{ bar})$.

From Appendix I, we get the Antoine equation constants as:

Benzene (a) $A = 4.72583$, $B = 1660.652$, and $C = -1.461$
Toluene (b) $A = 4.08245$, $B = 1346.382$, and $C = -53.508$

$$\log p_a^{\,\cdot} = 4.72583 - \frac{1660.652}{T - 1.461}; \; \log p_b^{\,\cdot} = 4.08245 - \frac{1346.382}{T - 53.508}$$

Given the total pressure 1.013 bar (equivalent to 1 atm), for $x = 0$, we assume T to be 373 K. Substituting these into the above equation we get $\log p_a^{\circ} = 0.256$, $p_a^{\circ} = 10^{0.256} = 1.804$. Similarly, $\log p_b^{\circ} = -0.132$, $p_b^{\circ} = 10^{-0.132} = 0.738$.

For $x = 0.0$, $p_a = xp_a^{\circ} = (0.0)(1.804) = 0.0$, $p_b = (1-x)p_b^{\circ} = (1.0)(0.738) = 0.738$. We get the sum of the two partial pressures $(p_a + p_b) = 0.738$, which is still far from the 1.013 bar we want to obtain.

We increase the temperature to 383 K and repeat the process. Substituting the new temperature into the equation we get $\log p_a^{\circ} = 0.373$, $p_a^{\circ} = 10^{0.373} = 2.362$. Similarly, $\log p_b^{\circ} = -.004$, $p_b^{\circ} = 10^{-.004} = 0.991$. The new value for p_a is 0.0 and p_b is 0.991. We get the sum of the two partial pressures $(p_a + p_b) = 0.991$, which is now near the 1.013 bar.

Increasing the temperature further to 383.75 K, we get $(p_a + p_b) = 1.013$ bar, which is now equal to the base pressure. From here we derive the value of y at this temperature:

$$y = p_a/(p_a + p_b) = 0.0/(0.0 + 1.013) = 0.0.$$

Applying the same process with other values of x we get the following values of y.

Table 3.1: Tabulation of x and corresponding y values.

x	T	p_a°	p_b°	p_a	p_b	$(p_a + p_b)$	y
0.000	383.750	2.409	1.013	0.000	1.013	1.013	0.000
0.100	379.200	2.136	0.888	0.214	0.799	1.013	0.211
0.200	375.150	1.914	0.788	0.383	0.630	1.013	0.378
0.300	371.520	1.731	0.706	0.519	0.494	1.013	0.512
0.400	368.200	1.576	0.637	0.631	0.382	1.013	0.623
0.500	365.200	1.447	0.579	0.723	0.290	1.013	0.714
0.600	362.450	1.335	0.530	0.801	0.212	1.013	0.791
0.700	359.900	1.238	0.488	0.867	0.146	1.013	0.856
0.800	357.550	1.154	0.451	0.923	0.090	1.013	0.911
0.900	355.350	1.080	0.419	0.972	0.042	1.013	0.959
1.000	353.270	1.013	0.390	1.013	0.000	1.013	1.000

We see that the process is very tedious, and with an infinite value of x we cannot derive all values of y. However, an equation can be approximated to represent the relationship, and by using polynomial regression we can derive that equation.

Benzene-Toluene Equilibrium System

Example Problem 3.2. Determine the polynomial curve equation that best represents the benzene-toluene equilibrium diagram given the following mol fraction of benzene in liquid and vapor phase:

Table 3.2: Mol fraction of benzene in liquid and vapor phase.

Mol fraction of Benzene in Liquid (x)	Mol fraction of Benzene in Vapor (y)
0.000	0.000
0.100	0.211
0.200	0.378
0.300	0.512
0.400	0.623
0.500	0.714
0.600	0.791
0.700	0.856
0.800	0.911
0.900	0.959
1.000	1.000

Using the polynomial regression program (Program Listing 2.5) we developed, we take the 3rd, 4th, and 5th polynomial orders. The following results were obtained:

3rd order polynomial:

$$y = 0.004406 + 2.196888x - 1.882051x^2 + 0.684149x^3$$

4th order polynomial:

$$y = 0.000587 + 2.329464x - 2.544930x^2 + 1.744755x^3 - 0.530303x^4$$

5th order polynomial:

$$y = 0.000068 + 2.370166x - 2.880267x^2 + 2.689467x^3 - 1.612034x^4 + 0.432692x^5$$

Table 3.3: Values of y_e using different polynomial orders.

x	y (given)	y_e (3rd order)	y_e (4th order)	y_e (5th order)
0.000	0.000	0.004406	0.000587	0.000068
0.100	0.211	0.205958	0.209776	0.210815
0.200	0.378	0.373975	0.377792	0.377965
0.300	0.512	0.512560	0.513195	0.512503
0.400	0.623	0.625819	0.623272	0.622580
0.500	0.714	0.717856	0.714037	0.714037
0.600	0.791	0.792777	0.790230	0.790923
0.700	0.856	0.854686	0.855321	0.856014
0.800	0.911	0.907688	0.911505	0.911332
0.900	0.959	0.955889	0.959706	0.958667
1.000	1.000	1.003392	0.999573	1.000092

Ave	0.632273		
$\mathbf{S}_t$	1.058036		
$\mathbf{S}_r \mathbf{s}$			
$\mathbf{r^2}$	0.999885	0.999995	0.999999
r	0.999943	0.999997	1.000000
υ	7	6	5
$\mathbf{S}_{yx}$	0.004161	0.000947	0.000376

Substituting back the value of x in Table 3.2 we get the values of y as shown in Table 3.3.

To choose among the three equations that best fit the data points in Table 3.2, we determine the coefficient of determination r^2 by solving both S_t and S_r. Based on the result in Table 3.3, the 5th order polynomial gives the highest coefficient of determination $r^2 = 0.999999$ (closest to 1). Likewise, the smallest standard error of estimate is the 5th order polynomial, so this equation will be used to represent the benzene-toluene curve.

$$y = 0.000068 + 2.370166x - 2.880267x^2 + 2.689467x^3 - 1.612034x^4 + 0.432692x^5.$$

Consequently,

$$x = -0.000053 + 0.473581y - 0.276536y^2 + 1.749841y^3 - 2.133077y^4 + 1.185847y^5.$$

NOTE ▶ *We can derive the second equation by treating all values of y in Table 3.2 as x, and x as y, and run the regression program again.*

Develop the header file "`benz_tol.h`" using the 5th order polynomial equation:

```
#include <stdio.h>
#include <conio.h>
#include <math.h>
float benzenetoluene(float xy1,int xory)
{
float ans;
if(xory==2)
{
  ans=0.000068+(2.370166*xy1)-(2.880267*pow(xy1,2))
  +  (2.689467*pow(xy1,3))-(1.612034*pow(xy1,4))  +
  (0.432692*pow(xy1,5));
}
else
{
```

```
ans=-0.000053+(0.473581*xy1)-(0.276536*pow(xy1,2))+
(1.749841*pow(xy1,3))-(2.133077*pow(xy1,4))  +
(1.185847*pow(xy1,5));
}
return(ans);
}
```

Program Listing 3.1: Benzene-Toluene equilibrium system program

```
 1  #include <stdio.h>
 2  #include <stdlib.h>
 3  #include <conio.h>
 4  #include "benz_tol.h"
 5  int main()
 6  {
 7   char try;
 8   int lv;
 9   double val,res;
10   do{
11    do{
12    system("cls");
13    printf("This Program will provide data for Benzene-
           Toluene Equilibrium System\n");
14    printf("\nWhich fraction of benzene do you want to
           determine? \n1)Liquid 2)Vapor Select number: ");
15    scanf("%d",&lv);
16    }while(lv<1 || lv>2);
17   if(lv==1)
18   {
19    printf("\nEnter the mole fraction of vapor benzene: ");
20    scanf("%lf",&val);
21    res=benzenetoluene(val,lv);
22    printf("\nThe corresponding liquid fraction is %lf",
           res);
23   }
24   else
25   {
26    printf("\nEnter the mole fraction of liquid benzene: ");
27    scanf("%lf",&val);
28    res=benzenetoluene(val,lv);
29    printf("\nThe corresponding vapor fraction is
           %lf",res);
30   }
```

```
31  printf("\nDo you want to try again? (Y/N)");
32  try=getch();
33  }while(try=='y'||try=='Y');
34 return 0;
35 }
```

Line 4: The `benz_tol.h` header is linked in the program so that the `benzenetoluene()` function can be used in the program.

Line 14: This line asks the user to select which phase of the benzene mass fraction to consider.

Lines 17 to 30: Use the `if - else` to perform the corresponding task based on the user's choice in Line 14.

Lines 21, 28: Use the `benzenetoluene()` function to get the required information.

Lines 22, 29: The results are then printed on the screen.

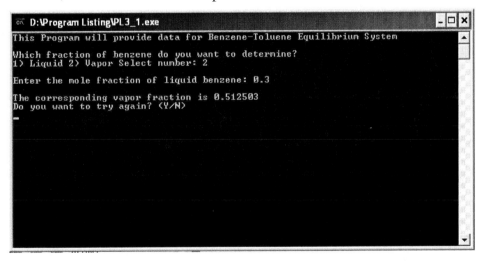

FIGURE 3.2: Output of Program Listing 3.1.

SATURATED STEAM DATA

There are many studies conducted on the thermodynamic properties of water and steam. In fact, through the years, different formulations have been introduced that approximate steam/water properties. Below are some of the recognized formulations:

- The International Association for Properties of Water and Steam Formulation for Industrial Use (IAPWS-IF97)

- The International Association for Properties of Water and Steam Formulation for General and Scientific Use (IAPWS-95)
- The National Bureau of Standards and National Research Council of Canada Formulation (NBS/NRC-84)
- Keenan, Keyes, Hill, and Moore "Thermodynamic Properties of Water Including Vapor, Liquid, and Solid Phase" (1969)
- The Industrial Formulation of 1967 (IFC-67)

In this section we will focus on the thermodynamic properties of saturated steam valid from 273.15 K to 647.14 K. We will also go over the different formulas indicated in the *Revised Supplementary Release on Saturation Properties of Ordinary Water Substance by IAPWS* (September 1992).

Given the reference constants:
$$T_c = 647.096 \text{ K}$$
$$p_c = 22064000 \text{ Pa}$$
$$\rho_c = 322 \text{ kg/m}^3$$
$$\alpha_0 = 1000 \text{ J/kg}$$
$$\varphi_0 = \alpha_0 / T_c$$
$$\theta = T/T_c$$
$$\tau = 1 - \theta$$

Saturated Pressure

At a given temperature T (in K), the saturated vapor pressure can be approximated as

$$p = p_c e^{\frac{T_c}{T}\left[a_1\tau + a_2\tau^{1.5} + a_3\tau^3 + a_4\tau^{3.5} + a_5\tau^4 + a_6\tau^{7.5}\right]}. \tag{3.4}$$

We get the derivative of the vapor pressure in relation to the saturation temperature:

$$\frac{dp}{dT} = (-p/T)\left[7.5a_6\tau^{6.5} + 4a_5\tau^3 + 3.5a_4\tau^{2.5} + 3a_3\tau^2 + 1.5a_2\tau^{0.5} + a_1 + \ln\left(p/p_c\right)\right] \tag{3.5}$$

With the following constants:

$a_1 =$	-7.85951783	$a_4 =$	22.6807411
$a_2 =$	1.84408259	$a_5 =$	-15.9618719
$a_3 =$	-11.7866497	$a_6 =$	1.80122502

Densities (ρ) and Specific Volumes (ν)

At a given temperature T, the densities of the saturated liquid can be taken as:

$$\rho' = \rho_c \left[1 + b_1 \tau^{1/3} + b_2 \tau^{2/3} + b_3 \tau^{5/3} + b_4 \tau^{16/3} + b_5 \tau^{43/3} + b_6 \tau^{110/3} \right]. \tag{3.6}$$

With the following constants:

$$
\begin{array}{llll}
b_1 = & 1.99274064 & b_4 = & -1.75493479 \\
b_2 = & 1.09965342 & b_5 = & -45.5170352 \\
b_3 = & -0.510839303 & b_6 = & -674694.450
\end{array}
$$

For saturated vapor:

$$\rho'' = \rho_c \, e^{\left[c_1 \tau^{1/3} + c_2 \tau^{2/3} + c_3 \tau^{4/3} + c_4 \tau^{3} + c_5 \tau^{37/6} + c_6 \tau^{71/6} \right]}. \tag{3.7}$$

Where the constants are as follows:

$$
\begin{array}{llll}
c_1 = & -2.03150240 & c_4 = & -17.2991605 \\
c_2 = & -2.68302940 & c_5 = & -44.7586581 \\
c_3 = & -5.38626492 & c_6 = & -63.9201063
\end{array}
$$

Specific Enthalpy

At a given T, the specific enthalpy of saturated liquid can be solved through

$$h' = \alpha + \frac{T}{\rho'} \frac{dp}{dT} \tag{3.8}$$

$$\alpha = \alpha_0 \left[d_\alpha + d_1 \theta^{-19} + d_2 \theta + d_3 \theta^{4.5} + d_4 \theta^{5} + d_5 \theta^{54.5} \right]. \tag{3.9}$$

$$
\begin{array}{llll}
d_1 = & -0.0000000565134998 & d_4 = & -135.003439 \\
d_2 = & 2690.66631 & d_5 = & 0.981825814 \\
d_3 = & 127.287297 & & \\
d_\alpha & -1135.905627715 & d_\varphi & 2319.5246
\end{array}
$$

And the specific enthalpy of saturated vapor can be taken as

$$h'' = \alpha + \frac{T}{\rho''}\frac{dp}{dT}.$$ (3.10)

Specific Entropy

At a given T, the specific entropy of saturated liquid can likewise be solved through

$$s' = \phi + \frac{1}{\rho'}\frac{dp}{dT},$$ (3.11)

where

$$\phi = \phi_0\left[d_\phi + \frac{19}{20}d_1\theta^{-20} + d_2\ln\theta + \frac{9}{7}d_3\theta^{3.5} + \frac{5}{4}d_4\theta^4 + \frac{109}{107}d_5\theta^{53.5}\right].$$ (3.12)

For saturated vapor, specific entropy can be derived as:

$$s'' = \phi + \frac{1}{\rho''}\frac{dp}{dT}.$$ (3.13)

Through these equations, we can now develop a computer program to determine the thermodynamic properties of steam given a certain temperature.

```
/* steam.h */
#include <stdio.h>
#include <math.h>
#include <stdlib.h>

#define Tc 647.096 /* K */
#define pc 22064000 /* Pa */
#define rho_c 322 /* kg/cu_m */
#define alpha_0 1000 /* J/kg */

double phi_0 = 1000/647.096;
double a[6]={-7.85951783, 1.84408259,
             -11.7866497, 22.6807411,
             -15.9618719, 1.80122502};
double b[6]={1.99274064, 1.09965342,
             -0.510839303, -1.75493479,
             -45.5170352, -674694.45};
```

```
double c[6]={-2.03150240, -2.68302940,
             -5.38626492, -17.2991605,
             -44.7586581, -63.9201063};
double d[5]={-0.0000000565134998,
             2690.66631, 127.287297,
             -135.003439, 0.981825814};
double d_alpha = -1135.905627715;
double d_phi = 2319.5246;

double sat_pres(double sat_temp)
{
 double ans, temp_var, theta, tau;
 theta = sat_temp/Tc;
 tau = 1 - theta;
 temp_var = (Tc/sat_temp)*(
             a[0]*tau +
             a[1]*pow(tau,1.5) +
             a[2]*pow(tau,3) +
             a[3]*pow(tau,3.5) +
             a[4]*pow(tau,4) +
             a[5]*pow(tau,7.5));
 ans = pc*exp(temp_var);
 return ans;
}

double rho_l(double sat_temp)
{
 double ans, temp_var, theta, tau, lntau;
 theta = sat_temp/Tc;
 tau = 1 - theta;
 lntau=log(tau);
 temp_var = 1 +
             b[0]*exp(lntau/3.0) +
             b[1]*exp(2.0*lntau/3.0) +
             b[2]*exp(5.0*lntau/3.0) +
             b[3]*exp(16.0*lntau/3.0) +
             b[4]*exp(43.0*lntau/3.0) +
             b[5]*exp(110.0*lntau/3.0);
 ans = rho_c* (temp_var*1.0);
 return ans;
}
```

```
double rho_v(double sat_temp)
{
 double ans, temp_var, theta, tau, lntau;
 theta = sat_temp/Tc;
 tau = 1 - theta;
 lntau=log(tau);
 temp_var = c[0]*exp(lntau/3.0) +
            c[1]*exp(2.0*lntau/3.0) +
            c[2]*exp(4.0*lntau/3.0) +
            c[3]*exp(3.0*lntau) +
            c[4]*exp(37.0*lntau/6.0) +
            c[5]*exp(71.0*lntau/6.0);
 ans = rho_c* exp(temp_var*1.0);
 return ans;
}


double sp_vol_l(double sat_temp)
{
 double ans;
 ans = 1/rho_l(sat_temp);
 return ans;
}


double sp_vol_v(double sat_temp)
{
 double ans;
 ans = 1/rho_v(sat_temp);
 return ans;
}


double alpha(double sat_temp)
{
 double ans, temp_var, theta;
 theta = sat_temp/Tc;
 temp_var = d_alpha +
            d[0]*pow(theta,-19) +
            d[1]*pow(theta,1) +
            d[2]*pow(theta,4.5) +
            d[3]*pow(theta,5) +
            d[4]*pow(theta,54.5);
 ans = alpha_0*temp_var;
```

```
       return ans;
   }

   double phi(double sat_temp)
   {
    double ans, temp_var, theta, lntheta;
    theta = sat_temp/Tc;
    lntheta = log(theta);
    temp_var = d_phi +
               ((19.0/20.0)*d[0]*exp(-20.0*lntheta)) +
               d[1]*lntheta +
               ((9.0/7.0)*d[2]*exp(3.5*lntheta)) +
               ((5.0/4.0)*d[3]*exp(4.0*lntheta)) +
               ((109.0/107.0)*d[4]*exp(53.5*lntheta));
    ans = phi_0*(temp_var*1.0);
    return ans;
   }

   double dp_dT(double pres, double sat_temp)
   {
    double ans, temp_var, theta, tau, ln_tau0,ln_tau1, ln_
    tau2;
    theta = sat_temp/Tc;
    tau = 1 - theta;
    ln_tau0 = 6.5*log(tau);
    ln_tau1 = 2.5*log(tau);
    ln_tau2 = 0.5*log(tau);
    temp_var = (7.5*a[5]*exp(ln_tau0)) +
               (4*a[4]*pow(tau,3)) +
               (3.5*a[3]*exp(ln_tau1)) +
               (3*a[2]*pow(tau,2)) +
               (1.5*a[1]*exp(ln_tau2)) +
               a[0]+
               (log(pres/pc));
    ans = -1*(pres/sat_temp)*temp_var;
    return ans;
   }

   double enthalpy_l(double sat_temp)
   {
    double ans, pres;
```

```
pres=sat_pres(sat_temp);
ans= alpha(sat_temp)+((sat_temp/rho_l(sat_temp))*dp_
dT(pres,sat_temp));
return ans;
}


double enthalpy_v(double sat_temp)
{
 double ans, pres;
 pres=sat_pres(sat_temp);
 ans= alpha(sat_temp)+((sat_temp/rho_v(sat_temp))*dp_
dT(pres,sat_temp));
 return ans;
}


double entropy_l(double sat_temp)
{
 double ans, pres;
 pres=sat_pres(sat_temp);
 ans= phi(sat_temp)+((1.0/rho_l(sat_temp))*dp_dT(pres,sat_
temp));
 return ans;
}


double entropy_v(double sat_temp)
{
 double ans, pres;
 pres=sat_pres(sat_temp);
 ans= phi(sat_temp)+((1.0/rho_v(sat_temp))*dp_dT(pres,sat_
temp));
 return ans;
}
```

Example Problem 3.3. Create a C program that will provide the different thermodynamic properties of water including:

- saturated pressure
- density (liquid and vapor)
- specific volume (liquid and vapor)
- enthalpy (liquid and vapor)
- entropy (liquid and vapor)

given a certain temperature. Test the program using the temperature 373.15 K.

Program Listing 3.2: Steam property

```
/*SteamProp.c*/
1   #include <stdio.h>
2   #include <math.h>
3   #include <stdlib.h>
4   #include "steam.h"
5   int main()
6   {
7   double temp,pres,dens_l,dens_v,vol_l,vol_v,h_l,h_v,s_
    l,s_v;
8   printf("Steam Property Program for 273.15K > T <
    647.14  K\n");
9   printf("Enter temperature (K): ");
10  scanf("%lf",&temp);
11  pres = sat_pres(temp);
12  dens_l = rho_l(temp);
13  dens_v = rho_v(temp);
14  vol_l = sp_vol_l(temp);
15  vol_v = sp_vol_v(temp);
16  h_l = enthalpy_l(temp);
17  h_v = enthalpy_v(temp);
18  s_l = entropy_l(temp);
19  s_v = entropy_v(temp);
20  printf("\n The saturated pressure is %lf Pa\n",
    pres);
21  printf(" The density (liquid) is %lf kg/cu m\n",
    dens_l);
22  printf(" The density (vapor) is %lf kg/ cu m\n",
    dens_v);
23  printf(" The specific volume (liquid) is %lf cu m/kg\
    n", vol_l);
24  printf(" The specific volume (vapor) is %lf cu m/kg\n",
    vol_v);
25  printf(" The enthalpy (liquid) is %lf J/kg\n", h_l);
26  printf(" The enthalpy (vapor) is %lf J/kg\n", h_v);
27  printf(" The entropy (liquid) is %lf J/kg-K\n", s_l);
28  printf(" The entropy (vapor) is %lf J/kg-K\n", s_
    v);
29  system("pause");
```

```
30  return 0;
31 }
```

Line 4: The `steam.h` header file is linked in the program so that the `sat_pres()`, `rho_l()`, `rho_v()`, `sp_vol_l()`, `sp_vol_v()`, `h_l()`, `h_v()`, `s_l()`, and `s_v()` functions can be used in the program.

Lines 9 and 10: Ask the user for the temperature.

Lines 11 to 19: Call the functions in `steam.h` to determine the different thermodynamic properties of water at the given temperature.

Lines 20 to 28: Print out the different values.

FIGURE 3.3:   Output of Program Listing 3.2.

The Saturated Steam Pressure-Temperature Relationship

There are engineering problems that require determination of the saturated temperature, given the saturated pressure of steam. In the previous section, we learned how to compute for the pressure given the temperature; reversing the formula will be a complicated task. However, an easier way is to get the pressure-temperature relationship through regression.

Plotting the different values of the saturated temperature against the saturated vapor, we get

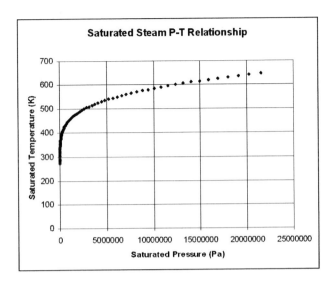

FIGURE 3.4: Saturated pressure -temperature relationship.

By doing regression analysis, the equation for the saturated steam P-T relationship can be established per section to obtain a high coefficient of determination.

Table 3.4: Saturated steam P-T relationship.

Pressure Range (Pa)	3rd Order Polynomial Equation (Temp in K)
615 - 2250	$y = 2.806712E\text{-}9\ x^3 - 1.54137E\text{-}5\ x^2 + 3.718675E\text{-}3\ x + 255.555589$
2251 - 12500	$y = 2.530233E\text{-}11\ x^3 - 7.262444E\text{-}7\ x^2 + 9.024094E\text{-}3\ x + 275.9391$
12501 - 68500	$y = 1.687943E\text{-}13\ x^3 - 2.8014759E\text{-}8\ x^2 + 2.011675E\text{-}3\ x + 302.6508$
68501 - 355000	$y = 1.348662E\text{-}15\ x^3 - 1.208023E\text{-}9\ x^2 + 4.775155E\text{-}4\ x + 335.5453$
355001 - 1700000	$y = 1.438385E\text{-}17\ x^3 - 6.381275E\text{-}11\ x^2 + 1.273902E\text{-}4\ x + 375.2343$
1700001 - 7350000	$y = 2.096919E\text{-}19\ x^3 - 4.182417E\text{-}12\ x^2 + 3.830535E\text{-}5\ x + 423.9684$
7350001 - 21500000	$y = 6.59654E\text{-}21\ x^3 - 4.54657E\text{-}13\ x^2 + 1.4498245E\text{-}5\ x + 477.9405$

A `sat_p2t()` function can then be developed and included in the `steam.h` header file.

```
double  sat_p2t(double pres)
{
```

```
double ans;
if(pres >614 && pres <=2250)
  ans = (2.806712*exp(-9*log(10))* pow(pres,3)) -
  (1.54137*exp(-5*log(10))*pow(pres,2)) + (3.718675*exp(-
  2*log(10))*pres) + 255.555589;
else if(pres >2250 && pres <=12500)
  ans = (2.530233*exp(-11*log(10))* pow(pres,3))
  - (7.262444*exp(-7*log(10))*pow(pres,2)) +
  (9.024094*exp(-3*log(10))*pres) + 275.9391;
else if(pres >12500 && pres <=68500)
  ans = (1.687943*exp(-13*log(10))* pow(pres,3))
  - (2.8014759*exp(-8*log(10))*pow(pres,2)) +
  (2.011675*exp(-3*log(10))*pres) + 302.6508;
else if(pres >68500 && pres <=355000)
  ans = (1.348662*exp(-15*log(10))* pow(pres,3))
  - (1.208023*exp(-9*log(10))*pow(pres,2)) +
  (4.775155*exp(-4*log(10))*pres) + 335.5453;
else if(pres >355000 && pres <=1700000)
  ans = (1.438385*exp(-17*log(10))* pow(pres,3))
  - (6.381275*exp(-11*log(10))*pow(pres,2)) +
  (1.273902*exp(-4*log(10))*pres) + 375.2343;
else if(pres >1700000 && pres <=7350000)
  ans = (2.096919*exp(-19*log(10))* pow(pres,3))
  - (4.182417*exp(-12*log(10))*pow(pres,2)) +
  (3.830535*exp(-5*log(10))*pres) + 423.9684;
else if(pres >7350000 && pres <=22070000)
  ans = (6.59654*exp(-21*log(10))* pow(pres,3))
  - (4.54657*exp(-13*log(10))*pow(pres,2)) +
  (1.4498245*exp(-5*log(10))*pres) + 477.9405;
else ans=0;
return ans;
}
```

HUMIDITY

Relative Humidity

For an air-water system, relative humidity (*%RH*) is defined as the ratio of the pressure exerted by the water vapor to the vapor pressure of the liquid water taken at that temperature (similar to air). Generally, relative humidity is expressed in terms of percentage.

Relative humidity, $\%RH = 100\% \dfrac{p_w}{P_w}$,   (3.14)

where p_w is the partial pressure of water vapor and P_w is the vapor pressure of water at the given temperature.

This can also be expressed as

$\%RH = 100\% \dfrac{p_{actual}}{p_{saturation}}$,   (3.15)

where p_{actual} is the actual vapor pressure and $p_{saturation}$ is the saturation vapor pressure at the given temperature. The $p_{saturation}$ can be determined using Eq 3.4.

Absolute Humidity

Similarly, for an air-water system, absolute humidity (H) is likewise defined as the mass of water vapor present per unit mass of vapor-free air. It depends solely on the pressure exerted by the vapor in the water vapor-air mixture when the total pressure (sum of the partial pressure of vapor and air) is constant.

Absolute humidity, $H = \dfrac{M_w p_w}{M_a p_a} = \dfrac{M_w p_a}{M_a (P_T - p_a)}$,   (3.16)

where M_a and M_w are mol fractions of air and water vapor, respectively, while p_a and p_w are partial pressures of air and water vapor, respectively. P_T is the total pressure and can be taken as atmospheric pressure.

Another way to compute for H is through the Ideal Gas Law equation. Basically, H can also be expressed as the density of water vapor in air. So, by using the Ideal Gas Law equation, we can present H through:

$H = {p_{actual}}\Big/{TR_w}$,   (3.17)

where $R_w = 461.512244565$ J/kg-K.

Dew Point (DP)

Dew point (DP) is the temperature at which the water vapor in a water-air mixture is saturated. It can be determined by cooling down the vapor-air mixture to the point where water vapor starts to condense, forming mist. That temperature will be the dew point.

To determine the dew point, we need to determine the saturation temperature of the given actual vapor pressure p_{actual}. To do that we can use the polynomial regression equation established for the saturated steam pressure-temperature relationship (see Figure 3.4).

With all these equations, we can now generate the program listing for solving the actual vapor pressure p_{actual}, absolute humidity H, and the dew point DP, given the relative humidity and the temperature (dry bulb).

```
/*humidity.h*/
#include <stdio.h>
#include <math.h>
#include "steam.h"

#define Rw 461.512244565
#define absT 273.15
double act_P(double RH, double Tdry)
{
   double ans;
   Tdry=Tdry+absT;
   ans =sat_pres(Tdry);
   ans =(RH/100)*ans;
  return ans;
}


double absH(double RH, double Tdry)
{
   double ans, p_act;
   Tdry=Tdry+absT;
   p_act =sat_pres(Tdry);
   p_act = (RH/100)*p_act;
   ans = p_act*1000/((Tdry)*Rw);
   return ans;
}


double dewpoint( double RH, double Tdry)
{
   double ans, p_act;
   Tdry=Tdry+absT;
   p_act =sat_pres(Tdry);
   p_act = (RH/100)*p_act;
   ans = sat_p2t(p_act);
```

```
        return (ans-absT);
}
```

Program Listing 3.3: Humidity program

```
1  #include <stdio.h>
2  #include <stdlib.h>
3  #include <conio.h>
4  #include "humidity.h"
5  int main()
6  {
7  double relhum, temp, pres,dp,H;
8  char res;
9  do
10 {
11  printf("Humidity Calculator\n");
12  printf("Enter percent relative humidity : ");
13  scanf("%lf",&relhum);
14  printf("Enter temperature (deg C) : ");
15  scanf("%lf",&temp);
16  pres = act_P(relhum,temp);
17  H = absH(relhum,temp);
18  dp = dewpoint(relhum,temp);
19  printf("\nActual vapor pressure : %lf Pa\n",pres);
20  printf("Absolute humidity : %lf g/cu m\n",H);
21  printf("Dew point : %f deg C\n",dp);
22  printf("\nDo you want to try again? (Y/N)\n");
23  res = getch();
24  system("cls");
25 }while (res =='y' || res == 'Y');
26  return 0;
27 }
```

Line 4: The humidity.h header file is linked in the program so that the act_P(), absH(), and dewpoint() functions can be used in the program.

Lines 12, 15: Request the relative humidity and temperature (dry bulb).

Lines 16 to 18: Call the functions in humidity.h to determine the actual pressure, humidity, and dew point.

Lines 19 to 22: Print out the different values.

Table 3.5 shows the summary of the output of Program Listing 3.3. This table indicates the absolute humidity (H) in g/m3 (upper row) and the dew point temperature (DP) of the air in oC (lower row) as a function of relative humidity (%RH).

FIGURE 3.5:   Output of Program Listing 3.3.

Table 3.5: Absolute humidity (H) and dew point (DP).

Relative Humidity (%RH)		10%	30%	50%	70%	90%
Dry Bulb [°C]						
50	H, g/m^3	8.28	24.85	41.41	57.98	74.54
	DP, °C	10.11	27.54	36.78	42.88	48.12
45	H, g/m^3	6.53	19.60	32.67	45.74	58.81
	DP, °C	6.38	23.35	32.16	38.30	42.85
40	H, g/m^3	5.11	15.33	25.55	35.77	46.00
	DP, °C	2.59	19.66	27.48	33.53	38.11
35	H, g/m^3		11.87	19.79	27.71	35.62
	DP, °C		14.76	23.00	28.62	33.16
30	H, g/m^3		9.11	15.18	21.25	27.32
	DP, °C		10.57	18.74	23.86	28.08
25	H, g/m^3		6.91	11.52	16.13	20.73
	DP, °C		6.24	13.80	19.69	23.21
20	H, g/m^3		5.19	8.65	12.10	15.56
	DP, °C		1.88	9.30	14.29	18.57
15	H, g/m^3			6.41	8.98	11.54
	DP, °C			4.65	9.61	13.32
10	H, g/m^3			4.70	6.58	8.46
	DP, °C			0.08	4.77	8.47
5	H, g/m^3					6.12
	DP, °C					3.46

Note that due to the program limitation, any value of a dew point lower than 0 can not be generated.

DUHRING'S RULE (BOILING POINT ELEVATION)

Generally, the vapor pressure of an aqueous solution is less than pure water at the same temperature. Therefore, for a given pressure, the boiling point of the solution is higher than water. The increase in the boiling point over that of water is known as the Boiling Point Elevation (BPE) of the solution.

Actually, there is no straightforward method of predicting the extent of the BPE in concentrated solutions. Many solutions have their boiling points at some concentrations tabulated in books, and these can be extended by the use of a relationship known as **Duhring's rule**. This rule states that the boiling point of a given solution is a linear function of the boiling point of pure water at the same pressure. The linear relationship as shown in the plotted diagram is called Duhring line.

Thus, if we take the vapor pressure-temperature relation of a reference liquid, usually water, and if we know two points on the vapor pressure–temperature curve of the solution that is being evaporated, the boiling points of the solution to be evaporated at various pressures can be obtained from a **Duhring line** diagram. The Duhring line will give the boiling point of solutions of various concentrations and pressures by interpolating along the line at constant concentration.

NaOH – Water System

In order to convert the diagram for a NaOH–water system—which is found mostly in textbooks—into a computer program, we first estimate the equation of the line to represent each given mass fraction in the diagram. From there, we use two pairs of x and y coordinates and the linear regression program (Program Listing 2.4) to generate the line equation. Table 3.6 shows the different linear equations that represent the boiling point of solutions based on the mass fraction of NaOH.

Table 3.6: NaOH concentration, boiling point data.

Mass Fraction, NaOH	(x_1,y_1)	(x_2,y_2)	Line Equation Generated
0.00	(50,50)	(350,350)	$y = x$
0.10	(50,55)	(350,355)	$y = 5 + x$
0.20	(50,60)	(350,370)	$y = 8.33 + 1.033x$
0.30	(50,70)	(350,382)	$y = 18 + 1.04x$
0.35	(50,80)	(350,392)	$y = 28 + 1.04x$
0.40	(50,90)	(350,408)	$y = 37 + 1.05x$

0.45		(50,110)	(350,425)	$y = 57.5 + 1.06x$
0.50		(50,117)	(350,438)	$y = 63.5 + 1.07x$
0.60		(50,140)	(325,440)	$y = 85.45 + 1.09x$
0.65		(50,160)	(300,435)	$y = 105 + 1.1x$
0.70		(50,173)	(250,400)	$y = 116.25 + 1.135x$

Data points were approximated and averaged from the Duhring line graph (NaOH–water system) from various sources. These are not results of a laboratory study and therefore may be inaccurate.

Program Listing 3.4 implements a linear interpolation routine for concentrations not specified in Table 3.6.

Program Listing 3.4:.NaOH–water Duhring line program.

```c
#include <stdio.h>
#include <stdlib.h>
#include <conio.h>
int main()
{
 double x[2],y[2],ansy,valx,boilwater,conc;
 char res;
 do{
  ansy=0;res='n';
  system("cls");
  printf(«DUHRING LINE Program for NaOH - Water System»);
  printf(«\n\nEnter boiling temperature of water in deg
  F: «);
  scanf(«%lf»,&boilwater);
  do{
   printf(«\nEnter mass fraction of NaOH in solution: «);
   scanf(«%lf»,&conc);
  }while (conc>1);
  if(conc <= 0.1)
  {
   x[0]=0;x[1]=0.1;
   y[0]=boilwater;y[1]=5+boilwater;
  }
  else if(conc <=0.2 && conc >0.1)
  {
   x[0]=.1;x[1]=0.2;
   y[0]=5+boilwater;y[1]=8.33+(1.033*boilwater);
  }
```

```
else if(conc <=0.3 && conc >0.2)
{
 x[0]=.2;x[1]=0.3;
 y[1]=18+(1.04*boilwater);y[0]=8.33+(1.033*boilwater);
}
else if(conc <=0.35 && conc >0.3)
{
 x[0]=.3;x[1]=0.35;
 y[0]=18+(1.04*boilwater);y[1]=28+(1.04*boilwater);
}
else if(conc <=0.4 && conc >0.35)
{
 x[0]=.35;x[1]=0.4;
 y[1]=37+(1.06*boilwater);y[0]=28+(1.04*boilwater);
}
else if(conc <=0.45 && conc >0.4)
{
 x[0]=.4;x[1]=0.45;
 y[0]=37+(1.06*boilwater);y[1]=57.5+(1.05*boilwater);
}
else if(conc <=0.5 && conc >0.45)
{
 x[0]=.45;x[1]=0.5;
 y[1]=63.5+(1.07*boilwater);y[0]=57.5+(1.05*boilwater);
}
else if(conc <=0.6 && conc >0.5)
{
 x[0]=.5;x[1]=0.6;
 y[0]=63.5+(1.07*boilwater);y[1]=85.45+(1.09*boilwater);
}
else if(conc <=0.65 && conc >0.6)
{
 x[0]=.6;x[1]=0.65;
 y[1]=105+(1.1*boilwater);y[0]=85.45+(1.09*boilwater);
}
else
{
 x[0]=.65;x[1]=0.7;
 y[0]=105+(1.1*boilwater);y[1]=116.25+(1.135*boilwater);
}
valx=conc;
```

```
ansy = y[0] + (((y[1]-y[0])/(x[1]-x[0]))*(valx-
x[0]));
printf(«\nThe boiling temp of NaOH Solution: %lf\
n»,ansy);
printf(«\nTry again (y/n)?\n»);
res=getch();
}while (res=='Y' || res=='y');
return 0;
}
```

Example Problem 3.4. Using Program Listing 3.4, determine the boiling point of the NaOH solution with 40% concentration where the boiling point of the solvent water is 150°F.

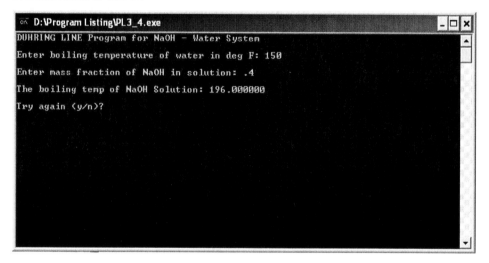

FIGURE 3.6:   Output of Program Listing 3.4.

ENTHALPY-CONCENTRATION

NaOH–Water System

Given the different enthalpy–concentration data for the NaOH water system in Table 3.7, we can easily convert this into a corresponding equation using polynomial regression (5th order).

Table 3.7: NaOH concentration-enthalpy.

Mass Fraction	Enthalpy (H) at Temperature						
	40°F	**50°F**	**100°F**	**150°F**	**200°F**	**300°F**	**400°F**
0.0	8	18	67	118	167	268	375
0.1	6	15.5	61	107	152.5	244	335
0.2	5	14	57	101	144	232	322
0.3	13	21	64	106.5	151	237	324
0.4	42	53	93	133	177	260	344
0.5	92.5	103	140	182	223	300	375
0.6	163	175	210	243	285	352	422
0.7	250	257	281	314	352	415	480

Data points were approximated (some were extrapolated) and averaged from enthalpy-concentration diagrams (NaOH water system) from various sources. These are not results of a laboratory study and therefore may be inaccurate.

At temperature = 40°F:
$$y = 7.931235 + 39.015734x - 847.741842x^2 + 3216.127623x^3 - 2093.531469x^4 + 176.282052x^5$$

At temperature = 50°F:
$$y = 17.941725 + 39.262821x - 923.615968x^2 + 3445.148602x^3 - 2099.358975x^4 - 80.128205x^5$$

At temperature = 100°F:
$$y = 67.042541 - 46.906177x - 258.406177x^2 + 890.297203x^3 + 2145.979021x^4 - 2660.256410x^5$$

At temperature = 150°F:
$$y = 117.885781 - 61.953963x - 748.120630x^2 + 3657.415502x^3 - 3116.258742x^4 + 592.948718x^5$$

At temperature = 200°F:
$$y = 167.022727 - 144.000000x - 218.560606x^2 + 2035.037879x^3 - 833.333333x^4 - 625.000000x^5$$

At temperature = 300°F:
$$y = 267.981352 - 263.466200x + 15.442890x^2 + 2624.417250x^3 - 3432.400932x^4 + 1474.358974x^5$$

At temperature = 400°F:
$$y = 374.96212 - 571.931818x + 2048.295454x^2 - 3414.772728x^3 + 4261.363637x^4 - 2083.333334x^5$$

Program Listing 3.5: Enthalpy-concentration program for NaOH solution.

```
#include <stdio.h>
#include <conio.h>
#include <stdlib.h>
#include <math.h>
int main()
{
 double x[2],y[2],ansy,valx,tempsol,conc;
 char res;
 do{
  system(«cls»);
  ansy=0;res='n';
  printf(«ENTHALPY-CONCENTRATION Program for NaOH - Water
  System»);
  do{
   printf(«\n\nEnter temperature of solution in deg F
   (Min 40F): «);
   scanf(«%lf»,&tempsol);
   printf(«\nEnter mass fraction of NaOH in solution: «);
   scanf("%lf",&conc);
  }while (conc >1 || tempsol <40);
  if(tempsol <=50)
  {
   x[0]=40;y[0]=7.931235+(39.015734*conc)-(847.741842*
   pow(conc,2))+(3216.127623*pow(conc,3))-(2093.531469*
   pow(conc,4))+(176.282052*pow(conc,5));
   x[1]=50;y[1]=17.941725+(39.262821*conc)-(923.615968*
   pow(conc,2))+(3445.148602*pow(conc,3))-(2099.358975*p
   ow(conc,4))-(80.128205*pow(conc,5));
  }
  else if(tempsol <=100 && tempsol >50)
  {
   x[1]=100;y[1]=67.042541-(46.906177*conc)-(258.406177*
   pow(conc,2))+(890.297203*pow(conc,3))+(2145.979021*
   pow(conc,4))-(2660.256410*pow(conc,5));
   x[0]=50;y[0]=17.941725+(39.262821*conc)-(923.615968*
   pow(conc,2))+(3445.148602*pow(conc,3))-(2099.358975*p
   ow(conc,4))-(80.128205*pow(conc,5));   }
  else if(tempsol <=150 && tempsol >100)
  {
   x[0]=100;y[0]=67.042541-(46.906177*conc)-(258.406177*
```

```
pow(conc,2))+(890.297203*pow(conc,3))+(2145.979021*
pow(conc,4))-(2660.256410*pow(conc,5));
x[1]=150;y[1]=117.885781-(61.953963*conc)-(748.120630*
pow(conc,2))+(3657.415502*pow(conc,3))-(3116.258742*
pow(conc,4))+(592.948718*pow(conc,5));
}
else if(tempsol <=200 && tempsol>150)
{
 x[1]=200;y[1]=167.022727-(144.000000*conc)-(218
 .560606*pow(conc,2))+(2035.037879*pow(conc,3))-
 (833.333333*pow(conc,4))-(625.000000*pow(conc,5));
 x[0]=150;y[0]=117.885781-(61.953963*conc)-(748.120630*
 pow(conc,2))+(3657.415502*pow(conc,3))-(3116.258742*
 pow(conc,4))+(592.948718*pow(conc,5));
}
else if(tempsol <=300 && tempsol>200)
{
 x[0]=200;y[0]=167.022727-(144.000000*conc)-(218
 .560606*pow(conc,2))+(2035.037879*pow(conc,3))-
 (833.333333*pow(conc,4))-(625.000000*pow(conc,5));
 x[1]=300;y[1]=267.981352-(263.466200*conc)+(15.442890
 *pow(conc,2))+(2624.417250*pow(conc,3))-(3432.400932*
 pow(conc,4))+(1474.358974*pow(conc,5));
}
else
{
 x[1]=400;y[1]=374.962121-(571.931818*conc)+(2048.2954
 54* pow(conc,2))-(3414.772728*pow(conc,3))+(4261.3636
 37*pow(conc,4))-(2083.333334*pow(conc,5));
 x[0]=300;y[0]=267.981352-(263.466200*conc)+(15.442890
 *pow(conc,2))+(2624.417250*pow(conc,3))-(3432.400932*
 pow(conc,4))+(1474.358974*pow(conc,5));
}
valx=tempsol;
ansy = y[0] + (((y[1]-y[0])/(x[1]-x[0]))*(valx-x[0]));
printf(«\nThe Enthalpy of NaOH Solution: %lf Btu/lb\
n»,ansy);
printf(«\nTry again (y/n)?\n»);
res=getch();
}while (res=='Y' || res=='y');
return 0;
}
```

Example Problem 3.5 Determine the enthalpy of a 50% concentration NaOH-water solution at 160oF.

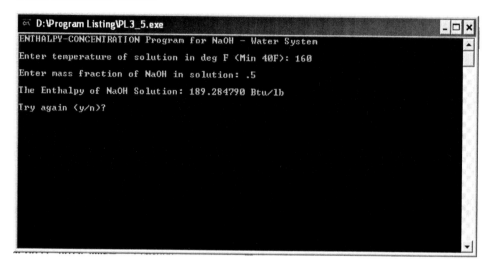

```
D:\Program Listing\PL3_5.exe
ENTHALPY-CONCENTRATION Program for NaOH - Water System
Enter temperature of solution in deg F (Min 40F): 160
Enter mass fraction of NaOH in solution: .5
The Enthalpy of NaOH Solution: 189.284790 Btu/lb
Try again (y/n)?
```

FIGURE 3.7:   Output of Program Listing 3.5.

FRICTION FACTOR

One of the common parameters we encounter when dealing with flow of fluid is the friction factor f. In both turbulent and laminar flow, the friction factor depends on the Reynolds number N_{Re}, which describes the nature of the flow. In the previous discussion, we mentioned that the Reynolds number can be computed as

$$N_{Re} = \frac{Dv\rho}{\mu},$$

(3.18)

where:

D = diameter of the circular pipe
v = velocity of fluid
ρ = density of fluid
μ = viscosity of fluid

In addition to the Reynolds number, the friction factor for turbulent flow is also dependent on the surface roughness of the pipe. However, in laminar flow the roughness has no significant effect. Many literary sources provide large numbers of experimental data on friction factors for smooth tubes and for pipes with varying degrees of roughness.

Once the friction factor is known, other information can be obtained. In both laminar and turbulent flow, the frictional pressure drop Δp_f can be derived through

$$\Delta p_f = 2 f \rho \frac{Lv^2}{Dg_c},$$ (3.19)

where

f = friction factor (dimensionless)
L = pipe length
g_c = proportionality constant, 1 $(kg)(m)/(N)(sec^2)$ or 32.174 $(lb_m)(ft)/(lb_f)(sec^2)$

Also, friction loss F_f can be determined through an empirical expression known as the Fanning equation:

$$F_f = \frac{\Delta p_f}{\rho} = 2 f \frac{Lv^2}{Dg_c}.$$ (3.20)

The values of these parameters are quite important in solving fluid flow problems.

Reynolds Number/Friction Factor Relationship

Example Problem 3.6. Determine an empirical equation to represent the relationship between the Reynolds number and the friction factor for a smooth pipe given the following data:

Table 3.8: Reynolds number (N_{Re}) –friction factor (f) relationship.

N_{Re}	f
8500	0.008
20000	0.0065
30000	0.006
60000	0.005
700000	0.003
1000000	0.0028
10000000	0.002

The N_{Re} –f can be best represented using a log-log plot. With the data provided in Table 3.8, we construct such a diagram where N_{Re} represents the x-axis and f the y-axis.

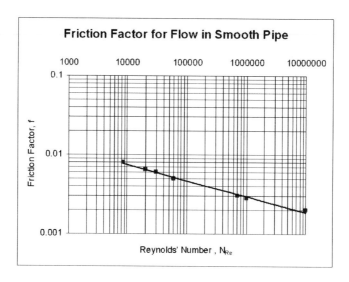

FIGURE 3.8: Friction factor for flow in smooth pipe.

We can see that at logarithmic scale, the x-y can be represented by a straight line. With this, we can use linear regression to determine the equation given both logarithmic values of x and y.

Table 3.9: Logarithmic values of N_{Re} and f.

Log (N_{Re})	Log (f)
3.929419	-2.09691
4.30103	-2.18709
4.477121	-2.22185
4.778151	-2.30103
5.845098	-2.52288
6	-2.55284
7	-2.69897

Using the Linear Regression Program (Program Listing 2.4), we get

$\log y = -1.330166 - 0.200117 \log x$.
From the anti-log, we obtain $y = 0.04675 \, x^{-0.20012}$. So the empirical formula to represent the $N_{Re} - f$ relationship for smooth pipe can be written as

$$f = 0.04675 \, N_{Re}^{-0.20012}. \tag{3.21}$$

 The regression method using the log values of x and y can also be referred to as a power equation (see Eq 2.12).

Through this equation, we can develop a function for such a relationship and store it in the `pipe.h` header file.

```
/*pipe.h*/
#include <stdio.h>
#include <math.h>

double friction(double Nre)
   {
   double ans;
   ans = 0.04675*exp(-0.20012*log(Nre));
   return(ans);
}
```

Program Listing 3.6: Friction factor program.

```
1   #include <stdio.h>
2   #include <stdlib.h>
3   #include <conio.h>
4   #include "pipe.h"
5   int main()
6   {
7     char try;
8     double Nre, f;
9     do{
10     system("cls");;
11     printf("Friction Factor for smooth pipe.\n");
12     printf("\nEnter the Reynolds Number ( >2100 ): ");
13     scanf("%lf",&Nre);
14     f=friction(Nre);
15     printf("\nThe friction factor is %lf\n",f);
16     printf("\nDo you want to try again? (Y/N)\n");
17     try=getch();
18   }while(try=='y'||try=='Y');
19   return 0;
20 }
```

Line 4: The `pipe.h` header is linked in the program so that the `friction()` function can be used in the program.

Line 12: Requests the Reynolds number (must be greater than 2100 to be considered non-laminar).

Line 14: Uses the `friction()` function to provide the equivalent friction factor for the given Reynolds number.

Line 15: Prints out the value of the friction factor on screen.

Example Problem 3.6. Determine the friction factor for flow in smooth pipe with a Reynolds number that equals 2×10^5.

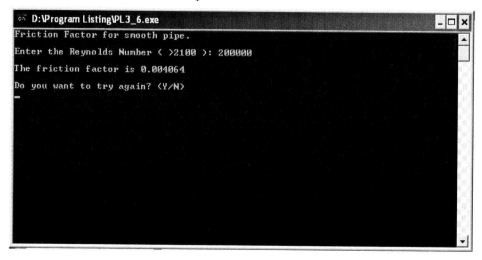

FIGURE 3.9:   Output of Program Listing 3.6.

LABORATORY EXERCISES

1) Create a computer program that can provide the boiling temperature of a sodium chloride (NaCl) salt solution given the mass concentration of salt and the boiling point of water (in °F and K). Use the chart provided below.

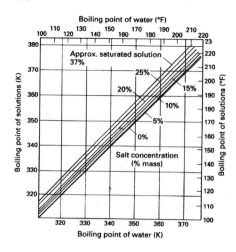

FIGURE 3E.1:   Duhring lines – NaCl water solution.

Reprint from *Chemical Engineering* Vol. 2 Issue 4, J.M. Coulson and J.F. Richardson, *Particle Technology and Separation Process*, Page No. 624, Copyright (1991), with permission from Elsevier, Inc..

2) Create a C program that will calculate the saturation vapor pressure (p_{sat}), the actual vapor pressure (p_{act}), the dew point (dp), and the absolute humidity (H), given the relative humidity $(\%RH)$ and the temperature (T) using the equations from Jensen et al. (1990):

$$p_{sat} = 0.611e^{(17.27T/(T+237.3))} \qquad \text{in kPa}$$
$$p_{act} = p_{sat} * (\%RH/100) \qquad \text{in kPa}$$
$$dp = (116.9+237.3*\ln(p_{act}))/(16.78-\ln(p_{act})) \qquad \text{in } ^\circ C$$
$$H = (p_{act}*1000)/((T+273.15)*R_w) \qquad \text{in kg/m}^3$$

where R_w (the gas constant for water vapor) = 461.512244565J/kg-K

3) Determine the friction factor equation for laminar flow, given the following data:

N_{Re}	f
100	0.16
200	0.08
400	0.04
800	0.02
1600	0.01
2000	0.008

Use linear regression analysis in deriving the equation.

4) Complete the missing values of y using the Antoine equation and determine the polynomial curve equation that can best represent the methanol-water equilibrium diagram given the following mol fraction of methanol in liquid (x) and vapor phase (y). Validate the derived equation by acquiring the coefficient of determination and standard error of estimate.

Mol fraction of Methanol in Liquid (x)	Mol fraction of Methanol in Vapor (y)
0	0.000
0.10	0.285
0.20	0.478
0.30	0.616
0.40	0.718
0.50	-
0.60	-
0.70	-

0.80   -
0.90   -
1.0   -

5) Develop a new function `sat_pres2()`, which can provide the pressure of saturated steam given the saturation temperature. But instead of using Eq 3.4, use polynomial regression in establishing the equation and validate through statistical analysis.

4 APPLICATIONS USING C

This chapter presents the core of Part I of this book. So far, we have reviewed the basics of the C language, learned numerical computation, and prepared functions and header files that predict physical properties of materials. In this chapter, we will use all of our experience from the previous activities to solve several chemical engineering problems we have encountered.

In this chapter we will develop complex code to solve tedious problems involving:

Material Balance
  Multiple Reactors
  Condensation
Energy Balance
  Multiple Effect Evaporators
Fluid Flow
  Pressure Drop Determination
  Pipe Diameter Calculation
Mass and Heat Transfer
  Temperature Determination in Two-dimensional Conduction
  Evaporation
  Distillation

Optimization
Process Design
Plant Operation

All program listings discussed in this chapter can be found on the CD-ROM.

\Program Listings\Chapter 4\PL4_1.c	Program Listing 4.1
\Program Listings\Chapter 4\PL4_2.c	Program Listing 4.2
\Program Listings\Chapter 4\PL4_3.c	Program Listing 4.3
\Program Listings\Chapter 4\PL4_4.c	Program Listing 4.4
\Program Listings\Chapter 4\PL4_5.c	Program Listing 4.5
\Program Listings\Chapter 4\benz_tol.h	Benzene-Toluene Header File
\Program Listings\Chapter 4\pipe.h	Pipe Header File
\Program Listings\Chapter 4\steam.h	Steam Header File

MATERIAL BALANCE

Material balance follows the law of mass conservation, whereby all materials entering a system either accumulate or leave the system. There can be neither loss nor gain of mass along the way. In this section, the example system follows such a law and exhibits the simple form of input equals output.

Multiple Reactors

Example Problem 4.1. In Figure 4.1, four reactor tanks are connected by pipes where directions of flow are depicted by means of arrows. The transfer rate of fluid in the tanks is taken as the product of the flow rate (Q, liters per second) and the concentration (C, grams per liter). If given that the incoming flow rate is equal to the outgoing rate, establish the material balance equations and solve for the concentration of fluid at each tank.

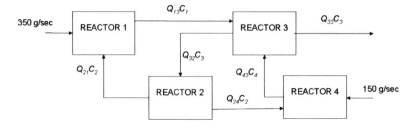

FIGURE 4.1: Multiple reactor tanks.

Q_{13} = 75 liters/sec Q_{24} = 20 liters/sec Q_{33} = 60 liters/sec

Q_{21} = 25 liters/sec Q_{32}= 45 liters/sec Q_{43} = 30 liters/sec

Performing material balance on each reactor, we get

Reactor 1

$$350 + Q_{21}C_2 = Q_{13}C_1$$
$$350 + 25C_2 = 75C_1$$
$$75C_1 - 25C_2 = 350 \tag{4.1}$$

Reactor 2

$$Q_{32}C_3 = Q_{21}C_2 + Q_{24}C_2$$
$$45C_3 = 25C_2 + 20C_2$$
$$45C_3 - 45C_2 = 0 \tag{4.2}$$

Reactor 3

$$Q_{13}C_1 + Q_{43}C_4 = Q_{32}C_3 + Q_{33}C_3$$
$$75C_1 + 30C_4 = 45C_3 + 60C_3$$
$$75C_1 + 30C_4 - 105C_3 = 0 \tag{4.3}$$

Reactor 4

$$150 + Q_{24}C_2 = Q_{43}C_4$$
$$150 + 20C_2 = 30C_4$$
$$30C_4 - 20C_2 = 150 \tag{4.4}$$

Summarizing Eqs 4.1, 4.2, 4.3, and 4.4:

$$
\begin{aligned}
75C_1 - 25C_2 & & & = 350 \\
-45C_2 + 45C_3 & & & = 0 \\
75C_1 - 105C_3 + 30C_4 & & & = 0 \\
-20C_2 + 30C_4 & & & = 150
\end{aligned}
$$

Since this can be presented in matrix form, we can use one of the programs developed in Chapter 2 to solve for simultaneous linear equations. In as much as we are dealing with a 4 by 4 matrix, we can use the Gauss-Jordan program (Program Listing 2.2) with some changes on the dimension of the array.

Program Listing 4.1: Revised Gauss-Jordan program

```
#include <stdio.h>
#include <stdlib.h>
int main()
```

```
{
 int r,c,i, numeq;
 double a[20][20],b[20];
 double d,x[40],e1;
 do{
  system("cls");
  printf("\nCan solve up to 20 Equations - 20 Unknowns\n");
  printf("\nEnter number of unknowns (max 20): ");
  scanf("%d",&numeq);
 }while(numeq>20);
  for (r=0; r<numeq; r++)
 {
  for (c=0; c<numeq; c++)
  {
   printf(" Enter a[%d][%d]: ",r,c);
   scanf("%lf",&a[r][c]);
  }
  printf(" Enter b[%d]: ",r);
  scanf("%lf",&b[r]);
  x[r]=0;
 }
 for(r=0;r<numeq;r++)
 {
  d=-1/a[r][r];
  for(c=0;c<numeq;c++)
  {
   if (c==r)(c++);
   a[r][c]=a[r][c]*d;
  }
  d=-d;
  for(i=0;i<numeq;i++)
  {
   if(i==r)(i++);
   e1=a[i][r];
   for(c=0;c<numeq;c++)
   {
    if(c==r)(a[i][r]=a[i][r]*d);
    else(a[i][c]=a[i][c]+a[r][c]*e1);
   }
  }
  a[r][r]=d;
```

```
    }
    for(i=0;i<numeq;i++)
    {
      for(c=0;c<numeq;c++)
      x[i]=x[i]+b[c]*a[i][c];
    }
    for(i=0;i<numeq;i++)
    printf("variable x[%d]= %lf\n",i,x[i]);
    system("pause");
    return 0;
    }
```

Using such a program, we derived the values of the following concentrations:

C_1 = 7.444444 g per liter C_2 = 8.333333 g per liter
C_3 = 8.333333 g per liter C_4 = 10.555556 g per liter

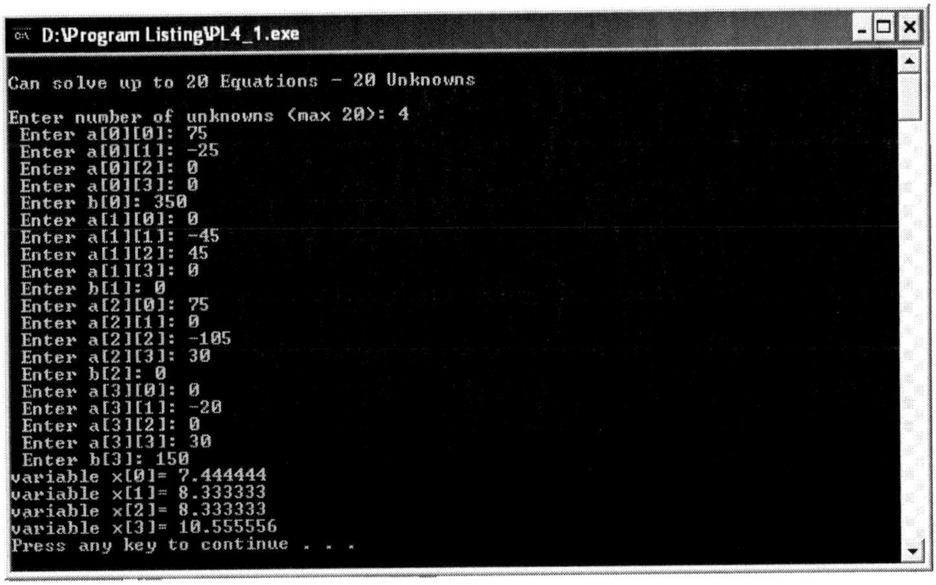

```
 D:\Program Listing\PL4_1.exe

Can solve up to 20 Equations - 20 Unknowns

Enter number of unknowns (max 20): 4
 Enter a[0][0]: 75
 Enter a[0][1]: -25
 Enter a[0][2]: 0
 Enter a[0][3]: 0
 Enter b[0]: 350
 Enter a[1][0]: 0
 Enter a[1][1]: -45
 Enter a[1][2]: 45
 Enter a[1][3]: 0
 Enter b[1]: 0
 Enter a[2][0]: 75
 Enter a[2][1]: 0
 Enter a[2][2]: -105
 Enter a[2][3]: 30
 Enter b[2]: 0
 Enter a[3][0]: 0
 Enter a[3][1]: -20
 Enter a[3][2]: 0
 Enter a[3][3]: 30
 Enter b[3]: 150
variable x[0]= 7.444444
variable x[1]= 8.333333
variable x[2]= 8.333333
variable x[3]= 10.555556
Press any key to continue . . .
```

FIGURE 4.2: Output of Program Listing 4.1.

Condensation

Air contains water vapor in varying amounts. As the temperature of air decreases, its relative humidity increases until it reaches its dew point whereby water vapor in air becomes saturated. Further cooling at this point will result in condensation and a decrease in the amount of water in the air.

In this section, the solution for a material balance problem involving condensation will be discussed. In Chapter 3, we developed a computer program for absolute humidity in relation to the temperature (dry bulb); we also determined the dew point with respect to relative humidity. Now we are going to use the program to solve the following problem.

Example Problem 4.2. Initially the air in a cold storage room has a temperature of 28°C and a relative humidity of 73%. As the day progresses, air inside the storage room has reduced its temperature to 16°C with a relative humidity of 55%. Assuming that the storage room was never opened the entire day, determine the amount of water condensed if the total air inside is 150 cubic meters.

Solution:
For the initial state:
Dry bulb temperature = 28°C and %RH = 73%.
Using the Absolute Humidity Program (Program Listing 3.3), we determined that the water content at this condition is 19.869656 g/m³.

For the final state:
Dry bulb temperature = 16°C and %RH = 55%.
Using the same program, we determined that the water content at the final state is 7.495775 g/m³.

From this we can determine the total water condensed is
(19.869656 – 7.495775)g/m³ (150 m³) = 1856.08 g or 1.856 kg.

ENERGY BALANCE

When determining energy balance, it is advisable to look at the system from a macro point of view. This helps to know what energy is being transferred in and out of the system. This principle is similar to that used in material balance, where we determine the energy being generated and consumed to be able to predict how much is accumulated as a result.

The following examples exhibit such a basic form of input equals output of energies.

Multiple-Effect Evaporators

In this section we will not go into evaporation problems just yet, but what we will focus on here is energy balance part in preparation for the problems dealing with evaporation that we will explore later in this chapter.

Example Problem 4.3. Consider the multiple effect evaporators as shown in Figure 4.3.

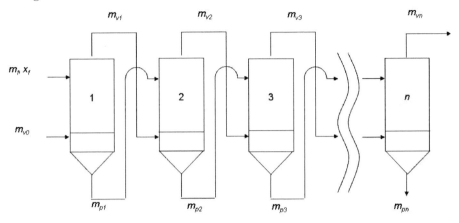

FIGURE 4.3: Multiple effect evaporators.

Steam (m_{v0}) is introduced to the first effect to heat the feed (m_f) with a known concentration (x_f). The solution leaving the first effect (m_{p1}) is fed to the second effect for further concentration using the vapor leaving the first effect (m_{v1}) as a heating medium. The process continues until the *nth* effect, where the desired concentration (x_p) of the product is reached.

We do the energy balance around this system, given n numbers of effect.

Around the first effect:

$$m_f h_f + m_{v0}(H_{v0}\text{-}h_{v0}) = m_{v1}H_{v1} + m_{p1}h_{p1} \quad \text{but } m_{p1} = m_f - m_{v1.}$$

By substituting and rearranging, we get

$$m_{v0}(H_{v0}\text{-}h_{v0}) + m_{v1}(h_{p1}\text{-}H_{v1}) = m_f(h_{p1}\text{-}h_f). \tag{4.5}$$

Around the second effect:

$$m_{p1}h_{p1} + m_{v1}(H_{v1}\text{-}h_{v1}) = m_{v2}H_{v2} - m_{p2}h_{p2} \quad \text{but } m_{p2} = m_{p1} - m_{v2} = m_f - m_{v1} - m_{v2.}$$

By substituting and rearranging, we get

$$m_{v1}(H_{v1}+h_{p2}\text{-}h_{p1}\text{-}h_{v1}) + m_{v2}(h_{p2}\text{-}H_{v2}) = m_f(h_{p2}\text{-}h_{p1}). \tag{4.6}$$

Around the third effect:
Similarly,

$$m_{v1}(h_{p3}-h_{p2}) + m_{v2}(H_{v2}+h_{p3}-h_{p2}-h_{v2}) + m_{v3}(h_{p3}-H_{v3}) = m_f(h_{p3}-h_{p2}). \tag{4.7}$$

Around the nth effect:

$$m_{v1}(H_{vn}-h_{pn-1})+m_{v2}(H_{vn}-h_{pn-1})+ \ldots + m_{vn}(H_{vn}+H_{vn-1}-h_{pn-1}-h_{vn-1}) = m_f[H_{vn}-h_{pn-1}+(x_f/x_p)(h_{pn}-H_{vn})], \tag{4.8}$$

where H_{vi} and h_{vi} are the enthalpies of vapor and condensate, respectively; in the ith effect
h_{pi} is the enthalpies of product in the ith effect.
Subscripts i includes all effects from 1 to n.

We can summarize all equations and present them in matrix form.

$$\tag{4.9}$$

$$[A] = \begin{bmatrix} H_{v0}-h_{v0} & h_{p1}-H_{v1} & 0 & 0 & 0 \\ 0 & H_{v1}+h_{p2}-h_{p1}-h_{v1} & h_{p2}-H_{v2} & 0 & 0 \\ 0 & h_{p3}-h_{p2} & H_{v2}+h_{p3}-h_{p2}-h_{v2} & h_{p3}-H_{v3} & 0 \\ \vdots & \vdots & \vdots & \vdots & \vdots \\ 0 & h_{pn-1}-h_{pn-2} & h_{pn-1}-h_{pn-2} & H_{vn-2}+h_{pn-1}-h_{pn-2}-h_{vn-2} & h_{pn-1}-H_{vn-1} \\ 0 & H_{vn}-h_{pn-1} & H_{vn}-h_{pn-1} & H_{vn}-h_{pn-1} & H_{vn}+H_{vn-1}-h_{pn-1}-h_{vn-1} \end{bmatrix}$$

$$[x] = \begin{bmatrix} m_{v0} \\ m_{v1} \\ m_{v2} \\ \vdots \\ m_{vn-2} \\ m_{vn-1} \end{bmatrix} \text{ and the right-hand side vector } [b] = \begin{bmatrix} m_f(h_{p1}-h_f) \\ m_f(h_{p2}-h_{p1}) \\ mf(h_{p3}-h_{p2}) \\ \vdots \\ mf(h_{pn-1}-h_{pn-2}) \\ mf\left[H_{vn}-h_{pn-1}+\dfrac{x_f}{x_p}(h_{pn}-H_{vn})\right] \end{bmatrix}$$

By solving the matrix we can determine the values of $m_{v0}, m_{v1}, m_{v2}, m_{v3}, \ldots m_{vn-1}$. We can then solve for m_{vn} through material balance $m_{vn} = [(x_f/x_p)m_f] - m_{v1} - m_{v2} \ldots - m_{vn-1}$.

If specific heats (Cp) are given instead of enthalpies, we can derive the corresponding h_f and h_l by using the boiling point of the solution as the base temperature (T_{base}).

$h_f = Cp_f(T_f - T_{base})$ *and* $h_p = Cp_p(T_{base} - T_{base}) = 0$,

where h_f and h_p are enthalpies of feed and product, respectively.

FLUID FLOW

Chemical engineers are sometimes involved in the design structure for transporting fluid from one location to another through pipes. This involves determining pressure drop, friction losses, and pipe dimensions, among other things. In this section, we will touch on some of these aspects and demonstrate how computers can help speed up computation.

Pressure-Drop Determination

Example Problem 4.4. Water is flowing at 21°C through smooth pipe with an internal diameter of 0.1 m at the velocity of 0.5 m/sec. The pipe is 200 m long. Calculate the pressure drop.

The properties of water at 21°C are: $\rho = 997.92$ kg/m³, $\mu = 0.000982$ kg/m-s.

Using Eq 3.18 to solve for Reynolds number, $N_{Re} = \dfrac{Dv\rho}{\mu} = 50{,}810.59$,

N_{Re} will be used to determine the friction factor from the Friction Factor Program in Program Listing 3.6.
Output: $f = 0.005347$

Finally, using Eq 3.19, we can determine the pressure drop as

$$\Delta p_f = 2f\rho \frac{Lv^2}{Dg_c} = 5335.88 \text{ N/m}^2.$$

Pipe Diameter Calculation

Example Problem 4.5. Water flowing through smooth pipe with a length of 250 m at a flow rate of 0.0085 m³/s has a temperature of 21°C. Water source is elevated to overcome the friction loss F_f of 27.5 J/kg. Calculate the pipe diameter.

The properties of water at 21°C are: ρ = 997.92 kg/m^3 and μ = 0.000982 kg/m-s.

The process for solving the pipe diameter will be trial-and-error based on the assumed pipe diameter.

The first assumption, D = 0.1 m, solving for area (cross-sectional) = $\pi D^2/4$ = 0.007854 m^2. Correspondingly, v = flow rate/ area = 0.0085/0.007854 = 1.08225 m/s.

Using Eq 3.18 to solve for Reynolds number, $N_{Re} = \dfrac{Dv\rho}{\mu}$ = 109,979.89,

N_{Re} will be used to determine the friction factor from the Friction Factor Program (Program Listing 3.6).

Output: f = 0.004581

With F_f given as 27.5 J/kg, using Eq 3.20 $F_f = 2f\dfrac{Lv^2}{Dg_c}$ we can solve for D with

f taken as 0.004581. The resulting D is 0.098 m, which is closer to the first D we assumed (0.1 m). We can say that the correct value of D is between these two values.

Implementing further trial-and-error we found that the correct pipe diameter is 0.0995 m or 9.95 cm.

Solving this kind of problem using manual computation is very tedious and a computer program would be very useful in these instances. Program Listing 4.2 will address fluid flow problems with unknown pipe diameter such as the one.

Program Listing 4.2: Pipe diameter calculation.

```
1    #include <stdio.h>
2    #include <stdlib.h>
3    #include <conio.h>
4    #include "pipe.h"
5    int main()
6    {
7      char try;
8      double density, visco, dia1, rate, Nre, f, length,
       area,Ff,dia2, velocity;
9      system("cls");
10     printf("Solving for diameter of smooth pipe.\nPlease
       ensure   all SI units are consistent.\n");
11     printf("\nEnter density of the fluid: ");
12     scanf("%lf",&density);
13     printf("Enter viscosity of the fluid: ");
14     scanf("%lf",&visco);
```

```
15  printf("Enter pipe length: ");
16  scanf("%lf",&length);
17  printf("Enter flow rate: ");
18  scanf("%lf",&rate);
19  printf("Enter friction loss: ");
20  scanf("%lf",&Ff);
21  do{
22   printf("\nEnter assumed pipe diameter: ");
23   scanf("%lf",&dia1);
24   area=3.1416*(dia1*dia1)/4;
25   velocity = rate/area;
26   Nre= dia1*velocity*density/visco;
27   f=friction(Nre);
28   dia2=2*f*length*velocity*velocity/(Ff);
29   printf("The derived diameter is %lf",dia2);
30   printf("\n\nDo you want to enter new diameter? (Y/
     N)\n");
31   try=getch();
32  }while(try=='y'||try=='Y');
33 return 0;
34 }
```

Line 4: The `pipe.h` header is linked in the program so that the `friction()` function can be used in the program.

Lines 11 to 20: This section asks the user to provide the needed information.

Lines 21 to 32: Utilize the `do - while` loop trial-and-error determination of the pipe diameter.

Line 24: Computes the cross-sectional area of the pipe.

Line 25: Computes the velocity based on the flow rate and computed area.

Line 26: Determines the Reynolds number.

Line 27: Utilizes the `friction()` function to generate the friction factor based on the computed Reynolds number.

Line 28: Determines the diameter of the pipe based on the friction loss and generated friction factor.

```
 D:\Program Listing\PL4_2.exe                                    _ □ ✕
Solving for diameter of smooth pipe.
Please ensure all SI units are consistent.

Enter density of the fluid: 997.92
Enter viscosity of the fluid: 0.000982
Enter pipe length: 250
Enter flow rate: 0.0085
Enter friction loss: 27.5

Enter assumed pipe diameter: .1
The derived diameter is 0.097546

Do you want to enter new diameter? (Y/N)

Enter assumed pipe diameter: .0995
The derived diameter is 0.099421

Do you want to enter new diameter? (Y/N)
‐
```

FIGURE 4.4:   Output of Program Listing 4.2.

MASS AND HEAT TRANSFER

Nearly all operations that are handled by chemical engineers involve transfer of mass and heat. This may include thermal conduction, evaporation of solvent, and distillation of mixed liquids. The driving force for the transfer is usually concentration for mass and temperature gradient for heat. In this section, we will solve mass and heat transfer problems using computer programs previously developed. We will even develop new program to address more complex problems.

Temperature Determination in Two-Dimensional Conduction

Example Problem 4.6. Consider a long, rectangular flue (Figure 4.5) with steady heat conduction along the x and y axes.

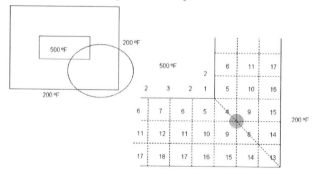

FIGURE 4.5:   Rectangular flue.

Since both ends are symmetrical, only 1/8 of the figure will be considered. From this section, a rectilinear network of nodes can be arranged. A heat balance on the volume on a sample node (4) gives

$$q_4 = \frac{kA}{\Delta x}(T_5 - T_4) + \frac{kA}{\Delta x}(T_9 - T_4) + \frac{kA}{\Delta y}(T_5 - T_4) + \frac{kA}{\Delta y}(T_9 - T_4).$$

When using a square net, Δx is equal to Δy, so by rearranging kA, the equation becomes

$$\frac{q_4}{kA} = 2T_5 + 2T_9 - 4T_4.$$

When the steady state is reached, the imaginary sink (q_4/kA) is equal to zero. So by completing the equations at various nodes, we yield

$$T_1 = T_2 = T_3 = 500°F$$
$$T_{13} = T_{14} = T_{15} = T_{16} = T_{17} = T_{18} = 200°F.$$

At node 4: $-4T_4 + 2T_5 + 2T_9 = 0$
At node 5: $T_4 - 4T_5 + T_6 + T_{10} = -500$
At node 6: $T_5 - 4T_6 + T_7 + T_{11} = -500$
At node 7: $2T_6 - 4T_7 + T_{12} = -500$
At node 8: $-4T_8 + 2T_9 = -400$
At node 9: $T_4 + T_8 - 4T_9 + T_{10} = -200$
At node 10: $T_5 + T_9 - 4T_{10} + T_{11} = -200$
At node 11: $T_6 + T_{10} - 4T_{11} + T_{12} = -200$
At node 12: $T_7 + 2T_{11} - 4T_{12} = -200$

Using the program developed for solving simultaneous linear equations, the results are presented below:

$T_4 = 312.006169°F$ $T_5 = 369.593072°F$ $T_6 = 387.904852°F$
$T_7 = 392.193617°F$ $T_8 = 227.209633°F$ $T_9 = 254.419267°F$
$T_{10} = 278.461265°F$ $T_{11} = 289.832720°F$ $T_{12} = 292.964765°F$

Evaporation

To concentrate a solution of a nonvolatile solute in a volatile solvent, usually water, evaporation is preferred. In an actual evaporator, there are many factors that need to be considered, which affect its performance. Entrainments of solids, pressure of non-condensable gases in the vapor steam, and the effect of the liquid head over the tubes are some considerations. But in preliminary design, all these effects are ignored. We further assume that all heat losses are negligible and superheating in steam, as well as subcooling in condensate does not occur.

Consider the evaporator shown in Figure 4.6.

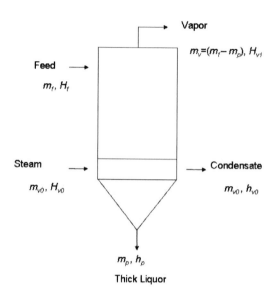

FIGURE 4.6: Single effect evaporator.

If we neglect steam superheat and condensate subcooling, the enthalpy balance for the steam is

$$q_{v0} = m_{v0}(h_{v0} - H_{v0}) = -m_{v0}\lambda_{v0}$$

h_{v0} = enthalpy of condensate,
H_{v0} = enthalpy of steam,
m_{v0} = steam flow rate,
q_{v0} = heat transfer rate through heating surface to steam, and
λ_{v0} = latent heat of condensation of steam.

For the liquid side, the enthalpy balance is

$$q = (m_f - m_p)H_v - m_f h_f + m_p h_p$$

h_f = enthalpy of feed liquor,
H_{v1} = enthalpy of vapor,
q = heat transfer rate from heating surface to liquid, and
h_p = enthalpy of thick liquor.

NOTE   *If the liquor has similar characteristics as the solvent and no BPE then $h_v = h_p$.*

If we assume no heat loss $q = -q_{v0}$ or $q = m_{v0}\lambda_{v0} = (m_f - m_p)H_{v1} - m_f h_f + m_p h_p$

The h_p and h_f will depend on both temperature and concentration (refer to Table 3.7 NaOH concentration-enthalpy diagram as an example). Once the enthalpy and material balances have been completed, we can note:

Heat Transfer Area $(A) = q/U(T_{v0} - T_{v1})$
Capacity $= m_f - m_p$
Steam Economy $= (m_f - m_p)/m_s$

Example Problem 4.7. A single vertical-tube evaporator is used to concentrate 29,000 lb/hr of organic colloid (60°F) from 25% to 60% solid. The solution is considered to have a negligible boiling point elevation, and its properties are similar to that of water at all concentrations. Saturated steam is introduced at 25 psia and the pressure in the condenser is 1.69 psia. The overall heat transfer coefficient is 300 Btu/ft²-h-deg F. Determine the heat transfer area in ft², and the steam consumption in lb per hr (see Figure 4.7).

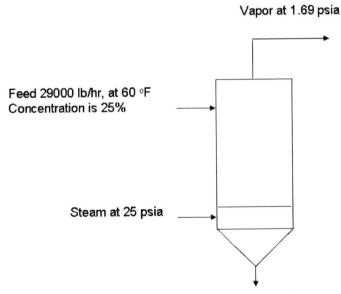

Vapor at 1.69 psia

Feed 29000 lb/hr, at 60 °F
Concentration is 25%

Steam at 25 psia

Product Liquor Concentration is 60%

FIGURE 4.7: Single-effect evaporator flow diagram of example problem 4.7

By conducting material balance, we can derive the amount of product liquor.

$$m_p = (x_f/x_p)(m_f) = (0.25/0.6)\,(29000) = 12083.33 \text{ lb/hr}$$
$$m_{vl} = m_f - m_p = 29000 - 12083.33 = 16916.67 \text{ lb/hr}$$

Find the saturation temperature of steam and vapor using Program Listing 4.3.

$$T_{v0} = 240.29°F \quad T_{vl} = 120.86°F$$

Since the solution fed into the system follows the characteristics of water, without BPE at any concentration, we can say $h_p = h_v$.

Find the values of the different enthalpies using Program Listing4.3.

$$H_{v0} = 1160.73 \text{ Btu/lb} \qquad H_{vl} = 1113.59 \text{ Btu/lb} \qquad h_f = 28.08 \text{ Btu/lb}$$
$$h_{v0} = 208.78 \quad \text{Btu/lb} \qquad h_{vl} = 88.86 \text{ Btu/lb}$$

From the enthalpy balance around this effect, we get

$$m_{v0}(H_{v0}-h_{v0}) = q = m_{vl}H_{vl} - m_f h_f + m_p h_{vl}$$
$$q = 16916.67(1113.59) - 29000(28.08) + 12083.33(88.86) = 19097639.25 \text{ Btu/hr}$$
$$m_{v0} = 19097639.25/(1160.73-208.78) = 20061.60 \text{ lb/hr}$$
$$A = 19097639.25/(300)(240.29-120.86) = 533.02 \text{ ft}^2$$

To distinguish symbols of enthalpies for saturated liquid from saturated vapor, use the small letter (h) for the former and capital letter (H) for the latter.

Program Listing 4.3: Steam enthalpy program

```c
#include <stdio.h>
#include <stdlib.h>
#include <conio.h>
#include "steam.h"
int main()
{
  double pres, temp, hl, hv;
  char res,opt;
  do
    {
      system("cls");
      printf("Input: Pressure or Temperature? (P/T)\n");
      opt=getch();
      if(opt=='p' || opt =='P')
      {
        printf("\nEnter pressure in psia: ");
        scanf("%lf", &pres);
```

```
      pres = pres * 6894.757;
      temp = sat_p2t(pres);
      }
    else
      {
      printf("\nEnter temperature in deg F: ");
      scanf("%lf", &temp);
      temp=((temp-32)/1.8)+273.15;
      }
    hl = enthalpy_l(temp);
    hl = hl/(1055.056*2.2046226);
    hv = enthalpy_v(temp);
    hv =hv/(1055.056*2.2046226);
    temp = ((temp-273.15)*1.8)+32;
    printf("\n");
    if(opt=='p' || opt =='P')
      printf("Saturation temperature is   %lf  deg  F\
      n",temp);
    printf("Enthalpy (liquid) is %lf Btu/lb\n",hl);
    printf("Enthalpy (vapor) is %lf Btu/lb\n",hv);
    printf("\nTry again? (Y/N)\n");
    res=getch();
   }while(res == 'y' || res == 'Y');
  return 0;
 }
```

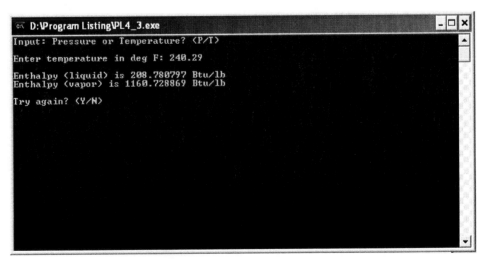

FIGURE 4.8: Output of Program Listing 4.3.

Multiple-Effect Evaporation

There are different types of multiple-effect evaporators based on the feeding method. However, in this section, we will only focus on the forward-feed type.

In this type, the feed is transferred into the first effect and passes on through the succeeding effects as shown earlier in Figure 4.3. The concentration increases as it goes through the different evaporators with the evaporation of the solvent from each effect. Steam is used in heating the first effect; however, the evaporated solvent or the vapor from the first effect will then be the source of heat in the second effect and similarly for the succeeding effects.

Heat transfer areas (A) are equal, which is common practice. Solving multiple-effect evaporators usually requires determining the heat transfer area, so with the area being equal, $A_1 = A_2 \ldots = A_n$ and the rate of heat transfer at a steady state, the temperature distribution can be assumed:

$$\Delta T_1 = \frac{1/U_1}{\dfrac{1}{U_1} + \dfrac{1}{U_2} + \ldots \dfrac{1}{U_n}} \left(\Delta T_{Total} \right)$$

$$\Delta T_n = \frac{U_1}{U_n} \Delta T_1$$

By performing the material and heat balances, the rate of heat transfer (q) from each effect can be obtained. And from there, the heat transfer area (A) is computed. If the area (A) is far from equal, the temperature is adjusted and the computation is repeated until the resulting area of each effect is almost the same. Let's consider an example problem.

Example Problem 4.8. An organic colloidal solution with a negligible boiling point rise is to be concentrated from 13% to 55% solids in a triple-effect forward-feed evaporator. Steam is available at 29.82 psia and the third effect is 1.27 psia. Feed enters at a rate of 45000 lb/hr at temperature of 60°F. The specific heat of the solution at all concentrations is the same as that of water. The overall coefficients are $U_1 = 530$, $U_2 = 350$, $U_3 = 195$ Btu/ft²-hr-°F. The area of each effect has to be the same. Calculate the area of the heating surface (A), the steam consumption (m_{v0}), and the steam economy.

First, conduct materials balance to solve for m_p and m_{vtotal}.

$$m_p = (x_f/x_p)(m_f) = (0.13/0.55)\,(45000) = 10636.36 \text{ lb/hr}$$
$$m_{vtotal} = m_f - m_p = 45000 - 10636.36 = 34363.64 \text{ lb/hr}$$

Find the saturation temperature of steam and vapor using Program Listing 4.3.

$$T_{v0} = 250.21°F \text{ (at P = 29.82 psi)} \qquad T_{v31} = 109.60°F \text{ (at P = 1.27 psi)}$$

The ΔT_{total} therefore is 140.61°F.

Estimating ΔT at each effect given $U_1 = 530$, $U_2 = 350$, $U_3 = 195$ Btu/ft²-hr-°F.

$$\Delta T_1 = \frac{140.61/530}{\dfrac{1}{530} + \dfrac{1}{350} + \dfrac{1}{195}} = 26.87 \qquad \Delta T_2 = \frac{140.61/350}{\dfrac{1}{530} + \dfrac{1}{350} + \dfrac{1}{195}} = 40.70$$

$$\Delta T_3 = \frac{140.61/195}{\dfrac{1}{150} + \dfrac{1}{350} + \dfrac{1}{195}} = 73.04$$

$$T_1 = 250.21 - 26.87 = 223.34°F$$
$$T_2 = 223.34 - 40.70 = 182.64°F$$
$$T_3 = 182.64 - 73.04 = 109.60°F$$

With all temperatures already determined, we now get the different enthalpies (for steam and water) from each effect using Program Listing 4.3. Since the solution fed into the system follows the characteristics of water, without BPE at any concentration, we can say $h_p = h_v$.

Enthalpies in (Btu/lb):

$hf = 28.08$	at 60°F	
$H_{v0} = 1164.19$	$h_{v0} = 218.84$	at 250.21°F
$H_{v1} = 1154.60$	$h_{v1} = 191.64$	at 223.34°F
$H_{v2} = 1138.99$	$h_{v2} = 150.69$	at 182.64°F
$H_{v3} = 1108.83$	$h_{v3} = 77.62$	at 109.60°F

And with the other given information $m_f = 45000$ lb/hr, $x_f = 0.13$, $x_p = 0.55$, we substitute this into the matrix below:

$$\begin{bmatrix} H_{v0} - h_{v0} & h_{v1} - H_{v1} & 0 \\ 0 & H_{v1} + h_{v2} - 2h_{v1} & h_{v2} - H_{v2} \\ 0 & H_{v3} - h_{v2} & H_{v3} + H_{v2} - 2h_{v2} \end{bmatrix} \begin{bmatrix} m_{v0} \\ m_{v1} \\ m_{v2} \end{bmatrix} = \begin{bmatrix} m_f(h_{v1} - h_f) \\ m_f(h_{v2} - h_{v1}) \\ m_f\left[H_{v3} - h_{v2} + \dfrac{x_f}{x_p}(h_{v3} - H_{v3}) \right] \end{bmatrix}$$

So substituting all values we get

$$\begin{bmatrix} 945.35 & -962.96 & 0 \\ 0 & 922.01 & -988.30 \\ 0 & 958.14 & 1946.44 \end{bmatrix} \begin{bmatrix} m_{v0} \\ m_{v1} \\ m_{v2} \end{bmatrix} = \begin{bmatrix} 7360200 \\ -1842750 \\ 32147975.45 \end{bmatrix}.$$

Solving for the simultaneous linear equation using Program Listing 2.2 (the Gauss-Jordan Program):

m_{v0} = 18257.83 lb/hr, m_{v1} = 10280.63 lb/hr, m_{v2} = 11455.63 lb/hr. Likewise, m_{v3} = 34363.64 − 10280.63 − 11455.63 = 12627.38 lb/hr.

Once we have the mass flow rate of steam and the different vapors in each effect, we can then solve for the rate of heat transfer q and the heating surface area A.

$$q_1 = m_{v0} (H_{v0} - h_{v0}) = 18257.83 (945.35) = 17260039.59 \text{ Btu/hr}$$
$$A_1 = q_1/[U_1(\Delta T_1)] = 17260039.59/[530(26.87)] = 1211.99 \text{ ft}^2$$

$$q_2 = m_{v1} (H_{v1} - h_{v1}) = 10283.63 (962.96) = 9902724.34 \text{ Btu/hr}$$
$$A_2 = q_2/[U_2(\Delta T_2)] = 9902724.34/[350(40.70)] = 695.17 \text{ ft}^2$$

$$q_3 = m_{v2} (H_{v2} - h_{v2}) = 11455.631 (988.30) = 11321599.13 \text{ Btu/hr}$$
$$A_3 = q_3/[U_3(\Delta T_3)] = 11321599.13/[195(73.04)] = 794.90 \text{ ft}^2$$

Since $A_1 \neq A_2 \neq A_3$, we need to adjust the temperature and repeat the process again. We do this by first getting the average A of the three effects.

$$A_{average} = (1211.99 + 695.17 + 794.90)/3 = 900.69 \text{ ft}^2$$

Adjusting the ΔTs we get

$$\Delta T_1 = 26.87 (1211.99/900.69) = 36.16$$
$$\Delta T_2 = 40.70 (695.17/900.69) = 31.41$$
$$\Delta T_3 = 73.07 (794.90/900.69) = 64.49$$

Find the total ΔT = 36.16 + 31.41 + 64.49 = 132.06, and since it is $\neq$ to 140.61, we need to further adjust it.

$$\Delta T_1 = 36.16 (140.61/132.06) = 38.50$$
$$\Delta T_2 = 31.41 (140.61/132.06) = 33.44$$
$$\Delta T_3 = 64.49 (140.61/132.06) = 68.67$$
$$\Delta T_{total} = 38.50 + 33.44 + 68.67 = 140.61$$

The estimated temperature at each effect will then be:

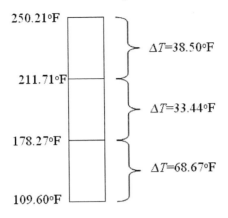

$$T_1 = 250.21 - 38.50 = 211.71°F$$
$$T_2 = 211.71 - 33.44 = 178.27°F$$
$$T_3 = 178.27 - 68.67 = 109.60°F$$

We then repeat the entire computation using the new temperatures. Once we obtain the values of the heating surface area, which are almost equal from each effect, we stop the iteration. For this problem, the heating surface area A for all effects is 846.25 ft^2, the steam consumption m_{v0} is 17968.4 lb/hr, and the steam economy is 1.91.

Take the algorithm and use it to develop the following program.

Program Listing 4.4: Evaporation program.

```
1    #include <stdio.h>
2    #include <conio.h>
3    #include <stdlib.h>
4    #include "steam.h"
5    int main()
6    {
7      int k,w,j,i;
8      double xmatrx[20][20], constant[20], d ,vapor[20],
         e1, area[20];
9      double hv[20],hl[20],hd[20],f,aream,tarea,heat,stmec
         on, temp_t;
10     double tv[20],tl[20],p,vt,totv,ut,dta,dtn,dtr,dt[20]
         ,u[20] ,conc[20],ps,pk,ts;
11     char flag;
12     hl[0]=0;
13     dt[0]=0;
14     hd[0]=0;
```

```
15   printf("Solving Problems on Multiple Effect Evaporators
     - Assuming no BPE\n");
16   printf("\nEnter number of evaporators: ");
17   scanf("%d",&k);
18   printf("Enter feed concentration in fraction: ");
19   scanf("%lf",&conc[0]);
20   printf("Enter feed temperature in deg F: ");
21   scanf("%lf",&tl[0]);
22   printf("Enter product concentration in fraction: ");
23   scanf("%lf",&conc[k]);
24   printf("Enter amount of feed in lb/hr: ");
25   scanf("%lf",&f);
26   printf("Enter steam pressure in psia: ");
27   scanf("%lf",&ps);
28   printf("Enter pressure (psia) at evaporator no. %d : ",k);
29   scanf("%lf",&pk);
30   for(j=1;j<=k;j++)
31   {
32    printf("Enter OHT Coefficient for evaporator no. %d : ",j);
33     scanf("%lf",&u[j]);
34     }
35     tv[0]=sat_p2t(ps*6894.757);
36     tv[0] =((tv[0]-273.15)*1.8)+32;
37     tv[k]=sat_p2t(pk*6894.757);
38     tv[k] =((tv[k]-273.15)*1.8)+32;
39     ts=tv[0];
40     p=f*conc[0]/conc[k];
41     vt=(f-p)/k;
42     for(j=1;j<k;j++)(conc[j]=f*conc[0]/(f-(vt*j)));
43     ut=0;
44     for(j=1;j<=k;j++)(ut=ut+(1/u[j]));
45     dta=tv[0]-tv[k];
46     dtn=0;
47     for(j=1;j<k;j++)
48     {
49      dt[j]=dta*((1/u[j])/ut);
50      dtn=dtn+dt[j];
51      }
52     dt[k]=dta-dtn;
53     tl[k]=tv[k];
54     dtr=tl[k]-tv[k];
```

```
55   for(j=1;j<k;j++)
56   {
57    tv[j]=tv[j-1]+dt[j];
58    tl[j]=tv[j];
59    dtr=dtr+(tl[j]-tv[j]);
60   }
61   dta=tv[0]-tv[k]-dtr;
62   dtn=0;
63   for(j=1;j<k;j++)
64   {
65    dt[j]=dta*((1/u[j])/ut);
66    dtn=dtn+dt[j];
67   }
68   dt[k]=dta-dtn;
69   flag='n';
70   do
71   {
72    for(j=k;j>0;j--)
73    {
74    tl[j]=tv[j];
75    tv[j-1]=tl[j]+dt[j];
76    }
77    tv[0]=ts;
78    system("cls");
79    for (j=0;j<=k;j++)
80    {
81     printf("temperature of vapor[%d] in deg F = %lf \
       r\n",j,tv[j]);
82     printf("temperature of liquid[%d] in deg F= %lf \
       r\n",j,tl[j]);
83    }
84    printf("product in lb/hr = %lf \r\n",p);
85    for (j=0;j<=k;j++)
86    {
87     temp_t=((tv[j]-32)/1.8)+273.15;
88     hv[j]=enthalpy_v(temp_t);
89     hv[j]=hv[j]/(1055.056*2.2046226);
90     temp_t=((tl[j]-32)/1.8)+273.15;
91     hl[j]=enthalpy_l(temp_t);
92     hl[j]=hl[j]/(1055.056*2.2046226);
93    }
```

```
94    for(j=0;j<k;j++)
95    {
96      temp_t =((tv[j]-32)/1.8)+273.15;
97      hd[j+1]=enthalpy_l(temp_t);
98      hd[j+1]=hd[j+1]/(1055.056*2.2046226);
99    }
100   for(j=0;j<k;j++)(vapor[j]=0);
101   for(j=0;j<=k;j++)
102   {
103     for(w=0;w<=k;w++)(xmatrx[j][w]=0);
104   }
105   constant[0]=f*(hl[0]-hl[1]);
106   for(j=1;j<k-1;j++)(constant[j]=f*(hl[j]-hl[j+1]));
107   constant[k-1]=f*(hl[k-1]-hv[k]+(conc[0]*(hv[k]-
      hl[k])/conc[k]));
108   xmatrx[k-1][0]=0;
109   for(j=1;j<(k-1);j++)
110   {
111     xmatrx[k-1][j]=hl[k-1]-hv[k];
112   }
113   xmatrx[k-1][k-1]=hd[k]+hl[k-1]-hv[k-1]-hv[k];
114   xmatrx[0][0]=hd[1]-hv[0];
115   for(j=0;j<(k-1);j++)(xmatrx[j][j+1]=hv[j+1]-
      hl[j+1]);
116   for(j=1;j<(k-1);j++)(xmatrx[j][j]=hd[j+1]+hl[j]-
      hv[j]-hl[j+1]);
117   for(j=2;j<(k-1);j++)
118   {
119     for(w=1;w<(k-2);w++)(xmatrx[j][w]=hl[j]+hl[j+1]);
120   }
121   for(w=0;w<k;w++)
122   {
123    d=-1/xmatrx[w][w];
124    for(j=0;j<k;j++)
125    {
126      if (j==w)(j++);
127      xmatrx[w][j]=xmatrx[w][j]*d;
128    }
129    d=-d;
130    for(i=0;i<k;i++)
131    {
```

```
132    if(i==w)(i++);
133    e1=xmatrx[i][w];
134    for(j=0;j<k;j++)
135    {
136      if(j==w)(xmatrx[i][w]=xmatrx[i][w]*d);
137      else(xmatrx[i][j]=xmatrx[i][j]+xmatrx[w][j]*e1);
138    }
139    }
140   xmatrx[w][w]=d;
141   }
142   for(i=0;i<k;i++)
143   {
144     for(j=0;j<k;j++)
145     vapor[i]=vapor[i]+constant[j]*xmatrx[i][j];
146   }
147   totv=0;
148   for(j=1;j<k;j++)(totv=totv+vapor[j]);
149   vapor[k]=f-p-totv;
150   for(i=0;i<=k;i++)( printf("vapor[%d]= %lf\r\
      n",i,vapor[i]));
151   for(i=0;i<k;i++)
152   {
153    heat=vapor[i]*(hv[i]-hd[i+1]);
154    printf("Rate of heat transfer at Evaporator %d is
       %lf BTU/hr\r\n",i+1,heat);
155   }
156   stmecon=(totv+vapor[k])/vapor[0];
157   printf("Steam economy = %lf\r\n",stmecon);
158   for(j=1;j<=k;j++)
159   {
160    area[j]=(vapor[j-1]*(hv[j-1]-hd[j]))/(u[j]*(tv[j-
       1]-tl[j]));
161    printf("Area for evaporator no. %d = %lf\r\
       n",j,area[j]);
162   }
163   if(k==1)
164   {
165    (flag='y');
166    getch();
167   }
168   else
```

```
169  {
170   printf("Are the areas OK ?(y/n)\n");
171   flag=getch();
172   if(flag=='n')
173   {
174    tarea =0;
175    for(j=1;j<=k;j++)(tarea=tarea+(dt[j]*area[j]));
176    aream=tarea/dta;
177    for(j=1;j<=k;j++)(dt[j]=dt[j]*area[j]/aream);
178   }
179  }
180  }while (flag=='n');
181  return 0;
182 }
```

Lines 1 and 3: Again, the #include statement tells the C compiler to use the <stdio.h>, <stdlib.h>, and <conio.h> header files, which contain the standard input/output functions printf() and scanf(), and the console input/output function getch(), and the system() function.

Line 4: The steam.h header is linked in the program so that the sat_p2t(), enthalphy_v(), and enthalpy_l() functions can be used in the program.

Line 5: The main() function is where the main operation of the program occurs.

Line 7: Declaration of variables k, w, j, and i as integers.

Lines 8 to 10: Declaration of variable xmatrx[][], constant[], d, vapor[], e1, area[], hv[], hl[], hf, f, aream, tarea, heat, stmecon, tv[], tl[], p, vt, totv, ut, dta, dtn, dtr, dt[], u[], conc[], ps, pk, and ts as doubles.

Line 11: Declaration of flag as character.

Lines 12 to 14: Initializes hl[], dt[], and hd[] as 0.

Lines 15 to 34: Request all the needed values using printf() and scanf() and the looping statement for loop.

Lines 35 to 39: Convert the given pressure of steam to saturation temperature (in oF) using sat_p2t() in steam.h.

Line 40: Solves for the mass flow of the product based on given mass flow of feed and concentrations.

Lines 41 to 54: Solve for the estimated T in each effect.

Lines 55 to 77: Solve for estimated temperature of liquid and vapor in each effect.

Lines 78 to 84: Clear the screen and then print the temperature of liquid and vapor, including the product mass flow rate.

Lines 85 to 99: Approximate the enthalpies of liquid and vapor based on the derived temperature using `enthalpy_v()` and `enthalpy_l()` functions.

Lines 100 to 120: Fill up the matrix Eq 4.9 (energy balance).

Lines 121 to 146: Perform Gauss-Jordan matrix inversion and multiplication to solve for the simultaneous linear equation and determine the mass flow rate of steam and vapor in each effect.

Lines 147 to 181: Compute for the heat transfer area and steam economy. The user will be asked if the areas provided are acceptable; if the areas are not acceptable, then the program will redo all the computation until the areas of all effects are almost equal.

```
D:\Program Listing\PL4_4.EXE
Solving Problems on Multiple Effect Evaporators - Assuming no BPE

Enter number of evaporators: 3
Enter feed concentration in fraction: .13
Enter feed temperature in deg F: 60
Enter product concentration in fraction: .55
Enter amount of feed in lb/hr: 45000
Enter steam pressure in psia: 29.82
Enter pressure (psia) at evaporator no. 3 : 1.27
Enter OHT Coefficient for evaporator no. 1 : 530
Enter OHT Coefficient for evaporator no. 2 : 350
Enter OHT Coefficient for evaporator no. 3 : 195
```

```
D:\Program Listing\PL4_4.EXE
temperature of vapor[0] in deg F = 250.212511
temperature of liquid[0] in deg F= 60.000000
temperature of vapor[1] in deg F = 223.338163
temperature of liquid[1] in deg F= 223.338163
temperature of vapor[2] in deg F = 182.642723
temperature of liquid[2] in deg F= 182.642723
temperature of vapor[3] in deg F = 109.599624
temperature of liquid[3] in deg F= 109.599624
product in lb/hr = 10636.363636
vapor[0]= 18257.960995
vapor[1]= 10280.631484
vapor[2]= 11455.639978
vapor[3]= 12627.364901
Rate of heat transfer at Evaporator 1 is 17260233.606461 BTU/hr
Rate of heat transfer at Evaporator 2 is 9899834.314032 BTU/hr
Rate of heat transfer at Evaporator 3 is 11321551.879358 BTU/hr
Steam economy = 1.882118
Area for evaporator no. 1 = 1211.805357
Area for evaporator no. 2 = 695.046923
Area for evaporator no. 3 = 794.862777
Are the areas OK ?(y/n)
```

After six iterations:

```
D:\Program Listing\PL4_4.EXE
temperature of vapor[0] in deg F = 250.212511
temperature of liquid[0] in deg F= 60.000000
temperature of vapor[1] in deg F = 212.339424
temperature of liquid[1] in deg F= 212.339424
temperature of vapor[2] in deg F = 178.154129
temperature of liquid[2] in deg F= 178.154129
temperature of vapor[3] in deg F = 109.599624
temperature of liquid[3] in deg F= 109.599624
product in lb/hr = 10636.363636
vapor[0]= 17968.375196
vapor[1]= 10438.917102
vapor[2]= 11415.344416
vapor[3]= 12509.374846
Rate of heat transfer at Evaporator 1 is 16986472.557856 BTU/hr
Rate of heat transfer at Evaporator 2 is 10125210.599038 BTU/hr
Rate of heat transfer at Evaporator 3 is 11312726.660275 BTU/hr
Steam economy = 1.912451
Area for evaporator no. 1 = 846.246001
Area for evaporator no. 2 = 846.246103
Area for evaporator no. 3 = 846.246105
Are the areas OK ?(y/n)
```

FIGURE 4.9:   Output of Program Listing 4.4.

Distillation

Generally, distillation includes evaporation and condensation. The difference between the evaporation processes previously discussed and distillation is that the former involves a volatile solvent and a non-volatile solute. Separation in this case is less complicated because the solute will not evaporate together with the solvent. Distillation, however, deals with the mixture of volatile substances where separation of the components is accomplished through partial vaporization. Separate recovery is done on the vapor and the residual liquid.

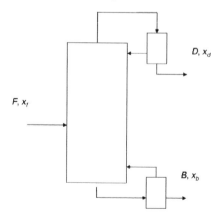

FIGURE 4.10:   Distillation column.

Material balances for distillation are shown below:

Above Feed $\quad y_{n+1} = (L_n / V_{n+1})\, x_n + (Dx_d / V_{n+1})$

Below Feed $\quad y_{m+1} = (L_m / V_{m+1})\, x_m + (Bx_b / V_{m+1})$

Use a total balance around the top of the column $D = V_{n+1} - L_n$ and around the bottom $B = L_m + V_{m+1}$. Substitute the above equations to get rid of V_{n+1} *and* V_{m+1} and the material balance becomes

$$y_{n+1} = L_n\, x_n /(L_n+D) + Dx_d /(L_n+D)$$

and

$$y_{m+1} = L_m\, x_n /(L_m-B) + Bx_b /(L_m+B).$$

If the molal overflow L_n and L_m can be assumed constant, the operating lines (results of the material balances) are straight.

Reflux ratio R_D can be equated as $R_D = L/D = (V-D)/D$.

From this, the above feed operating line can be taken as

$$y_{n+1} = R_D x_n /(R_D+1) + x_d /(R_D+1).$$

Since the operating line passes through the diagonal line ($y = x$) *at* $x = x_d$, the rectifying line (above feed) goes through the operating line point $y = x = x_d$ with a slope of $R_D/(R_D+1)$ and a y-intercept $x_d/(R_D+1)$.

$$y = Lx/(L-B) - Bx_b/(L-B)$$

where L is the liquid flow down the column below the feed point which passes through the point $y = x = x_b$.

The q factor determines numerically the effect of the feed on the column internal flows and is defined as the mols of liquid in the stripping section per mol of feed.

$q > 1$, cold feed
$q = 1$, saturated liquid
$0 < q < 1$, partially vapor
$q = 0$, saturated vapor
$q < 0$, supersaturated vapor

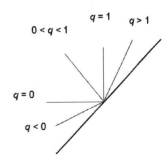

With this, the feed line can be obtained by the equation $y = qx/(q-1) + x_f(q-1)$.

This line passes through the diagonal at $y = x = x_f$ and the slope can be taken as $q/(q-1)$. Therefore, the rectifying, the stripping, and the feed line pass through a common point.

With these three operating lines, the number of theoretical stages may be determined using the McCabe-Thiele graphical method.

The binary distillation operating at a fixed pressure can be solved given the following parameters: x_d, x_f, x_b, q, and R_D. First, draw the equilibrium curve in an x, y diagram with the $y = x$ diagonal line. Map out the given x_d, x_f, and x_b. From the q factor, determine the slope of the feed line. Next, determine the y-intercept of the rectifying line. Plot both the feed and the rectifying lines, then the stripping line with one end at the intersection of the rectifying and feed lines and the other end on the diagonal line where $x = x_b$. Once the rectifying, feed, and stripping lines are plotted, start stepping down the stages. The objective is to get the smallest number of stages by shifting from rectifying to stripping as soon as it is possible (see Figure 4.11).

Example Problem 4.9. Given:

$x_d = 0.85$
$x_f = 0.45$
$x_b = 0.15$
$q = 1$
$R_D = 3.5$

Solve for the y-intercept of the rectifying line: y-intercept = 0.1889.
Answer: Five actual stages.

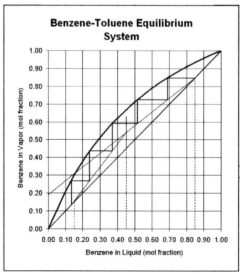

FIGURE 4.11: McCabe-Thiele method used to determine the actual number of plates.

The minimum number of plates or stages is obtained using the equilibrium curve and the diagonal line to generate the stages. When all the liquid from the condenser is returned to the column as reflux (reflux ratio approaches infinity), the slope of the rectifying line approaches zero coinciding with the diagonal line (see Figure 4.12).

Example Problem 4.10. Given:

$x_d = 0.85$
$x_f = 0.45$
$x_b = 0.15$
$q = 1$
$R_D = 3.5$

Answer: Four minimum stages.

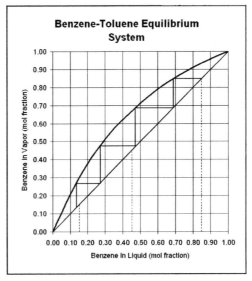

FIGURE 4.12: McCabe-Thiele method used to determine the minimum number of plates.

The minimum reflux ratio R_{Dmin} is obtained when the number of stages approaches infinity. In other words, the rectifying line approaches the intersection of the feed line with the equilibrium curve. From the slope or intercept of the rectifying line, we can compute for the minimum reflux ratio (see Figure 4.13).

Example Problem 4.11. Given:

$x_d = 0.85$
$x_f = 0.45$

$x_b = 0.15$
$q = 1$
$R_D = 3.5$

Answer: y-intercept is 0.47, therefore, $R_{Dmin} = 0.81$.

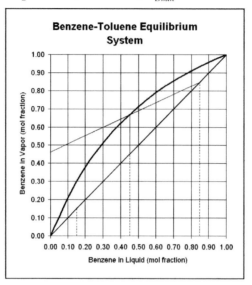

FIGURE 4.13: McCabe-Thiele method used to determine the minimum reflux ratio.

Program Listing 4.5: Distillation program.

```
 1  #include <stdio.h>
 2  #include <stdlib.h>
 3  #include "benz_tol.h"
 4  int main()
 5  {
 6  double xd=0,xf=0,xb=0,xcoor[100],ycoor[100],q,ycoorf
    ,xcoorr;
 7  double ycoorr,mr,mb,rmin,bmin,bact,dcoorr,ref,delm,
    delb;
 8  int j,n=0,fplate;
 9  printf("\nThis program will solve problems on
    Distillation of Benzene-Toluene\n");
10  printf("\nEnter fraction of benzene in distillate : ");
11  scanf("%lf",&xd);
12  printf("Enter fraction of benzene in feed : ");
```

```
13   scanf("%lf",&xf);
14   printf("Enter fraction of benzene in bottom : ");
15   scanf("%lf",&xb);
16   printf("Enter q value : ");
17   scanf("%lf",&q);
18   j=0;ycoor[0]=xd;
19   printf("Minimum Number of Plates\n");
20   do{
21    j++;
22    xcoor[j]=benzenetoluene(ycoor[j-1],1);
23    ycoor[j]=xcoor[j];
24    n=j;
25   }while(xcoor[j]>xb);
26   for(j=1;j<=n;j++)
27   printf("fraction    x=%lf    y=%lf       plate no.=%d\
     n\r",xcoor[j],ycoor[j-1],j);
28   /* for Minimum Reflux Ratio */
29   if (q==1)
30   {
31    xcoorr=xf;
32    ycoorr=benzenetoluene(xcoorr,2);
33   }
34   else if (q==0)
35   {
36   ycoorr=xf;
37   xcoorr=benzenetoluene(ycoorr,1);
38   }
39   else
40   {
41    if (q>1) xcoorr= xf;
42    else if (q < 0)xcoorr= benzenetoluene(xb,1);
43    else xcoorr= benzenetoluene(xf,1);
44    do
45    {
46     xcoorr=xcoorr+0.00000005;
47     ycoorf=(((-1*q)/(1-q))*xcoorr)+(xf/(1-q));
48     ycoorr= benzenetoluene(xcoorr,2);
49     dcoorr=ycoorf-ycoorr;
50    }while(dcoorr > 0.000001 || dcoorr< -0.000001);
51   }
52   mr=(xd-ycoorr)/(xd-xcoorr);
```

```
53  bmin=ycoorr-(mr*xcoorr);
54  rmin=(xd/bmin)-1;
55  printf("Minimum reflux is = %lf\r\n",rmin);
56  printf("Enter reflux ratio : ");
57  scanf("%lf",&ref);
58  /* for Actual number of Plates */
59  if(q==1)(xcoorr=xf);
60  else
61  {
62   delm=(xd/(ref+1))-(xf/(1-q));
63   delb=((-1*q)/(1-q))-(ref/(ref+1));
64   xcoorr=delm/delb;
65  }
66  ycoorr=((ref/(ref+1))*xcoorr)+(xd/(ref+1));
67  mb=(ycoorr-xb)/(xcoorr-xb);
68  bact=ycoorr-(mb*xcoorr);
69  j=0;ycoor[0]=xd;
70  do
71  {
72   j++;
73   xcoor[j]=benzenetoluene(ycoor[j-1],1);
74   if(xcoor[j] >= xcoorr)
75   {
76    ycoor[j]=((ref/(ref+1))*xcoor[j])+(xd/(ref+1));
77    fplate=j+1;
78   }
79   if(xcoor[j] < xcoorr)(ycoor[j]=(mb*xcoor[j])+bact);
80   n=j;
81  }while(xcoor[j]>xb);
82  printf("Actual Number of Plates\r\n");
83  for(j=1;j<n+1;j++)
84  printf("fraction  x=%lf     y=%lf      plate no.=%d\
    n\r",xcoor[j],ycoor[j-1],j);
85  printf("\nFeed is introduced in plate number %d\r\
    n",fplate);
86  system("pause");
87  return 0;
88 }
```

Lines 1 and 2: The #include statement tells the C compiler to use the <stdio.h> and <stdlib.h> header files, which contain the standard input/output functions `printf()` and `scanf()` and the `system()` function.

Line 3: The `benz_tol.h` header is linked in the program so that the `benzenetolune()` function can be used in the program. (Refer to Chapter 3, Phase Equilibrium (the benzene-toluene equilibrium system).)

Line 4: The `main()` function is where the main operation of the program occurs.

Lines 6 to 7: Declaration of variables `xd`, `xf`, `xb`, `xcoor[]`, `ycoor[]`, `q`, `ycoorf`, `xcoorr`, `ycoorr`, `mr`, `mb`, `rmin`, `bmin`, `bact`, `dcoorr`, `ref`, `delm`, and `delb` as doubles.

Line 8: Declaration of variables `j`, `n`, and `fplate` as integers.

Lines 9 to 17: Request all the needed values using `printf()` and `scanf()` and the looping statement `for` loop.

Lines 18 to 27: Solve for the minimum number of plates and then print the result on the screen, giving the concentration of benzene (for both liquid and vapor) at each stage.

Lines 29 to 55: Solve for the minimum reflux ratio and print the result.

Lines 56 to 57: Request the user to provide the reflux ratio and stores the value in `ref` variable.

Lines 59 to 85: Solve for the actual number of plates and the feed inlet location and then print out the result on the screen with the corresponding concentration of benzene (liquid and vapor) at each stage.

```
D:\Program Listing\PL4_5.exe                                    _ □ ×

This program will solve problems on Distillation of Benzene-Toluene

Enter fraction of benzene in distillate : .85
Enter fraction of benzene in feed : .45
Enter fraction of benzene in bottoms : .15
Enter q value : 1
Minimum Number of Plates
fraction    x=0.690002      y=0.850000      plate no.=1
fraction    x=0.471863      y=0.690002      plate no.=2
fraction    x=0.267675      y=0.471863      plate no.=3
fraction    x=0.131138      y=0.267675      plate no.=4
Minimum reflux is = 0.815317
Enter reflux ratio : 3.5
Actual Number of Plates
fraction    x=0.690002      y=0.850000      plate no.=1
fraction    x=0.513644      y=0.725557      plate no.=2
fraction    x=0.362272      y=0.588390      plate no.=3
fraction    x=0.237294      y=0.431649      plate no.=4
fraction    x=0.128579      y=0.263159      plate no.=5

Feed is introduced in plate number 3
Press any key to continue . . .
```

FIGURE 4.14: Output of Program Listing 4.5.

OPTIMIZATION

There are two areas in chemical engineering where optimization may be applied. These areas involve solving problems dealing with process design and plant operation. Optimization requires selecting the best option among different solutions. By optimizing the process design, the result will be maximum production and profit, while minimizing cost and energy usage. Using the Simplex Method Program (Program Listing 2.8) calculations involving these two areas become feasible and convenient.

Process Design

Example Problem 4.12. Two streams of unfiltered sugar solution will be processed and split into two streams (Figure 4.15). The distribution of each feed in the two filtration equipment is presented in Table 4.1.

Table 4.1: Percent (%) distribution of feed sugar solution

Feed	Percent (%) Distribution		Profit after selling (US $/liter)
	Plate and Frame		Candle Filter
F1	60	40	1.70
F2	25	75	2.00

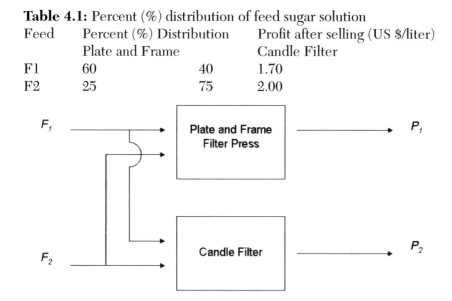

FIGURE 4.15:   Filter Press and Candle Filter Data

The distribution ensures both outputs (P_1 and P_2) will have the same sugar color base on reference base unit (RBU). Likewise, the profit from processing each feed and selling the resulting products is indicated in the same table.

Due to supply constraints, the filtration of F_1 will not exceed 8000 liters per day and F_2 will not exceed 6000 liters per day. On the other hand, equipment limitations allow the plate and frame filter press to process only up to 6000 liters per day and the candle filter up to 12000 liters per day. Design a process setup that can maximize the profit.

To answer this, we first write the objective function:

$$\text{Maximize} \quad z = 1.7x_1 + 2.0x_2$$
$$\text{Constraints} \quad x_1 \le 8000$$
$$x_2 \le 6000$$
$$0.6x_1 + 0.25x_2 \le 6000$$
$$0.4x_1 + 0.75x_2 \le 12000$$
$$x_1 \ge 0, x_2 \ge 0$$

where x_1 is the amount of F_1 to be clarified and x_2 is the amount of F_2 to be clarified.

After which we can then use the simplex program developed in Chapter 2. The result gives us

$$x_1 = 7500 \text{ liters per day}$$
$$x_2 = 6000 \text{ liters per day}$$
$$\text{Total Profit} = \$ 24,750.00$$

Plant Operation

Example Problem 4.13. A sugar mill has two batches of sugar cane that can produce various grades of sugar as shown in Table 4.2. Because of equipment and warehouse limitations, the production of bottler's grade, premium, and standard sugar must be limited, also shown in Table 4.2.

The profit on processing Batch #1 is US $2.00/sack and on Batch #2, $1.60/sack. Find the approximate optimum weekly milling rate of the two batches of sugar cane.

Table 4.2: Grades of sugar.

Sugar Grade	Batch #1 (% distribution)	Batch #2 (% distribution)	Max. Production Rate (sacks/week)
Bottler's Grade	8	11	1500
Premium	29	54	5500
Standard	63	35	11000

where x_1 = no. of sacks of sugar from Batch 1
x_2 = no. of sacks of sugar from Batch 2

Write the objective function and the constraints:

Maximize $z = 2.0x_1 + 1.6x_2$

Constraints $0.08x_1 + 0.11x_2 \leq 1500$

$0.29x_1 + 0.54x_2 \leq 5500$

$0.63x_1 + 0.35x_2 \leq 11000$

From Program Listing 2.8, using the Simplex Method Program we determine that in maximizing production rate, 16820 sacks of sugar per week should be produced from Batch #1 and 1152 sacks per week should be derived from Batch #2 sugar cane. This will result in a total profit of $35,483.87 per week.

LABORATORY EXERCISES

1) Determine the concentration (g/liter) in each reactor tank based on the setup provided,

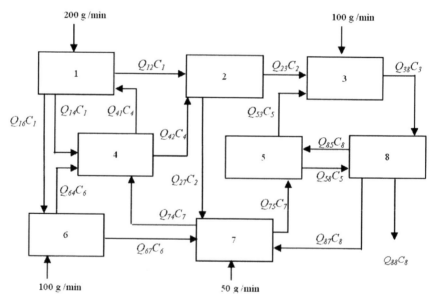

FIGURE 4E.1: Multiple reactor tanks.

where the following Qs are provided in liters per min:

$Q_{41} = 50$	$Q_{16} = 70$	$Q_{14} = 40$	$Q_{12} = 40$	$Q_{64} = 80$
$Q_{67} = 90$	$Q_{74} = 30$	$Q_{42} = 100$	$Q_{27} = 70$	$Q_{23} = 70$
$Q_{87} = 10$	$Q_{75} = 190$	$Q_{85} = 20$	$Q_{53} = 110$	$Q_{58} = 100$
$Q_{38} = 230$	$Q_{88} = 300$			

2) The new operations manager of a fertilizer plant wants to increase the plant's profit by optimizing the production of their three grades of fertilizers. The mixture components, composition, and the corresponding profit of each grade type are presented in the table below:

Grade	Nitrate (%)	Phosphate (%)	Potash (%)	Profit(%)
1	46.54	28.40	25.06	$1050/ton
2	23.42	44.58	32.00	$900/ton
3	25.00	45.37	29.63	$950/ton

Review of the current inventory indicates that phosphate and potash stocks can still reach 40 and 25 tons, respectively, but the nitrate is only half of the phosphate's quantity. If you are the new manager, determine how each stock should be utilized to maximize profit.

3) Determine the temperature of each node in a long rectangular flue with steady heat conduction along x and y axes as shown in the figure.

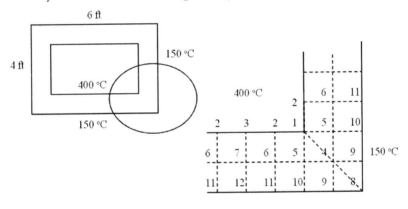

FIGURE 4E.2: Rectangular flue.

4) A plant will distill a water-acetic acid mixture containing 0.4 mol faction water. The distillate is to contain 0.8 mol of fraction water and the bottom product contains 0.2 mol fraction. The feed is 50% vapor and the reflux ratio is 3.0. The equilibrium data are given as: (x , y) = (0.0,0.0), (0.1,0.168), (0.2,0.312), (0.3,0.437), (0.4,0.546), (0.5,0.643), (0.6,0.729), (0.7,0.807), (0.8,0.877), (0.9,0.941), (1.0,1.0), where x is the mol fraction of water in liquid and y is the mol fraction of water in vapor. Calculate: (a) the minimum number of plates; (b) the minimum reflux ratio; and (c) the actual number of plates. Modify Program Listing 4.5, the Distillation Program, and the corresponding header file to solve this problem.

5) Modify Program Listing 4.4, the Evaporation Program, to solve the problem below:

An 80,000 lb/hr NaOH solution with 7% concentration is to be fed into a triple-effect forward-feed evaporator. The solution enters at 75°F and is to be concentrated to 45% NaOH. The overall coefficients of the first, second, and third effects are 750, 500, and 300 Btu/hr ft^2°F, respectively. The steam is available at 40 psia and the vapor space in the third effect is maintained at 1 psia. The heating surface areas are equal. Determine the heating area for each effect, the steam consumption, and the corresponding steam economy.

Convert Program Listing 3.4 (NaOH Water Duhring Line Program) and Program Listing 3.5 (Enthalpy-Concentration Program) from Chapter 3 into functions and store in a separate header file to be linked to your program.

Chapter 5   OVERVIEW OF C++

C++ is an improved version of C in terms of features and programming style capabilities, which is why we see an increment symbol ++ after the C. Since C++ is a superset of C, we can use the C++ compiler to compile existing C programs (but not the other way around). In this way, we can improve the C program structure to pattern it to C++.

The best way to learn C++ syntax is to compare it with the language we already know–C. We will be reviewing the C program listings presented previously and comparing them with the way C++ programs are developed with respect to the same problems. The transition from C to C++ will be much easier, as we can always relate it to the language we are familiar.

In this chapter we will cover:

C and C++ Syntax
    Basic Output and Input Stream
    Conditional Statements
    Looping Statements
    Arrays
    Functions

Object-Oriented Programming (OOP) in C++
    Class

Distillation Program Listing Revisited

All program listings discussed in this chapter can be found on the CD-ROM.

C AND C++ SYNTAX

Basic Output and Input Stream

Like the `printf()` and `scanf()` functions in C, C++ also has equivalent yet more powerful input and output streams. These are the `cout` for the output statement and the `cin` for the input statement. To see application of these statements, we revisit programs in Chapter 1 and convert them to C++ format. All C++ code will be compared with the original C program listing.

Example Problem 5.1. Make a C++ program that can calculate the product of two numbers.

Program Listing 5.1: Product of two numbers
```
1   #include <iostream.h>
2   #include <stdlib.h>
```

```
3   int main()
4   {
5    int product,n1,n2;
6    cout<<"Enter two whole numbers\n";
7    cout<<"\n First Number: ";
8    cin>>n1;
9    cout<<"\n Second Number: ";
10   cin>>n2;
11   product=n1*n2;
12   cout<<"\n The product is "<<product<<".\n"<<endl;
13   system("pause");
14   return 0;
15  }
```

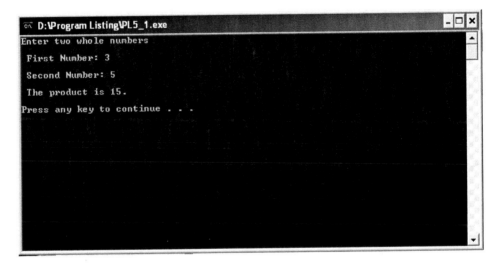

FIGURE 5.1: Output of Program Listing 5.1.

The new standard library header file `<iostream.h>` introduced contains prototypes for the standard input and output functions. The `cout` statement displays data to the standard output device, and the `cin` statement reads data from the standard input device. The `endl` stream manipulator (abbreviation for end line) outputs a new line and then clears out the output buffer.

A new operator (`<<`) is used together with `cout` to display a message on the screen. Similarly, the operator (`>>`) is also used together with `cin` to accept values from the program user. If these symbols are unfamiliar, it is because these are only present in C++ and not in C. Table 5.1 provides some other operators found only in C++.

Furthermore, %d, %f, and other data format types will no longer be necessary in the quoted string (Line 12 of Program Listing 5.1).

Table 5.1: Additonal operators in C++

Symbol	Meaning
::	unary or binary scope resolution
<<	bitwise left shift
>>	bitwise right shift
.*	pointer to member via object
->	pointer to member via pointer

Example Problem 5.2. Make a C++ program that can compute the average of three input numbers.

Program Listing 5.2: Average of three numbers

```
1    #include<iostream.h>
2    #include <stdlib.h>
3    int main()
4    {
5    double num1,num2,num3;
6    double ave;
7    cout<<"Enter three numbers\n";
8    cout<<"\n First Number: ";
9    cin>>num1;
10   cout<<"\n Second number: ";
11   cin>>num2;
12   cout<<"\n Third number: ";
13   cin>>num3;
14   ave=(num1+num2+num3)/3;
15   cout<<"\nThe average is "<<ave<<"."<<endl;
16   system("pause");
17   return 0;
18   }
```

Lines 1 and 2: The #include statement tells the C++ compiler to use the <iostream.h> and <stdlib.h> header files, which contain the standard input/output cout and cin and the console input/output function system() from C.

Line 3: The main() function is where the main operation of the program occurs.

Lines 5, 6: Declaration of variables num1, num2, num3, and ave as doubles.

Line 7: The `cout` prints out the statement on the screen. The << operator indicates that the statement on the right will be printed out on the screen. The control character \n ensures that the statement occupies a new line before being printed.

Lines 9,11,13: The `cin` requests three values from the user and stores these values in the `num1`, `num2`, and `num3` variables. The >> operator indicates to the variable where to store the inputted values.

Line 14: The three variables are processed using addition (+) and division (/) operation to give a resulting value, which in turn assigns this to variable `ave`.

Line 15: The `ave` is printed out on the screen using `cout`.

Line 16: The program pauses while waiting for the user to hit any key before terminating.

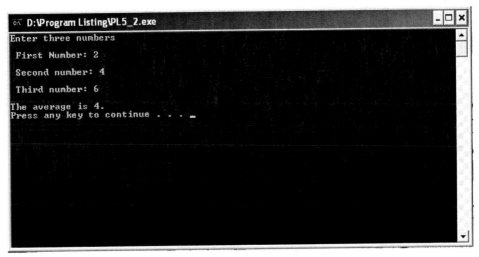

FIGURE 5.2: Output of Program Listing 5.2.

Conditional Statements

Conditional statements available in C are also available for use in C++. Revisiting previous example problems on conditional statements in Chapter 1 will illustrate how equivalent C++ programs can be developed.

Example Problem 5.3. Write a program that determines if the input Reynolds number indicates laminar flow (less than 2100) or not (greater than or equal to 2100). Use the `if - else` statement in C++.

Program Listing 5.3: Using the `if - else` statement

```
1    #include <iostream.h>
2    #include <stdlib.h>
3    int main()
4    {
5     int nre;
6     cout<<"\nEnter Reynolds number: ";
7     cin>>nre;
8     if(nre<2100)
9       cout<<"\nThe flow is laminar.\n"<<endl;
10    else
11      cout<<"\nFlow is not laminar.\n"<<endl;
12    system("pause");
13    return 0;
14   }
```

Line 5: Declaration of variable `nre` as an integer.

Line 7: The `cin` requests a value from the user and stores this in the `nre` variable through the `>>` operator.

Lines 8 to 11: These lines cover the `if - else` statements. When conditions are met, the corresponding `cout <<` stream is initiated to print the desired output statement.

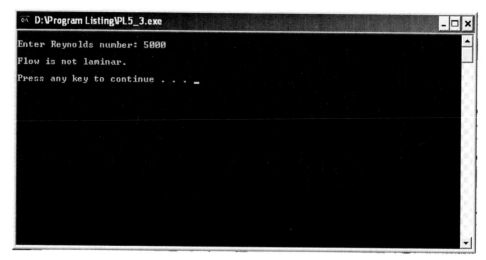

FIGURE 5.3: Output of Program Listing 5.3.

Example Problem 5.4. Write a C++ program that accepts a temperature in °C and then determines the phase of water (solid, liquid, or gas). Use the nested `if - else if` statement.

Program Listing 5.4: Using the nested `if - else if` statement

```
1   #include <iostream.h>
2   #include <stdlib.h>
3   int main()
4   {
5    int temp;
6    cout<<"\nEnter the temperature in deg C: ";
7    cin>>temp;
8    if(temp>100)
9     cout<<"\nThe water is in gas phase.\n"<<endl;
10   else if(temp <=100 && temp>0)
11    cout<<"\nThe water is in liquid phase.\n"<<endl;
12   else
13    cout<<"\nWater is in solid phase.\n"<<endl;
14   system("pause");
15  return 0;
16  }
```

Line 5: Declaration of variable `temp` as an integer.

Line 7: The `cin` input stream requests a value from the user and stores this in the `temp` variable.

Lines 8 to 13: Cover the nested `if - else if` statements. When conditions are met, the corresponding `cout<<` stream is initiated to print the desired output statement.

Line 10: Uses the logical operator (`&&`), which tells the computer that the condition will only be true when `temp` is less than or equal to 100 and greater than 0.

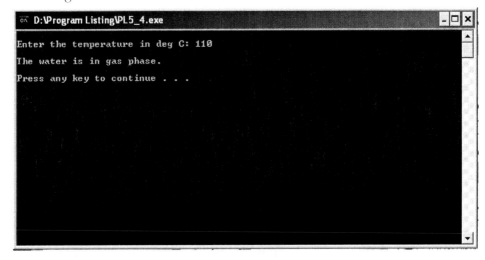

FIGURE 5.4: Output of Program Listing 5.4.

Example Problem 5.5. Make a C++ program that accepts a value to convert and allows the user to choose from a given unit of temperature to another unit of temperature. Use the `switch/case` statement in C++.

Program Listing 5.5: Using the `switch/case` statement

```
1   #include <iostream.h>
2   #include <stdlib.h>
3   int main()
4   {
5   double init_temp, final_temp;
6   char unit;
7   cout<<"\n Enter value to be converted: ";
8   cin>>init_temp;
9   cout<<"\n Select initial and final units:";
10  cout<<"\n A) deg C to F";
11  cout<<"\n B) deg C to K";
12  cout<<"\n C) deg F to R";
13  cout<<"\n D) deg F to C";
14  cout<<"\n E) K to deg C";
15  cout<<"\n F) R to deg F \n";
16  cin>>unit;
17  switch(unit){
18    case 'A':case 'a':final_temp= (init_temp*1.8)+32;break;
19    case 'B': case 'b':final_temp= (init_temp+273);break;
20    case 'C': case 'c':final_temp= (init_temp+460);break;
21    case 'D': case 'd':final_temp= (init_temp-32)/1.8;break;
22    case 'E': case 'e':final_temp= (init_temp - 273); break;
23    case 'F': case 'f':final_temp= (init_temp - 460);break;
24    default: cout<<"\n Choice unavailable";break;
25  }
26  cout<<"\n The converted value is: "<<final_temp<<" "<<endl;
27  system("pause");
28  return 0;
29  }
```

Lines 5, 6: Declaration of variable `init_temp` and `final_temp` as an integer, and `unit` as a character.

Line 8: The input stream `cin>>` requests a value from the user and stores this in the `init_temp` variable.

Lines 9 to 15: Print out the selection menu.

Line 16: The input stream `cin>>` gets the selected option and stores it in the `unit` character variable.

Lines 17 to 25: Cover the switch statements. When conditions are met, the corresponding calculation is done. The break command is used to skip the other conditions.

Line 26: The result is printed using the cout<< output stream.

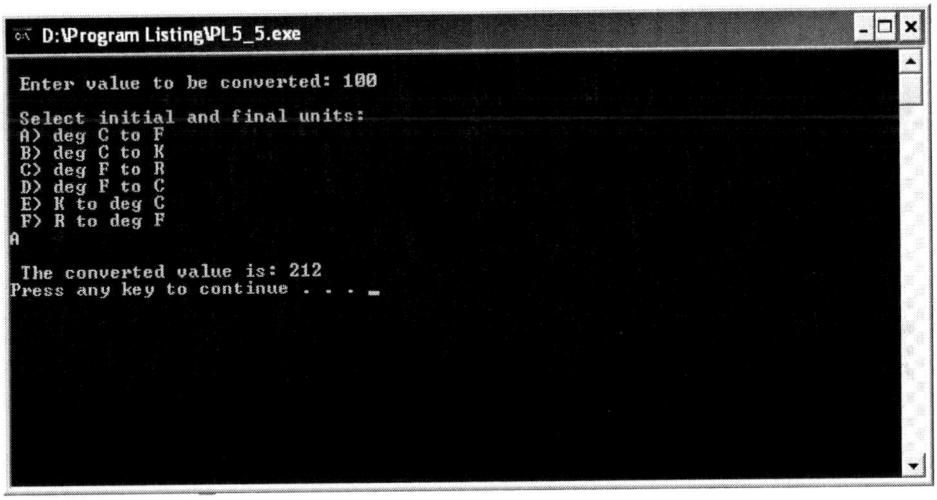

```
ⒸⓍ D:\Program Listing\PL5_5.exe                            _ □ ×

Enter value to be converted: 100

Select initial and final units:
A> deg C to F
B> deg C to K
C> deg F to R
D> deg F to C
E> K to deg C
F> R to deg F
A

The converted value is: 212
Press any key to continue . . . _
```

FIGURE 5.5:   Output of Program Listing 5.5.

Looping Statements

Looping statements, like Conditional statements, are common to both C and C++. Dealing with the same example problems in Chapter 1 can provide a good comparison between C and C++ programs.

Example Problem 5.6. Write a C++ program that will compute for the concentration of A in a zero order A→ B reaction from start (0 sec) up to one minute for every 10 seconds given Ca_0 15 g mol/L and k is 0.0567 g mol/L-sec.

Using the for loop statement in C++.

Program Listing 5.6: Using the for loop statement

```
1   #include <iostream.h>
2   #include <stdlib.h>
3   int main()
4   {
5     double conc, init_con,k;
6     int time;
```

```
 7   init_con = 15.00;
 8   k = 0.0567;
 9   cout<<»\nTIME(sec)   CONC.(g mol/L)»;
10   for (time=0; time<=60; time = time+10)
11   {
12    conc = init_con - (k*time);
13    cout<<"\n   "<<time<<"\t\t"<<conc<<"\n";
14   }
15   cout<<endl;
16   system("pause");
17   return 0;
18 }
```

Lines 5, 6: Declaration of variables k, init_con, and conc as a double, and time as an integer.

Lines 7, 8: Initialize init_con with a value of 15.00 and k with a value of 0.0567.

Lines 10 to 14: These lines cover the for loop statement. In the loop, cout<< is utilized to print the resulting conc. The loop stops when the value of time reaches 60.

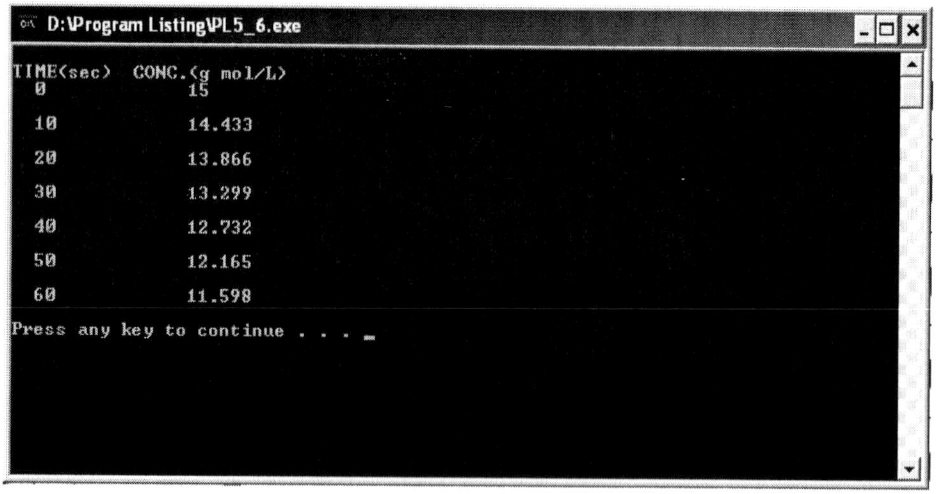

FIGURE 5.6:   Output of Program Listing 5.6.

Addressing the same problem in Example Problem 5.6 but using the while statement in C++ we developed the following program.

Program Listing 5.7: Using the `while` statement

```
1  #include <iostream.h>
2  #include <stdlib.h>
3  int main()
4  {
5  double conc, init_con,k;
6  int time;
7  init_con = 15.00;
8  k = 0.0567;
9  time =0;
10 cout<<»\nTIME(sec)  CONC.(g mol/L)»;
11 while (time<=60)
12 {
13  conc = init_con - (k*time);
14  cout<<"\n  "<<time<<"\t\t"<<conc;
15  time=time+10;
16 }
17 cout<<endl;
18 system("pause");
19 return 0;
20 }
```

Lines 12 to 17: Similarly, these lines cover the `while` loop statement. In the loop, `cout<<` is utilized to print the resulting `conc`. The loop stops when the value of `time` reaches 60 as indicated in the condition under the `while()` function.

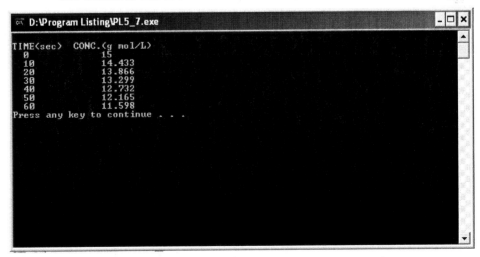

FIGURE 5.7: Output of Program Listing 5.7.

Address the same problem in Example Problem 5.6 using the do - while statement in C++.

Program Listing 5.8: Using the do - while statement

```
1   #include <iostream.h>
2   #include <stdlib.h>
3   int main()
4   {
5    double conc, init_con,k;
6    int time;
7    init_con = 15.00;
8    k = 0.0567;
9    time =0;
10   cout<<»\nTIME(sec)   CONC.(g mol/L)»;
11   do
12   {
13    conc = init_con - (k*time);
14    cout<<"\n   "<<time<<"\t\t"<<conc;
15    time=time+10;
16   }while (time<70);
17   cout<<endl;
18   system("pause");
19   return 0;
20  }
```

Lines 11 to 16: Cover the do - while statement. Note that looping is done as long as time is less than 70. The statements within the curly braces { } will be done at least once regardless of the condition.

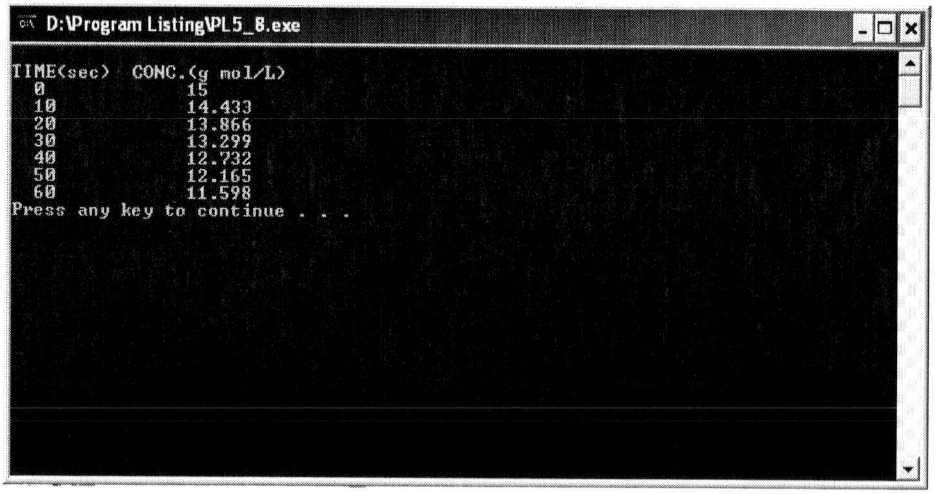

FIGURE 5.8: Output of Program Listing 5.8.

Arrays

In C++, arrays are handled similarly to C. We can use the same example problem in Chapter 1 and develop an equivalent C++ program to see the similarities.

Example Problem 5.7. Write a C++ program that will print out onscreen a temperature conversion table. This table should present the given values of temperature in degree Centigrade (°C) with the corresponding equivalents in degrees Fahrenheit (°F), Kelvin, and Rankine.

Program Listing 5.9: Using Arrays

```
1   #include <iostream.h>
2   #include <stdlib.h>
3   int main()
4   {
5    double temp[4][4];
6    int i;
7    for(i=0;i<4;i++)
8    {
9     temp[i][0] = (i*50)+100;
10    temp[i][1] = (temp[i][0]*1.8)+32;
11    temp[i][2] = temp[i][0]+273;
12    temp[i][3] = temp[i][1]+460;
13   }
14   cout<<"\nTable of Temperature Conversion";
15   cout<<"\n deg C    deg F       K          R";
16   for(i=0;i<4;i++)
17   cout<<"\n   "<<temp[i][0]<<"         "<<temp[i][1]<<"
     "<<temp[i][2]<<"        "<<temp[i][3];
18   cout<<endl;
19   system("pause");
20   return 0;
21  }
```

Lines 5 and 6: Declaration of variable `i` as an integer and `temp[][]` as an array double.

Lines 7 to 13: The `for` loop is used to initialized the `temp[][]` array with different values based on the mathematical operation performed.

Lines 16 to 17: Prints out the values of the array `temp[][]` using again the `for` loop statement.

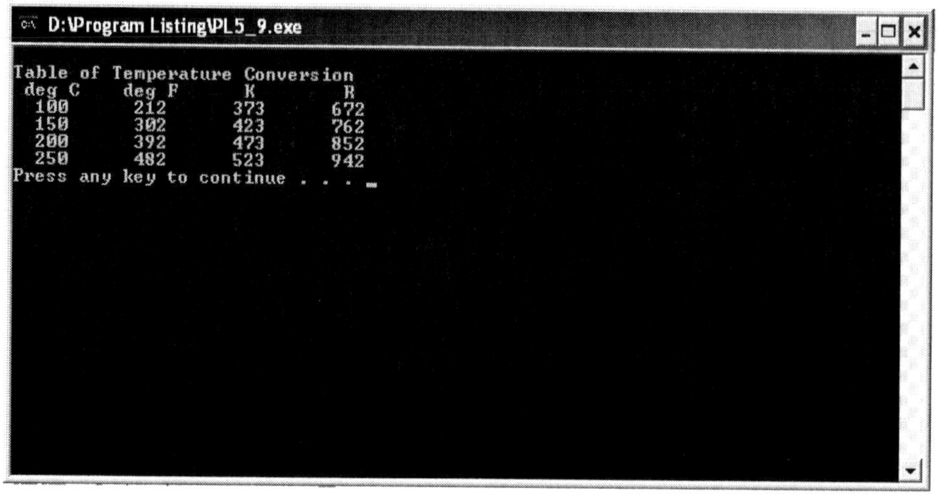

FIGURE 5.9:   Output of Program Listing 5.9.

Functions

Like C, we can also develop user defined functions in C++, but there are some differences when it comes to its form and its use of preprocessor directives as observed in the following example problem and program listings.

Example Problem 5.8. Revise Program Listing 5.1 to make use of a new function `multiply()`, which is included in a separate `MULTIPLY.hpp` header file. Place the function `multiply()` in class `MULTIPLY`.

Program Listing 5.10: Multiply header file

```
1   /*MULTIPLY.hpp*/
2   #ifndef MULTIPLY_H
3   #define MULTIPLY_H
4   class MULTIPLY{
5    public:
6    int multiply(int, int);
7   };
8   #endif
```

Lines 2, 3, 8: The `#ifndef`, `#define`, and `#endif` preprocessors are used to prevent the `MULTIPLY.hpp` header file from being included more than once in the program that will use it.

It is a good practice to use the name of the header file with the period replaced by an underscore (_) in the #ifndef *and* #define *preprocessor directives of a header file.*

Program Listing 5.11: Multiply (Member) Function

```
1  /*MULTIPLY.cpp*/
2  #include "MULTIPLY.hpp"
3  int MULTIPLY::multiply(int n1, int n2)
4  {
5  int ans;
6  ans = n1*n2;
7  return(ans);
8  }
```

Line 3: The use of binary scope resolution (::) in the member function links the member name to the class name MULTIPLY to uniquely identify the member function multiply().

Program Listing 5.12: Main Program

```
1  #include <iostream.h>
2  #include <stdlib.h>
3  #include "MULTIPLY.hpp"
4  int main()
5  {
6  MULTIPLY M;
7  int product,num1,num2;
8  cout<<"\n Enter two whole numbers.\n";
9  cout<<"\n First number: ";
10 cin>>num1;
11 cout<<"\n Second number: ";
12 cin>>num2;
13 product=M.multiply(num1,num2);
14 cout<<"\n The product is "<<product<<"."<<endl;
15 system("pause");
16 return 0;
17 }
```

Line 3: The #include statement tells the C++ compiler to include the "MULTIPLY.hpp" header file, which contains the function multiply().

Line 6: Declares M as a user-defined type MULTIPLY (class).

Line 12: Uses the M prototype function multiply().

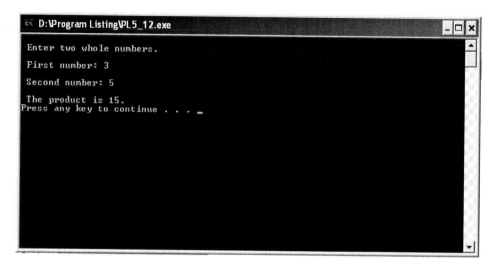

FIGURE 5.10: Output of Program Listing 5.12.

Through these examples, we have learned how to make C++ programs in their simplest and basic form. We also saw the similarities and differences of its programming style from C. Furthermore, Example Problem 5.8 touched on the concept of object-oriented programming. We will, in fact, discuss OOP further in the next section.

OBJECT-ORIENTED PROGRAMMING (OOP) IN C++

As we all know, the main purpose of developing a computer program is to help solve real-world problems that involve materials, equipment, or ideas. In fact, the problems discussed so far tackle real situations. Similarly, in object-oriented programming, programs use objects to represent real materials, equipment, or ideas. Since materials and ideas can be used again and again, the concept used for these objects is reusable. In this way we can easily develop other programs by reusing existing codes. Also, with reuse and rigorous application, these objects will become highly reliable and easy to maintain.

Aside from objects, other conceptual elements associated with object-oriented design include:

- Classes – A representation of an entity or object. The essential core of a class is the set of data that collectively describes the object. These sets of data are accessible through a set of interfaces that controls the operation.

- Encapsulation – Enclosing data and information within the object. By encapsulating information, it can only be accessed by procedures that have been defined as part of the class to which the object belongs. Encapsulation keeps related objects together.
- Information Hiding – Takes encapsulation one step further. Not only can the object never be accessed directly, its information can only be known by the one implementing the class.
- Message Passing – A necessary result of encapsulating information, wherein one object affects the condition of another object by sending a message that requests the second object to execute one of its procedures.
- Inheritance – A mechanism by which a new class can be readily derived from an existing base class, inheriting all its attributes and methods. The new class has the option of changing, as well as adding new methods and attributes. This somewhat exemplifies software reuse.

With the complexity of an object, using a single variable or even an array will not suffice. The data of an object that describes it cannot be stored in a variable represented by only one data type. That is why C is not designed for object-oriented programming, but C++ is. C++ allows us to package related variables into a class.

Class

A class can contain many variables with varying data types, allowing the program to represent closely a real-world object. Even though class can still be considered a data type, it can be defined by the programmer, unlike the existing data type we learned in Chapter 1 (i.e., float and integer).

Below is an example of class declaration from Program Listing 5.10:

```
1   class MULTIPLY{
2     public:
3     int multiply(int, int);
4   };
```

Line 1: The declaration starts with the syntax `class`, followed by the name of the class, in this case "`MULTIPLY`".

Lines 2 to 4: This is the body of the class, enclosed in curly braces. Note that it should always end with a semicolon (;).

The class declaration is usually placed in a header file. This header file is included in the program that will be using the class. This forms the class's public interface that provides the program using it with the function prototype it needs to be able to call the member function.

Now it is time to deal with more complex source code.

DISTILLATION PROGRAM LISTING REVISITED

We break up the original large Distillation Program in C into various parts, each of which performs a specific and separate task such as

-   determining the minimum number of stages in the distillation process,
-   determining the minimum reflux ratio, and
-   determining the actual number of stages.

Actually, we improved the C program for the distillation problem (Program Listing 4.5) and added features to handle water-acetic acid and methanol water-in addition to the benzene-toluene system. By applying similar regression analysis and a validation procedure, we get the following polynomial equation for the acetic acid water system:

$$y = 0.001713 + 1.751865x - 1.113520x^2 + 0.361305x^3$$
$$x = -0.000698 + 0.581927y + 0.101547y^2 + 0.316126y^3$$

And the following for the methanol water system:

$$y = 0.000388 + 3.399670x - 6.406920x^2 + 7.910111x^3 - 5.473485x^4 + 1.570513x^5$$
$$x = -0.000030 + 0.713269y - 3.256688y^2 + 9.906105y^3 - 11.799846y^4 + 5.434529y^5$$

We then developed the program for each binary equilibrium system. See Program Listings 5.13 to 5.18 for the acetic acid water system, benzene-toluene, and the methanol water system, respectively. Each of which is represented in separate classes: `WAT_ACET`, `BENZ_TOL`, and `METH_WAT`.

Program Listing 5.13: WAT_ACET Header File

```
/*WAT_ACET.hpp*/
#ifndef WAT_ACET_H
#define WAT_ACET_H
class WAT_ACET{
   public:
        double wateracetic(double, int);
};
#endif
```

Program Listing 5.14 WAT_ACET Function

```
/*WAT_ACET.cpp*/
#include <math.h>
```

```
#include "WAT_ACET.h"
double WAT_ACET::wateracetic(double xy1,int xory)
{
 double ans;
 if(xory==2)
  ans=0.001713+(1.751865*xy1)-(1.113520*pow(xy1,2))+(0.3
  61305*pow(xy1,3));
 else
  ans=-0.000698+(0.581927*xy1)+ (0.101547*pow(xy1,2))+(0
  .316126*pow(xy1,3));
 return ans;
}
```

Program Listing 5.15 BENZ_TOL Header File

```
/*BENZ_TOL.hpp*/
#ifndef BENZ_TOL_H
#define BENZ_TOL_H

class BENZ_TOL{
    public:
         double benzenetoluene(double, int);
};
#endif
```

Program Listing 5.16 BENZ_TOL Function

```
/*BENZ_TOL.cpp*/
#include "BENZ_TOL.hpp"
#include <math.h>
double BENZ_TOL::benzenetoluene(double xy1,int xory)
{
 double ans;
 if(xory==2)
  ans=0.000068+(2.370166*xy1)-(2.880267*pow(xy1,2))
  + (2.689467*pow(xy1,3))-(1.612034*pow(xy1,4))  +
  (0.432692*pow(xy1,5));
 else
  ans=-0.000053+(0.473581*xy1)-(0.276536*pow(xy1,2))+
  (1.749841*pow(xy1,3))-(2.133077*pow(xy1,4))  +
  (1.185847*pow(xy1,5));
 return ans;
}
```

Program Listing 5.17 METH_WAT Header File

```
/*METH_WAT.hpp*/
#ifndef METH_WAT_H
#define METH_WAT_H

class METH_WAT{
    public:
            double methanolwater(double, int);
};
#endif
```

Program Listing 5.18 METH_WAT Function

```
/*METH_WAT.cpp*/
#include "METH_WAT.hpp"
#include <math.h>

double METH_WAT::methanolwater(double xy1,int xory)
{
 double ans;
 if(xory==2)
  ans=0.000388+(3.399670*xy1)-(6.406920*pow(xy1,2))-
  (7.910111*pow(xy1,3))-(5.473485*pow(xy1,4))  +
  (1.570513*pow(xy1,5));
 else
  ans=-0.000030+(0.713269*xy1)-(3.256688*pow(xy1,2))+
  (9.906105*pow(xy1,3))-(11.799846*pow(xy1,4))+(5.43452
  9*pow(xy1,5));
 return ans;
}
```

The next step is to come up with functions to address the major tasks performed in the distillation program. These functions will then be part of a class to be developed. In making the functions, we copy the algorithm from the previous C program. Here are the functions that are now part of the DISTILL class found in the DISTILL.h header file (Program Listing 5.19).

Table 5.2: Additonal functions in DISTILL.h Header File

Function Name	Operation
minimumPlates()	Solve for minimum number of plates.
minimumReflux()	Determine the minimum reflux ratio.
actualPlates()	Solve for the actual number of plates required.

Program Listing 5.19: DISTILL Header File

```
/*DISTILL.hpp*/
#ifndef DISTILL_H
#define DISTILL_H
class DISTILL{
 public:
   DISTILL(char, double, double, double, double);
   double compute(double, int);
   void minimumPlates();
   void minimumReflux();
   void actualPlates(double);
 private:
   char choice;
   double xb, xd, xf, q;
};
#endif
```

Program Listing 5.20: DISTILL Functions

```
/* DISTILL.cpp */
#include <iostream.h>
#include <stdlib.h>
#include «DISTILL.hpp»
#include «WAT_ACET.hpp»
#include «BENZ_TOL.hpp»
#include «METH_WAT.hpp»
DISTILL::DISTILL(char choice, double xb, double xd, double
xf, double q){
 this->choice=choice;
 this->xb=xb;
 this->xd=xd;
 this->xf=xf;
 this->q=q;
 }

double DISTILL::compute(double fValue, int nValue){
 double result;
 switch(choice){
  case 'A':WAT_ACET WAT;
    result=WAT.wateracetic(fValue,nValue);break;
  case 'B':BENZ_TOL BENZ;
    result=BENZ.benzenetoluene(fValue,nValue);break;
```

```
  case 'C':METH_WAT METH;
   result=METH.methanolwater(fValue,nValue);break;
 }
 return result;
}


void DISTILL::minimumPlates(){
 int j=0,n;
 double ycoor[100],xcoor[100];
 ycoor[0]=xd;
 cout<<»MINIMUM NUMBER OF PLATES»<<endl;
 do{
  j++;
  xcoor[j]=compute(ycoor[j-1],1);
  ycoor[j]=xcoor[j];
  n=j;
  }while(xcoor[j]>xb);
   for(j=1;j<=n;j++)
   cout<<»fraction    x=»<<xcoor[j]<<»        y=»<<ycoor[j-
   1]<<»          plate no.=»<<j<<endl;
  }


void DISTILL::minimumReflux(){
 double xcoorr,ycoorr,dcoorr,mr,bmin,rmin,ycoorf;
 if (q==1)
 {
  xcoorr=xf;
  ycoorr=compute(xcoorr,2);
 }
 else if (q==0)
 {
  ycoorr=xf;
  xcoorr=compute(ycoorr,1);
 }
 else
 {
  if (q>1)
   xcoorr= xf;
  else if (q < 0)
   xcoorr= compute(xb,1);
  else
```

```
  xcoorr= compute(xf,1);
 do
 {
  xcoorr=xcoorr+0.00000005;
  ycoorf=(((-1*q)/(1-q))*xcoorr)+(xf/(1-q));
  ycoorr= compute(xcoorr,2);
  dcoorr=ycoorf-ycoorr;
 }while(dcoorr > 0.000001 || dcoorr< -0.000001);
 }
 mr=(xd-ycoorr)/(xd-xcoorr);
 bmin=ycoorr-(mr*xcoorr);
 rmin=(xd/bmin)-1;
 cout<<»Minimum reflux is = «<<rmin<<endl;
 }

void DISTILL::actualPlates(double ref){
 double xcoorr,delm,delb,ycoorr,mb,bact,ycoor[100],xcoor
 [100];
 int j,fplate,n;
 if(q==1)
  (xcoorr=xf);
 else
 {
  delm=(xd/(ref+1))-(xf/(1-q));
  delb=((-1*q)/(1-q))-(ref/(ref+1));
  xcoorr=delm/delb;
 }
 ycoorr=((ref/(ref+1))*xcoorr)+(xd/(ref+1));
 mb=(ycoorr-xb)/(xcoorr-xb);
 bact=ycoorr-(mb*xcoorr);
 j=0;ycoor[0]=xd;
 do
 {
  j++;
  xcoor[j]=compute(ycoor[j-1],1);
  if(xcoor[j] >= xcoorr)
  {
   ycoor[j]=((ref/(ref+1))*xcoor[j])+(xd/(ref+1));
   fplate=j+1;
  }
  if(xcoor[j] < xcoorr) (ycoor[j]=(mb*xcoor[j])+bact);
```

```
      n=j;
    }while(xcoor[j]>xb);
    cout<<»ACTUAL NUMBER OF PLATES\r\n»;
    for(j=1;j<n+1;j++)
    cout<<»fraction   x=»<<xcoor[j]<<»        y=»<<ycoor[j-1]
    <<»       plate no.=»<<j<<endl;
    cout<<»\nFeed is introduce in plate number
    «<<fplate<<endl;
    }
```

Program Listing 5.21: DISTILL Main Program

```
1   #include <iostream.h>
2   #include <stdlib.h>
3   #include <ctype.h>
4   #include <string.h>
5   #include "DISTILL.hpp"
6   void whatType(char, char*);
7   void menu();
8   int main()
9   {
10  double xd=0,xf=0,xb=0,q,ref;
11  char choice,type[10];
12  menu();
13  do{
14   choice=toupper(getch());
15  }while(choice<'A'||choice>'C');
16  whatType(choice, type);
17  cout<<choice<<"\n\nEnter fraction of "<<type<<" in
    distillate : ";
18  cin>>xd;
19  cout<<"Enter fraction of "<<type<<" in feed : ";
20  cin>>xf;
21  cout<<"Enter fraction of "<<type<<" in bottom : ";
22  cin>>xb;
23  cout<<"Enter q value : ";
24  cin>>q;
25  DISTILL D(choice, xb, xd, xf, q);
26  /* for Minimum number of Plates */
27  D.minimumPlates();
28  /* for Minimum Reflux Ratio */
29  D.minimumReflux();
30  cout<<"Enter reflux ratio : ";
```

```
31   cin>>ref;
32   /* for Actual number of Plates */
33   D.actualPlates(ref);
34   system("pause");
35   return 0;
36  }
37  void whatType(char choice, char *type){
38   switch(choice){
39    case 'A':strcpy(type,"water"); break;
40    case 'B':strcpy(type,"benzene"); break;
41    case 'C':strcpy(type,"methanol"); break;
42   }
43  }
44  void menu(){
45   cout<<"        [A] WATER - ACETIC ACID"<<endl;
46   cout<<"        [B] BENZENE - TOLUENE"<<endl;
47   cout<<"        [C] METHANOL - WATER"<<endl;
48   cout<<"Select the equilibrium system to use: ";
49  }
```

Lines 1, 2: The #include statement tells the C++ compiler to use the <iostream.h> and <stdlib.h> header files, which contain the cout, cin and function system() from C.

Lines 3 and 4: The #include statement tells the C++ compiler to use the <ctype.h> and <string.h> header files, which contain the strcpy() and toupper() functions.

Line 5: Contains the #include DISTILL.hpp header file, which contains the functions listed in Table 5.2.

Line 6: Calls on the whatType() function, which is in Lines 37 to 43 and informs the compiler that this function will be used in the main program.

Line 7: Calls on the menu() function in Lines 44 to 49 and informs the compiler that this function will also be used in the main program.

Line 8: The main() function required in every C program is also required in every C++ program. This is where the main operation is done.

Line 12: Uses the function menu() to print the selection on screen.

Line 16: Uses the function whatType() to determine the user's selected binary equilibrium system and stores the keyword (water, benzene, or methanol) in the character variable type.

Lines 19 to 24: Requests the needed information, which includes the mass fractions at distillate, feed, and bottom, as well as the q value.

Line 25: Calls on the DISTILL class prototype D and transfers the inputted information for use of the different functions in the class.

Line 27: Calls on the minimumPlates() function to determine the minimum number of plates.

Line 29: Performs the routine to determine the minimum reflux ratio.

Lines 30 to 31: Requests the reflux ration needed to perform the task in function actualPlates().

Line 33: Calls on the actualPlate() function to determine the actual number of plates.

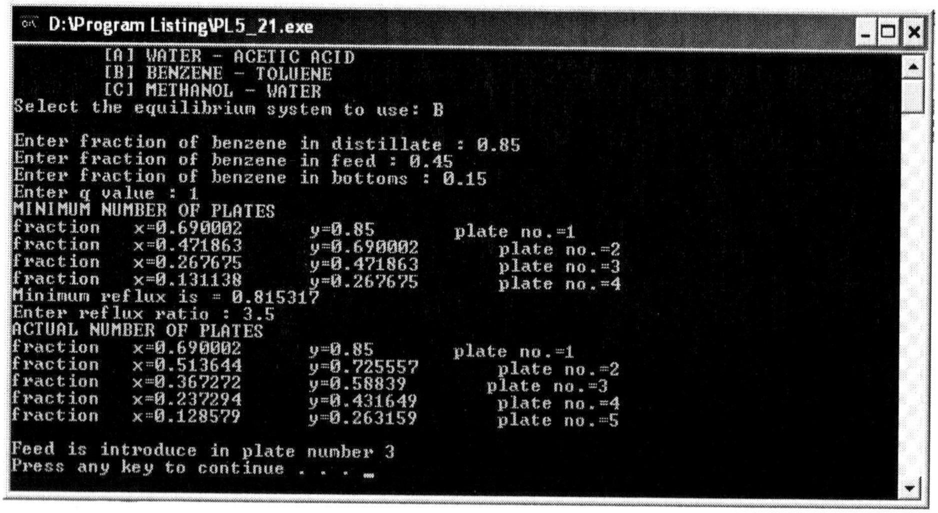

FIGURE 5.11: Output of Program Listing 5.21.

LABORATORY EXERCISES

1) Convert the Evaporation Program (Program Listing 4.2) from C to C++ using OOP design. Try solving Example Problem 4.8 and compare the answers of the newly developed C++ program from the solution provided.

2) Write a C++ program that can determine the root of a 3rd order polynomial equation using the Müller method. Provided below is the section of the program that performs the routine. The user needs to input the initial estimated root value and store it in variable xr, as well as provide values for ax[0], ax[1], ax[2], and ax[3] as the coefficients of the polynomial equation $0 = a_0 + a_1 x + a_2 x^2 + a_3 x^3$.

```
x[2]=xr;
x[1]=xr+(0.1*xr);
```

```
x[0]=xr-(0.1*xr);
do{
  fx[2]=0;fx[1]=0;fx[0]=0;
  for(i=0;i<3;i++)
  {
    for(j=3;j>=0;j--)  fx[i]=fx[i]+(ax[j]*pow(x[i],j));
  }
  h[0]=x[1]-x[0];
  h[1]=x[2]-x[1];
  d[0]=(fx[1]-fx[0])/h[0];
  d[1]=(fx[2]-fx[1])/h[1];
  a=(d[1]-d[0])/(h[1]+h[0]);
  b=a*h[1]+d[1];
  c=fx[2];
  rad2=(b*b)-(4*a*c);
  rad=sqrt(rad2);
  if(fabs(b+rad)>fabs(b-rad))  den=b+rad;
  else den=b-rad;
  dxr=-2*c/den;
  xr=x[2]+dxr;
  x[0]=x[1];x[1]=x[2];x[2]=xr;
}while(fabs(dxr)>0.0001);
```

The final root is stored in variable `xr`, which must be printed on the screen at the end of the `do-while` loop statement.

3) Convert the Simplex Method Program (Program Listing 2.8) into C++ using the object-oriented programming style.

4) Encode the following source codes into a text editor, debug for any errors until successfully compiled. Take note of all errors identified.

a) Linear Regression Program

```
#include<iostream.h>
#include <stdlib.h>

int main()
{
  int n,i;
  double x,y,sumx,sum_sqx,sumy,xy,sumxy,b,a;
  clrscr();
  cout<<"\n Enter number of x and y pairs: ";
  cin>>\n;
  sumx=0;sum_sqx=0;sumy=0;sumxy=0;
```

```
    for(i=0;i<n;i++)
    {
     cout<<"Enter x: ";
     cin>>x;
     cout<<"Enter y: ";
     cin>>y;
     sumx=sumx+x;
     sum_sqx=sum_sqx+(x*x);
     sumy=sumy+y;
     sumxy=sumxy+(x*y);
    }
    b=(sumxy-(sumx*sumy)/n)/(sum_sqx-(sumx*sumx)/n);
    a=(sumy/n)-(b*sumx/n);
    cout<<" Equation  of  line  is  Y  =  "<<a<<"  +
"<<b<<"X"<<endl
    system("pause");
    }
```

b) Lagrange Program

```
    #include <iostream.h>
    #include <stdlib.h>

    int main()
    {
     int i,j,n;
     double xval, yval, x[20], y[20],L[20];
     char res;
     res='n';
     system("cls");
     cout<<"\nEnter number of x and y pairs: ";
     cin>>n;
     for(i=0;i<n;i++)
     {
      cout<<"\nEnter x["<<i<<"]: ";
      cin>>x[i];
      cout<<"Enter y["<<i<<"]: ";
      cin>>y[i];
     }
    do{
     cout<<"\nEnter value of x for unknown y: ";
     cin>>xval;
     yval=0;
```

```
for(i=0;i<n;i++);
{
 L[i]=1;
 for(j=0;j<n;j++)
 {
   if (j!=i) L[i]=L[i]*((xval-x[j])/(x[i]-x[j]));
 }
 yval=yval+(L[i]*y[i]);
 }
 cout<<"\nThe value of y is "<<yval<<" ";
 cout<<"\nTry again (Y/N)?";
 res==getch();
}while(res=='Y' || res=='y')
 return 0;
}
```

c) Gauss-Seidel Program

```
#include<iostream.h>
#include <conio.h>

main
{
 int r,c,num;
 double a[20][20],b[20];
 double x[20],x_temp[20],x_prev[20],flag;
 do{
  clrscr();
  cout<<"\n Enter the number of unknowns (20 max):";
  cin>>num;
 }while(num>=20);
 for (r=0; r<num; c++)
 {
  for (c=0; c<num; c++)
   {
    cout<<" Enter a["<<r+1<<"]["<<c+1<<"]:";
    cin>>a[r][c];
   }
   cout<<" Enter b["<<r+1<<"]:";
   cin>>b[r];
 }
 for (r=0; r<num; r++)
 {
```

```
 x[r]=0;
 x_temp[r]=0;
}
for (r=0; r<num; r++)
{
 for (c=0; c<num; c++)
 {
  x_temp[r]=x_temp[r]+(a[r][c]*x[c]);
 }
 x[r]=(b[r]-x_temp[r])/a[r][r];
 x_temp[r]=0;
 x_prev[r]=x[r];
}
x_temp[r]=0;
do{
  for (r=0; r<num; r++)
  {
   for (c=0; c<num; c++)
   {
   if(c==r) x_temp[r]=x_temp[r]+(a[r][c]*0);
   else x_temp[r]=x_temp[r]+(a[r][c]*x[c]);
   }
   x[r]=(b[r]-x_temp[r])/a[r][r];
   x_temp[r]=0;
   flag=fabs(x_prev[r]-x[r]);
   x_prev[r]=x[r];
  }
}while(flag>0.00000000001);
for (r=0;r<num; r++);
cout<<"| X["<<r+1<<"] = "<<x[r]<<" |"<<endl;
system("pause");
return 0;
}
```

PART II

USING
MATLAB

Chapter 6

INTRODUCTION TO MATLAB®

There are many approaches to teaching the basics of MATLAB. This book will introduce MATLAB through a C programmer's point of view. In this way, the transition from the first half of the book (learning C/C++ programming) to the second half of the book (understanding MATLAB programming) will be much easier.

Like C/C++ and other programming languages, MATLAB, has arithmetic, logical, conditional, and looping operations but with much, much more.

This chapter will introduce MATLAB and explore the following concepts:

MATLAB Environment
>   MATLAB Desktop
>   Command Window
>   Command History Window
>   Editor/Debugger
>   Workspace
>   Current Directory Window
>   Array Editor
>   Help Window

Developing M-Files
Variable and Constant Name Declaration
Basic Output and Input Statements
Output Statements
Input Statements

Operators
Arithmetic and Assignment Operators
Relational Operators
Logical Operators

Conditional Statements
`if-else` Statement
`if-elseif` Statement
`switch/case` Statement

Looping Statements
`for` Loop Statement
`while` Loop Statement

Scalars, Vectors, and Matrices
Initialization
Manipulating Arrays
Arithmetic Operation

Creating Functions
Plotting
Basic 2D Plotting
Axis Command Functions
Titles and Labels
Adding Lines to Graphs
Setting of Line Style and Width
Basic 3D Plotting

All program listings discussed in this chapter can be found on the CD-ROM.

```
\Program Listings\Chapter 6\flue.mat      Flue Temp
                                          Distribution
\Program Listings\Chapter 6\PL6_1.m       Program Listing 6.1
\Program Listings\Chapter 6\PL6_2.m       Program Listing 6.2
\Program Listings\Chapter 6\PL6_3.m       Program Listing 6.3
\Program Listings\Chapter 6\PL6_4.m       Program Listing 6.4
```

THE MATLAB ENVIRONMENT

Through the years MATLAB has gained popularity not only because of the accuracy of its results but also due to its interface, which provides a friendly and interactive environment where numerical calculations can be easily performed.

MATLAB utilizes windows to support all the activities and programming being done. Each window has its own function, which will be discussed briefly in the following sections.

MATLAB Desktop

The main window is called the MATLAB desktop. It is the primary space where most interactions in MATLAB happen. The MATLAB desktop window manages sub-windows associated with it. These sub-windows may be located and visible inside the main desktop window depending on how MATLAB is set up.

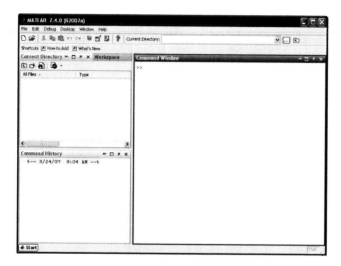

FIGURE 6.1: MATLAB desktop window.

Command Window

The MATLAB command window is a sub-window where a user can issue commands or instructions for MATLAB to process. The command window will display the prompt >> to signal that it is ready to accept instructions. Once the instruction is typed and the enter key is pressed, the instruction is immediately executed.

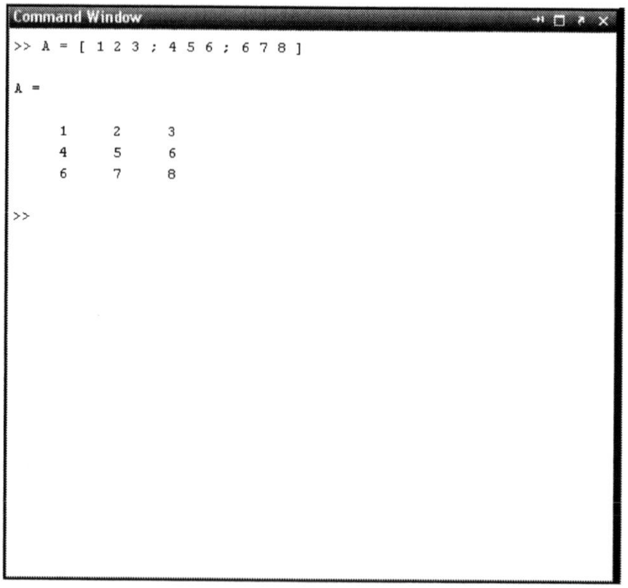

```
>> A = [ 1 2 3 ; 4 5 6 ; 6 7 8 ]

A =

     1     2     3
     4     5     6
     6     7     8

>>
```

FIGURE 6.2: MATLAB command window.

Command History Window

The command history window contains all the previous commands or statements declared in the command window. The list may extend even to the previous executions. It is also possible to delete one or more commands in the history list. This window may be handy during back-tracking of commands.

FIGURE 6.3: MATLAB command history window.

Editor/Debugger

The editor/debugger window provides an area where M-files can be created and edited prior to execution. Existing files can also be loaded for re-editing and debugging.

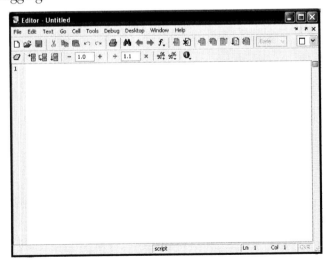

FIGURE 6.4: Editor/debugger window.

Workspace

The workspace is a user interface where loading, viewing, and saving of MATLAB variables are done.

Current Directory Window

The current directory window is an interface for viewing file directories associated with MATLAB from the current window.

Array Editor

The array editor is an interface for editing the values of MATLAB variables. Creating a variable and manually typing each value is possible in this window.

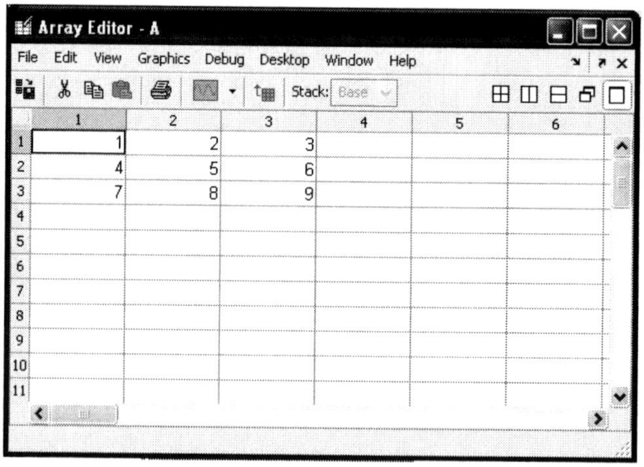

FIGURE 6.5: MATLAB array editor.

Help Window

The help window provides access to existing online help guides. The help windows consist of two sections: the display section and the help navigator. All information can be viewed through the display section. Under the help navigator, we can find the following components:

Search: We can use this if we want to look for a specific keyword.

Contents tab: Here we can view the titles and contents of the online help guide.

Index tab: Used to find relevant keywords arranged in alphabetical order.

Search results: Displays the results of the search command.

Demos tab: Here we can view and run demonstration programs included in the MATLAB package. It can also show the program listings of some of the demos.

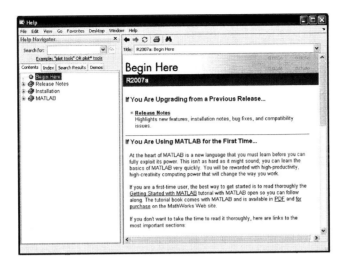

FIGURE 6.6: MATLAB help window.

DEVELOPING M-FILES

Entering a small set of instructions in the Command Window is easy and convenient. However, as the number of instructions increases, to deal with more complex problems, it is no longer practical to type all instructions in the command prompt (**>>**). A better approach is to store these instructions first in a text file, where editing is much easier, then once final, instruct MATLAB to load this file and read the instructions just as if it were typed in the command prompt. In this way, it is easier for the user to edit and change the programs if there are errors identified during execution. This type of text file (containing MATLAB instructions) is called an M-file, simply because the extension of the file name must be '.m' for MATLAB to recognize the file.

As mentioned earlier, all editing and debugging of M-files are done in the Editor/Debugger Window, and when this file is executed, the set of instructions in the M-file and the resulting action, can be seen in the Command Window.

We will deal with M-file samples in the next part of this chapter.

VARIABLE AND CONSTANT NAME DECLARATION

Like C or C++, MATLAB also has some rules for declaring constant and variable names. The limitations that must be considered include the following:

- Constant and variable names must be composed of letters, numbers, and/or underscores. Also, names must always start with a letter.

- Names cannot have more than two words with spaces in between. It is better to use an underscore if a two-word name cannot be avoided.
- MATLAB is case-sensitive so `VarName` is different from `varname`.
- Only the first 63 characters of the name are significant. Beyond that, MATLAB can no longer distinguish it.
- Existing keywords cannot be used as variable names.

The common way to assign values to or initialize variables and constants is with the use of the equal sign (=) assignment operator.

```
cons = <constant value>
var = <equation result or value>
```

Note that MATLAB no longer requires any data type formatting statements. The numeric format is automatically identified, and allocates appropriate storage memory once a value is assigned to a variable. The default format is short and presented in four decimal places (in the Command Window), but it can be modified to other formats by typing the following format commands in the Command Window (see Table 6.1).

 *Variables in MATLAB are **dynamically typed**. Therefore, the variables can be assigned without the need to declare them. Furthermore, their type can be modified while running the program.*

Table 6.1: Format commands.

Command	Display
>> format short	4 decimal places
>> format short e	4 decimal places with exponent
>> format short g	shortest decimal places based on the significant number
>> format long	14 decimal places
>> format long e	14 decimal places with exponent
>> format long g	shortest decimal places based on the significant number
>> format bank	2 decimal places

However, variables assigned with integer numbers will not be affected by the format command. When running instructions using M-files, floating numbers may be presented in 6 decimal places.

BASIC OUTPUT AND INPUT STATEMENTS

Output Statements

The common output statements use the `fprintf()` function. This allows printing of one or more variables in a string. This can be expressed as:

```
fprintf (<format string>, <variable>)
```

The format string is any string or character, variable, equation result, or other information. Printing the value of a variable can be done by placing the data type format followed by the variable name within the format string:

```
fprintf ('The temperature of the room is %5.3f.\n')
```

The use of the format specifier is similar to that of C. In the example, the format specifier, in this case is `%5.3f`, and can be expressed in terms of

- **marker (`%`):** This is required, meaning it should always be included during use of a format specifier.
- **field width (5 or any number):** This optional part specifies the total length of the value to display including the decimal point.
- **precision (3 or any number):** This is an optional part used to specify the number of decimal places in which the value will be presented or displayed.
- **data type** (d, f, e, g, c, or s): This is a required part where letters indicate the data type of the variable as indicated in Table 6.2.

Table 6.2: Data type specifier.

Specifier	Display
%d	integer/ whole number
%f	floating point
%e	exponential
%g	shortest format possible (insignificant zeroes are neglected)
%c	character
%s	string

To have more control of the output, control characters are added and inserted in the quoted string format. Presented in Table 6.3 are some of the control characters available in MATLAB.

Table 6.3: MATLAB control characters.

Characters	Meaning
\b	backspace
\n	new line/next line
\r	carriage return/enter key
\t	tab
\\	print backslash
%%	print percent character
' '	print single quote character

Another output statement is the `disp()` function. Unlike `fprintf()`, it can only display a string. So, for printing pure string statements, this function is preferred.

```
disp('The temperature of the room is 25.0 deg C')
```

Other useful commands include:

`pause` –allows the program to pause until a key is pressed
`pause(t)` – allows the program to pause for t seconds
`clc` – clears the screen of the Command Window

Input Statements

The common MATLAB input statement is the `input()` function. This function allows the printing of string statements and accepts values to store in a variable. This can be expressed as:

```
var = input('<format string>')
```

For example:

```
pressure = input('Enter the steam pressure :')
```

Once a value is input, it will be assigned to the variable `pressure`.

The input of string can be done through this format:

```
string = input('<format string>','s')
```

This returns the entered string as a text variable rather than as a numerical value.

OPERATORS

After receiving values through the use of input statements, the next step is to process the values using the various operators.

Table 6.4: MATLAB operators.

Operator	Symbol	Meaning
Arithmetic	+	addition
	−	subtraction
	*	multiplication
	/	right division
	\	left division
	^	power/exponentiation
	( )	evaluation rule specifier
Assignment	=	equal
Relational	>	greater than
	>=	greater than or equal to
	<	less than
	<=	less than or equal to
	==	equal to
	~=	not equal to
Logical	&	and
	&&	and (with shortcut)
	|	or
	||	or (with shortcut)
	~	not

Arithmetic and Assignment Operators

MATLAB supports the common arithmetic and assignment operators. Like any programming language, it uses the common arithmetic and equal signs.

However, the modulus % we learned from the previous chapter is no longer similar when applied to MATLAB. The % sign marks the beginning of a comment and should not be used to get the remainder of a division.

The power operator (^) is a convenient alternative to the function pow() used in C/C++. With MATLAB, exponentiation of numbers is much easier.

The common arithmetic precedence rule also applies to these operators; operations are usually recognized based on the following sequence: ^, *, / or \, +, − . However, if complex equations are created, it is better to use () to easily specify the evaluation order.

Relational Operator

In addition to arithmetic operations, MATLAB also supports relational operations. The relational operators compare the two values/arrays and return a result of either 1 (true) or 0 (false). When an instruction is directly encoded in the Command Window, the expression 8 < 2 will return a zero (0) value. The zero value signifies that the expression is false or incorrect.

```
>> 8 < 2
ans = 0
```

Relational operators are frequently used in conditional statements as part of a decision factor to determine what step to take given a certain condition.

Logical Operators

Logical operators are sometimes applied together with relational expressions to work on a logical value called Boolean. That is why logical operators are also called Boolean operators.

Table 6.4 provides some of the logical operators available in MATLAB. The difference between the & (*and* operator) and the && is the way it come up with a decision, by taking into consideration all the relational expressions provided. For example,

```
>> (8 < 2) & (6 > 2)
>> (8 < 2) && (6 > 2)
```

Both expressions will result to zero, but the expression using && will return the value first because it only requires the first relational expression (8 < 2) to come up with a decision that it is false, while the expression using & still evaluates the second relational expressions to come up with a decision. A similar situation can be observed for | and || logical operators.

⚠**WARNING** *The shortcut* && *and* || *operators are only available in MATLAB 7 and higher.*

Now let's try some MATLAB programming examples by applying what we have learned so far.

Example Problem 6.1. Create a MATLAB program that can calculate the product of two numbers.

Program Listing 6.1: Product of two numbers

```
1  clc;
2  fprintf('Enter two whole numbers.\n');
3  num1 = input('\n First Number: ');
4  num2 = input('\n Second Number: ');
5  product = num1*num2;
6  fprintf('\nThe product is %d.',product);
```

Line 1: The `clc` statement clears out the Command Window screen.

Line 2: The `fprintf()` function prints the statement on the screen. The control character \n ensures that the next statement occupies a new line before being printed.

Lines 3 and 4: The `input()` function requests two values from the user and stores these values in the `num1` and `num2` variables.

Line 5: The two variables are processed using the multiplication (`*`) operation to give a resulting value, which in turn is assigned to a variable `product`.

Line 6: The `product` is printed out on the screen using `fprintf()`.

 It is also possible to use `disp()` in lieu of `fprintf()`for basic text screen printing.

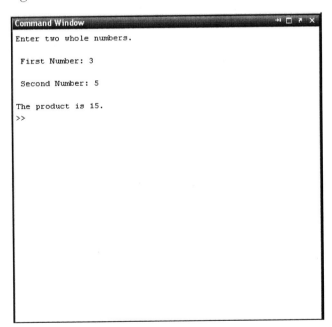

FIGURE 6.7: Output of Program Listing 6.1.

Example Problem 6.2. Create a MATLAB program that can compute the average of three input numbers.

Program Listing 6.2: Average of three numbers

```
1  clc;
2  fprintf('Enter three numbers.\n');
3  num1 = input('\n First Number: ');
4  num2 = input('\n Second Number: ');
5  num3 = input('\n Third Number: ');
6  ave = (num1+num2+num3)/3;
7  fprintf('\nThe average is %f.',ave);
```

Line 1: The `clc` statement clears out the Command Window screen to remove the clutter.

Line 2: The `fprintf()` function prints out the statement on the screen. The control character \n ensures that the next statement occupies a new line before being printed.

Lines 3, 4, 5: The `input()` function requests three values from the user and stores these values in the `num1`, `num2`, and `num3` variables.

Line 6: The three variables are processed using the addition (+) and division (/) operations to give a resulting value, which in turn assigns the result to variable `ave`.

Line 7: The `ave` is printed out on the screen using `fprintf()`.

FIGURE 6.8: Output of Program Listing 6.2.

CONDITIONAL STATEMENTS

Conditional statements provide the program with a selection mechanism to process a certain set of instructions. For MATLAB, there are three available conditional statements available; these are the `if-else`, the `if-elseif`, and the `switch/case` statements.

The `if-else` Statement

This conditional statement follows the format:

```
if condition1
    statement1
else
    statement2
end
```

Example Problem 6.3. Create a MATLAB program that determines if the inputted Reynolds number indicates laminar flow (less than 2100) or not (greater than or equal to 2100).

Program Listing 6.3: Using the `if - else` Statement

```
1  clc;
2  LIMIT = 2100;
3  nre = input('Enter Reynolds number: ');
4  if nre < LIMIT
5  fprintf('\n The flow is laminar.');
6  else
7  fprintf('\n The flow is not laminar');
8  end
```

Line 2: An assignment operator is used to provide a value for the variable `LIMIT`.

Line 3: The `input()` function requests the Reynolds number from the user and stores this in the `nre` variable.

Lines 4 to 8: Cover the `if-else` statements. When conditions are met, the corresponding `fprintf()` function is initiated to print the desired output message.

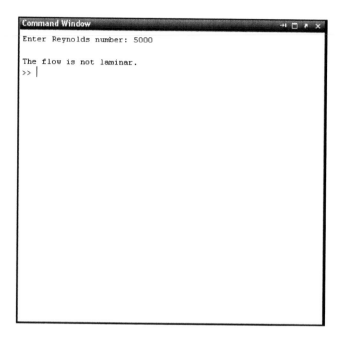

FIGURE 6.9: Output of Program Listing 6.3.

The `if-elseif` Statement

This conditional state follows the format:

```
if condition1
    statement1
elseif condition2
    statement2
elseif condition3
    statement3
else
    statement4
end
```

Example Problem 6.4. Write a MATLAB program that accepts a temperature in °C then determines the phase of water (solid, liquid, or gas).

Program Listing 6.4: Using the Nested `if - elseif` Statement

```
1   clc;
2   temp = input('Enter the temperature in deg C: ');
3   if temp > 100
```

```
4    fprintf('\n The water is in gas phase.');
5    elseif temp <= 100 & temp > 0
6     fprintf('\n The water is in liquid phase.');
7    else
8     fprintf('\n The water is in solid phase.');
9    end
```

Line 2:   The `input()` function requests a value from the user and stores the value in the `temp` variable.

Lines 3 to 9:   These lines cover the nested `if-elseif` statements. When conditions are met, the corresponding `fprintf()` function is initiated to print the desired output statement.

Line 5:   This line uses the logical operator (`&`). This instructs the computer that the condition will only be true when `temp` is less than or equal to 100 and greater than 0.

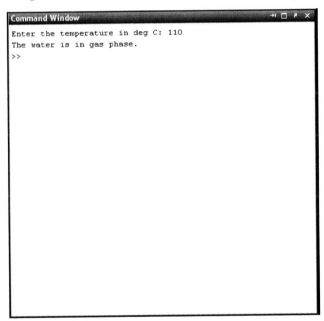

FIGURE 6.10:   Output of Program Listing 6.4.

The `switch/case` Statement

This conditional statement executes sets of instructions based on the value of a variable or expression.

```
switch   variable_or_expression
   case value1
        statement1
   case value2
        statement2
   case value3
        statement3
   otherwise
        statement4
end
```

Example Problem 6.5. Create a MATLAB Program that accepts a temperature that is entered by the user and converts the temperature from the given unit into another unit chosen by the user.

Program Listing 6.5: Using the `switch/case` Statement

```
 1  clc;
 2  init_temp = input('Enter the value to be converted: ');
 3  disp('Select initial and final units: ');
 4  disp('A) deg C to F');
 5  disp('B) deg C to K');
 6  disp('C) deg F to R');
 7  disp('D) deg F to C');
 8  disp('E) K to deg C');
 9  disp('F) R to deg F');
10  unit = input('Enter selection: ','s');
11  unit = lower(unit);
12  final_temp =0;
13  switch unit
14   case ('a')
15    final_temp   = (init_temp*1.8)+32;
16   case ('b')
17    final_temp = init_temp+273;
18   case ('c')
19    final_temp = init_temp + 460;
20   case ('d')
21    final_temp = (init_temp-32)/1.8;
22   case ('e')
23    final_temp = init_temp-273;
24   case ('f')
25    final_temp = init_temp-460;
26   otherwise
```

```
27   disp('Choice unavailable')
28 end
29 fprintf('\n The converted value is: %f',final_temp);
```

Line 2: The input() function requests a value from the user and stores it in the init_temp variable.

Lines 3 to 9: Print out the selection menu using disp().

Line 10: The input() function gets the selected option and stores it in the unit character variable.

Line 11: The lower() function converts the inputted character into lower case.

Line 12: The final_temp variable is set to have an initial value of zero.

Lines 13 to 28: Cover the switch/case statements. When conditions are met, the corresponding fprintf() function is initiated to print the desired output message.

Line 29: The result is then printed on screen using the fprintf() function.

⚠ WARNING   *Unlike in C, the MATLAB* switch *does not continue through. That is,* switch *stops when it finds its first match and subsequent matching will no longer be executed. Therefore,* break *statements in MATLAB are no longer necessary.*

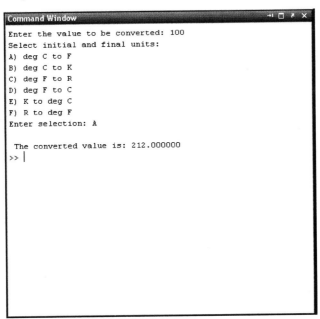

```
Command Window                                    ⊣ ⊡ ⊼ ✕
Enter the value to be converted: 100
Select initial and final units:
A)  deg C to F
B)  deg C to K
C)  deg F to R
D)  deg F to C
E)  K to deg C
F)  R to deg F
Enter selection: A

  The converted value is: 212.000000
>> |
```

FIGURE 6.11:   Output of Program Listing 6.5.

LOOPING STATEMENTS

The `for` Loop Statement

This loop statement repeats a set of instructions for a given number of times. This follows the format:

```
for var = init:increment/decrement:final
    statements
end
```

This can be shortened if the increment is one (1).

```
for var = init:final
    statements
end
```

Example Problem 6.6. Write a MATLAB program that will compute for the concentration of A in a zero order reaction, A -> B from start (0 sec) up to one minute for every 10 seconds given Ca_0 15 g mol/l and k is 0.0567 g mol/l-sec.

Program Listing 6.6: Using the `for` Loop Statement

```
1   init_conc = 15.00;
2   k = 0.0567;
3   clc
4   fprintf('\nTIME(sec)\t CONC.(g mol/L)\n');
5   for time = 0:10:60
6     conc = init_conc - (k*time);
7     fprintf(' %d\t\t\t\t\t\t%f\n',time,conc);
8   end
```

Lines 1 to 2: Initialize `init_con` with a value of 15.00 and `k` with 0.0567.

Lines 5 to 8: These lines cover the `for` loop statement. In the loop, `fprintf()` is utilized to print the resulting `conc`. The loop stops when value of `time` reaches beyond 60.

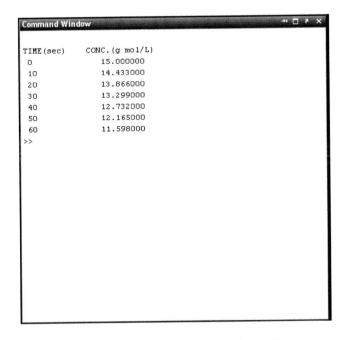

Command Window

```
TIME(sec)    CONC.(g mol/L)
  0            15.000000
 10            14.433000
 20            13.866000
 30            13.299000
 40            12.732000
 50            12.165000
 60            11.598000
>>
```

FIGURE 6.12:   Output of Program Listing 6.6.

The `while` Loop Statement

This loop statement repeats a set of instructions an indefinite number of times as long as the condition statement is still being met. This follows a format:

```
initialization
while condition
    statement
    increment/decrement
end
```

The next program listing will address the same problem (Example Problem 6.6) but using the `while` loop statement.

Program Listing 6.7: Using the `while` Loop Statement

```
1   init_conc = 15.00;
2   k = 0.0567;
3   time = 0;
4   clc
5   fprintf('\nTIME(sec)\t CONC.(g mol/L)\n');
6   while time<=60
```

```
 7   conc = init_conc - (k*time);
 8   fprintf(' %d\t\t\t\t%f\n',time,conc);
 9   time=time+10;
10 end
```

Lines 6 to 10: Similarly, these lines cover the `while` loop statement. In the loop, `fprintf()` is utilized to print the resulting `conc`. The loop stops when the value of `time` reaches more than 60.

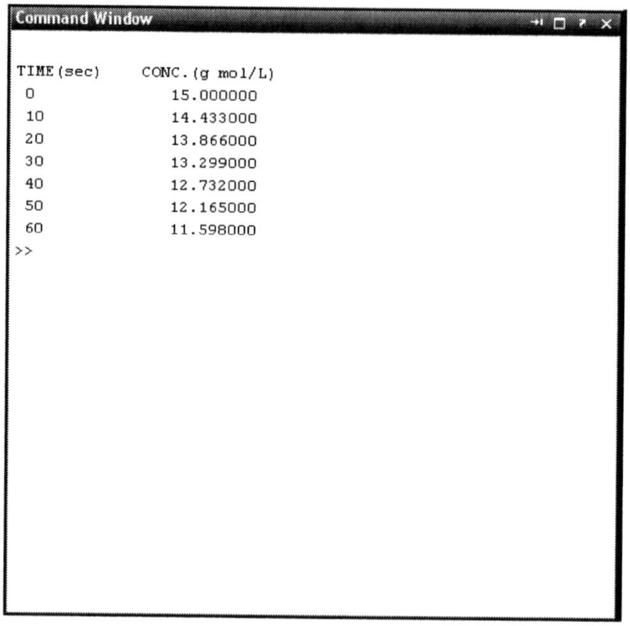

```
Command Window                                      ⇥ □ ⤢ ×

TIME(sec)      CONC.(g mol/L)
  0                15.000000
  10               14.433000
  20               13.866000
  30               13.299000
  40               12.732000
  50               12.165000
  60               11.598000
>>
```

FIGURE 6.13:   Output of Program Listing 6.7.

SCALARS, VECTORS, AND MATRICES

Initialization

MATLAB, which is short for MATrix LABoratory, was initially designed solely to perform matrix calculation. After reading the section, we will better appreciate this technical programming language specifically for handling arrays.

Initializing a scalar (1 × 1 array), a vector (one-dimensional array), and a matrix (two or more dimensional array) can be done through

```
array_name = expression
```

Unlike C or C++, there is no need to declare the data type. The data assigned to the array variable determines the type of data that is created. The following expression summarizes the format for declaring scalars, vectors, and matrices.

```
scalar_array = 10 or scalar_array = [10]
```

This creates a scalar (1 × 1 array), assigning the value 10 to `scalar_array`. Brackets are optional. We have been using this expression since the beginning of this chapter.

```
r_vector_array = [10 20 30]
```

This creates a 1 × 3 array with row vector `[10 20 30]`.

```
c_vector_array = [10 ; 20 ; 30]
```

This creates a 3 × 1 array containing the column vector $\begin{bmatrix} 10 \\ 20 \\ 30 \end{bmatrix}$.

```
matrix_array = [10 20 30 ; 40 50 60 ; 70 80 90]
```

This creates a 3 × 3 array containing the square matrix $\begin{bmatrix} 10 & 20 & 30 \\ 40 & 50 & 60 \\ 70 & 80 & 90 \end{bmatrix}$.

Manipulating Arrays

MATLAB provides the means to manipulate an existing array. We can modify existing values, extract, or rearrange them to suit our needs without complex programming. For example,

```
>> matrix_array = [10 20 30;40 50 60;70 80 90]
matrix_array =
    10    20    30
    40    50    60
    70    80    90
>> matrix_array(2)=100
matrix_array =
    10    20    30
   100    50    60
    70    80    90
```

This command changes the value in row 2 from 40 to 100.

```
>> matrix_array(1,2)=100
matrix_array =
     10    100     30
    100     50     60
     70     80     90
```

This command replaces the value in row 1 column 2 from 20 to 100.

```
>> matrix_array(:,3)=100
matrix_array =
     10    100    100
    100     50    100
     70     80    100
```

This command assigns the value 100 in all rows (:) in column 3.

Let us consider another set of arrays.
```
>> A = [9 8 7 ; 6 5 4 ; 3 2 1]
A =
     9      8      7
     6      5      4
     3      2      1
>> B = A(3:-1:1,3:-1:1)
B =
     1      2      3
     4      5      6
     7      8      9
```

This is an example of implementing a looping technique in an array. The 3:-1:1 expression in the row side of A(row,col) signifies that initially, the value of the row is 3 and will decrement by one (as expressed by the negative sign) until it reaches 1. This goes together with the column side. By running this, the B array will now have values similar to A but in reverse order.

The following functions are also available to initialize values for an array:

```
>> A = zeros(3,3)
A =
     0      0      0
```

```
        0      0      0
        0      0      0
>>  B  =  ones (3,5)


B  =
        1      1      1      1      1
        1      1      1      1      1
        1      1      1      1      1
>>  C  =  3.5*ones (4,4)


C  =
        3.5000      3.5000      3.5000      3.5000
        3.5000      3.5000      3.5000      3.5000
        3.5000      3.5000      3.5000      3.5000
        3.5000      3.5000      3.5000      3.5000
>>  D  =  rand (2,5)


D  =
        0.8147      0.1270      0.6324      0.2785      0.9575
        0.9058      0.9134      0.0975      0.5469      0.9649
>>  E  =  randn (3,2)


E  =
       -0.4326      0.2877
       -1.6656     -1.1465
        0.1253      1.1909
```

The rand () (based on uniform distribution) and the randn () (based on normal
distribution) functions generate random numbers to be assigned in the array.
    It is also possible to join matrices to form a bigger one. This method is called
concatenation and can be done using [ ] , the concatenation operator.

```
>>  A  =  ones (2,2);
>>  B  =  2*ones (2,2);
>>  C  =  3*ones (2,2);
>>  D  =  4*ones (2,2);
>>  E  =  [A B ; C D]


E  =
        1      1      2      2
        1      1      2      2
```

3	3	4	4
3	3	4	4

By concatenating A, B, C, and D, a bigger matrix E is formed. Note that the sequence of the smaller matrix, as to its location in the bigger matrix, can also be specified.

```
>> E(2,:)=[]
```

E =

1	1	2	2
3	3	4	4
3	3	4	4

Using an empty bracket [ ], we can delete an entire row or column of the matrix, as long as it is within dimension.

The transpose of a matrix, on the other hand, is generated by creating a new matrix, where the row of the original matrix is now the column of the new matrix. For example,

$$A = \begin{bmatrix} 10 & 20 & 30 \\ 40 & 50 & 60 \\ 70 & 80 & 90 \end{bmatrix} \qquad A' = \begin{bmatrix} 10 & 40 & 70 \\ 20 & 50 & 80 \\ 30 & 60 & 90 \end{bmatrix}$$

where A' is denoted as the transpose of matrix A. In MATLAB, the transpose can be derived using the symbol apostrophe (').

```
>> A = [10 20 30 ;40 50 60 ;70 80 90]

A =
       10       20       30
       40       50       60
       70       80       90
>> B = A'

B =
       10       40       70
       20       50       80
       30       60       90
```

There are also other means of manipulating arrays. A list of references found in Appendix K provides further information.

Arithmetic Operation

Like handling scalars, arithmetic operations can also be applied to vectors and matrices.

```
>> A = [2 4 ; 6 8]

A =

        2        4
        6        8
>> B = [1 3 ; 5 7]

B =

        1        3
        5        7
>> A + B
ans =

        3        7
       11       15
>> A - B
ans =

        1        1
        1        1
```

Implementing matrix addition and subtraction is quite easy with MATLAB. The tricky part, however, is the distinction between implementing multiplication and division between elements in a matrix, and from implementing multiplication and division between two matrices. To clarify further let us consider the same arrays, A and B.

```
>> A .* B
ans =
        2       12
       30       56
```

The .* operator implements element-by-element multiplication between A and B, as shown in the result. Note, however, that both arrays must have the same dimension.

```
>> A * B
ans =
       22        34
       46        74
```

The * operator, on the other hand, implements matrix multiplication. It is also required that the column in A must be equal to the number of rows in B. The same goes with division operation.

```
>> A ./ B
ans =
      2.0000      1.3333
      1.2000      1.1429
```

This operation is similar to A(r,c)/B(r,c). Division is done per element.

```
>> A / B
ans =
      0.7500      0.2500
     -0.2500      1.2500
```

This operation can also be expressed as A* inv(B), where inv(B) is the value of the inverse matrix of B. The same with A\B, which can be expressed as inv(A)*B. The difference between the operation of the matrix left division and the inv() function, is that matrix left division uses Gaussian elimination while inv() uses matrix inversion.

Let us now apply what we have learned so far by considering a problem.

Example Problem 6.7. Create a MATLAB program that allows input of two matrices (A and B) to generate a new matrix (C) as a result of arithmetic operation. Design the program to be flexible for the user to choose the operation to implement based on the following:

- addition
- subtraction
- element-by-element multiplication
- matrix multiplication
- element-by-element division
- matrix division

Program Listing 6.8: Arithmetic operations with arrays

```
1  clc
2  clear
3  dim = input ('Enter the dimension of both square
   matrices: ');
4  disp('For the first matrix.');
5  for r = 1:dim
6   for c = 1:dim
7    fprintf ('Enter A(%d,%d): ',r,c);
8    A(r,c) = input ('');
9   end
10 end
11 disp('For the second matrix.');
12 for r = 1:dim
13  for c = 1:dim
14   fprintf ('Enter B(%d,%d): ',r,c);
15   B(r,c) = input ('');
16  end
17 end
18 disp('_____');
19 disp('a) Addition');
20 disp('b) Subtraction');
21 disp('c) Element-by-element Multiplication');
22 disp('d) Matrix Multiplication');
23 disp('e) Element-by-element Division');
24 disp('f) Matrix Division');
25 oper=input('Select operation: ','s');
26 oper=lower(oper);
27 switch oper
28  case ('a')
29   C = A+B;
30  case ('b')
31   C = A-B;
32  case ('c')
33   C = A.*B;
34  case ('d')
35   C = A*B;
36  case ('e')
37   C = A./B;
38  case ('f')
39   C = A/B;
```

40 otherwise
41   disp('Choice unavailable');
42 end
43 C

Line 2: The clear statement clears all existing arrays in the memory.
Lines 4 to 17: Utilize the for loop statements to collect all values needed to assign to A and B arrays.
Lines 18 to 24: Display the selection menu using disp().
Lines 27 to 42: Utilize the switch statement to select a particular operation to perform based on the input of the user.
Line 43: Displays the C matrix in the Command Window.

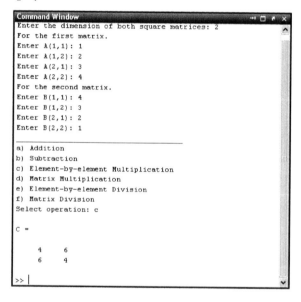

FIGURE 6.14:   Output of Program Listing 6.8.

CREATING FUNCTIONS

Like any programming language, MATLAB also allows the creation of functions to supplement existing functions. These serve as additional building blocks for larger programs. The created function in MATLAB can also accept input arguments and return output arguments. However, creating a function and saving it as an M-file requires that it have the same name.
This follows the format:

```
function [output] = function_name (input)
```

Let us consider again Program Listing 6.1 and create a separate function called `multiply`.

```
% save as multiply.m
function prod = multiply(num1,num2);
prod = num1*num1;
```

Program Listing 6.9: Using the created `multiply()` function

```
1   clc;
2   fprintf('Enter two whole numbers.\n');
3   num1 = input('\n First Number: ');
4   num2 = input('\n Second Number: ');
5   product = multiply(num1,num2);
6   fprintf('\nThe product is %d.\n',product);
```

By keeping the `multiply.m` in the current address, the `multiply()` function is made available to other programs.

Line 5: The `multiply()` function is utilized, processing the values of `num1` and `num2`. In turn, it receives a value after the mathematical operation and stores it in the `product` variable.

Line 6: Prints out the resulting `product`.

FIGURE 6.15:   Output of Program Listing 6.9.

The variable name used in the function `multiply()` may or may not be the one used when the function is called in the main program.

The output variable is also not limited to one value alone. It can be a set of values or an array.

PLOTTING

Basic 2D Plotting

Plotting a two-dimensional graph in MATLAB can be done using the `plot()` function. The basic format can be expressed as

```
plot(x,y)
```

where x contains the values of the x-coordinate and y contains values of the y-coordinate. But in order for this function to perform, it requires that the variable x and y already have values, therefore initialization is required.

Let us consider plotting the Benzene-Toluene equilibrium curve.

Program Listing 6.10: Plotting Benzene-Toluene equilibrium curve

```
x = [0.0 0.1 0.2 0.3 0.4 0.5 0.6 0.7 0.8 0.9 1.0];
y =[0.000 0.211 0.378 0.512 0.623 0.714 0.791 0.856 0.911
    0.959 1.000];
plot(x,y)
```

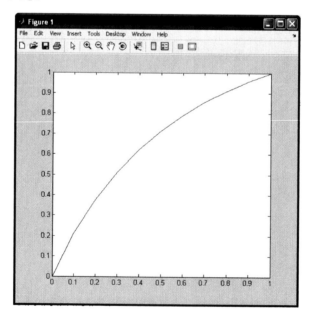

FIGURE 6.16:   Output of Program Listing 6.10.

It is also possible to do multiple plots with a single call of the plot function. This will result in multiple data in one graph.

Program Listing 6.11: Benzene-Toluene equilibrium curve with diagonal line

```
x = [0.0 0.1 0.2 0.3 0.4 0.5 0.6 0.7 0.8 0.9 1.0];
y =[0.000 0.211 0.378 0.512 0.623 0.714 0.791 0.856 0.911
    0.959 1.000];
diagonal =(0:1);
plot(x,y,diagonal,diagonal)
```

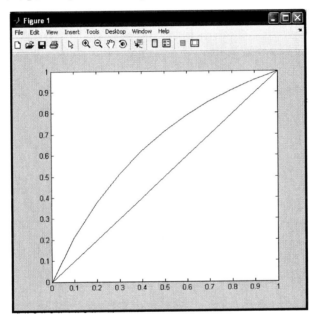

FIGURE 6.17:   Output of Program Listing 6.11.

There are other options for plotting, which can be done by adding the following marking symbols:

```
o   circle        *     star
d   diamond       <     triangle (pointing left)
h   hexagram      >     triangle (pointing right)
p   pentagram     ^     triangle (pointing up)
+   plus          v     triangle (pointing down)
.   point         x     cross mark
s   square
```

Express the new plot command as

```
plot(x,y,'*',diagonal,diagonal,'.')
```

Axis Command Functions

There are available axis command functions, which provide the user with options to control the scale, aspect ratio, axis limit display, as well as axis and grid visibility.

Program Listing 6.12: Using grid lines

```
x = [0.0 0.1 0.2 0.3 0.4 0.5 0.6 0.7 0.8 0.9 1.0];
y =[0.000 0.211 0.378 0.512 0.623 0.714 0.791 0.856 0.911
    0.959 1.000];
diagonal =(0:1);
plot(x,y,diagonal,diagonal)
grid on
axis square
axis equal
```

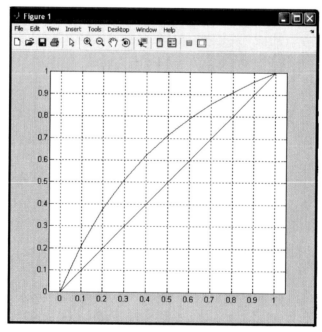

FIGURE 6.18:   Output of Program Listing 6.12.

Titles and Labels

It is also possible to incorporate text within the graph. The functions available to do this are:

title()	Places a title on the upper part of the graph
xlabel()	Provides a label for the x-axis
ylabel()	Provides a label for the y-axis
gtext()	Prints text in the graph (location provided through the mouse)

Program Listing 6.13: Using titles and labels

```
x = [0.0 0.1 0.2 0.3 0.4 0.5 0.6 0.7 0.8 0.9 1.0];
y =[0.000 0.211 0.378 0.512 0.623 0.714 0.791 0.856 0.911
    0.959 1.000];
diagonal =(0:1);
plot(x,y,diagonal,diagonal)
grid on
axis square
axis equal
xlabel('Benzene in Liquid (mole fraction)');
ylabel('Benzene in vapor (mole fraction)');
title('Benzene - Toluene Equilibrium System');
```

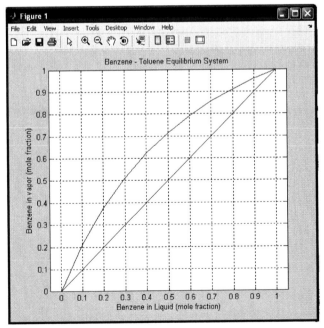

FIGURE 6.19: Ouput of Program Listing 6.13.

Adding Lines to Graphs

Lines can be added anywhere in the graph to improve its presentation. This can be done through the MATLAB line() function. By adding the line(), we modified further the previous MATLAB program.

Program Listing 6.14: Benzene-Toluene equilibrium curve with feed line

```
x = [0.0 0.1 0.2 0.3 0.4 0.5 0.6 0.7 0.8 0.9 1.0];
y =[0.000 0.211 0.378 0.512 0.623 0.714 0.791 0.856 0.911
   0.959 1.000];
diagonal =(0:1);
plot(x,y,diagonal,diagonal)
grid on
axis square
axis equal
xlabel('Benzene in Liquid (mole fraction)');
ylabel('Benzene in vapor (mole fraction)');
title('Benzene - Toluene Equilibrium System');
xd =[0.85 0.85];
yd =[0 0.85];
line(xd,yd);
xf =[0.45 0.45];
yf = [0 0.65];
line(xf,yf);
xb =[0.15 0.15];
yb = [0 0.15];
line(xb,yb);
```

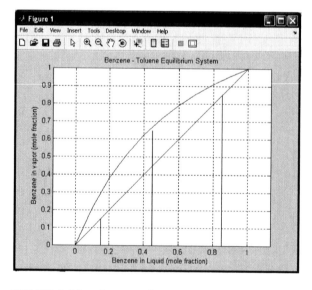

FIGURE 6.20: Output of Program Listing 6.14.

Setting Line Style and Width

It is also possible to specify the line style based on the available options:

- –   solid line
- :   dotted line
- –. dot and dash
- –– dashed lines

including the line width based on the pixel size with the use of the set() function.

Program Listing 6.15: Using line styles

```
x = [0.0 0.1 0.2 0.3 0.4 0.5 0.6 0.7 0.8 0.9 1.0];
y =[0.000 0.211 0.378 0.512 0.623 0.714 0.791 0.856 0.911
    0.959 1.000];
diagonal =(0:1);
plot(x,y,diagonal,diagonal)
grid on
axis square
axis equal
xlabel('Benzene in Liquid (mole fraction)');
ylabel('Benzene in vapor (mole fraction)');
title('Benzene - Toluene Equilibrium System');
xd =[0.85 0.85];
yd =[0 0.85];
set(line(xd,yd),'LineStyle',':');
xf =[0.45 0.45];
yf = [0 0.65];
set(line(xf,yf),'LineWidth',2);
xb =[0.15 0.15];
yb = [0 0.15];
set(line(xb,yb),'LineStyle',':');
```

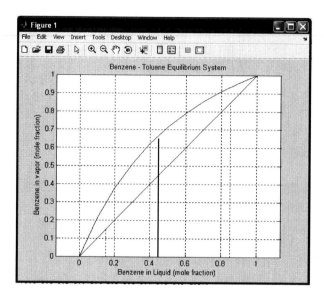

FIGURE 6.21:   Output of Program Listing 6.15.

Basic 3D Plotting

MATLAB is also capable of plotting three-dimensional graphs and the best way to go over these functions is through an example.

Let us revisit Example Problem 4.6 involving temperature determination in two-dimensional conduction.

Completing the temperature distribution (°F) for the entire cross-section of the rectangular flue, we have the following values based on their respective node location.

200	200	200	200	200	200	200	200	200	200	200	200
200	227.2	254.4	279.5	289.8	293	293	289.8	279.5	254.4	227.2	200
200	254.4	312	369.6	387.9	392.2	392.2	387.9	369.6	312	254.4	200
200	279.5	369.6	500	500	500	500	500	500	369.6	279.5	200
200	289.8	387.9	500	500	500	500	500	500	387.9	289.8	200
200	293	392.2	500	500	500	500	500	500	392.2	293	200
200	293	392.2	500	500	500	500	500	500	392.2	293	200
200	289.8	387.9	500	500	500	500	500	500	387.9	289.8	200
200	279.5	369.6	500	500	500	500	500	500	369.6	279.5	200
200	254.4	312	369.6	387.9	392.2	392.2	387.9	369.6	312	254.4	200
200	227.2	254.4	279.5	289.8	293	293	289.8	279.5	254.4	227.2	200
200	200	200	200	200	200	200	200	200	200	200	200

We first create an array T by typing in the Command Window

>> T = [];

Once the array T is created, we then copy all the values into the Array Editor to make a 12×12 matrix as shown in Figure 6.22.

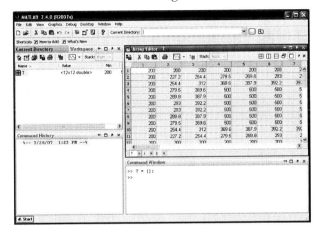

FIGURE 6.22:   Array editor.

There are various functions available in MATLAB to plot data in 3D graphs, an example of which is function `surf()`. Using `surf()`, we can generate a 3D image representation of the flue temperature profile, where the x- and y-axis represent the node location, while the z-axis represents the temperature in °F,

```
>>surf(T),xlabel('X'),ylabel('Y'),zlabel('Temperature in deg F')
```

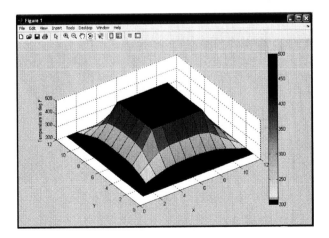

FIGURE 6.23:   3D plot.

There are additional basic plotting functions available, which can be accessed by selecting the Graphics Menu and clicking on the "More Plots…" option to display the Plot Catalog window.

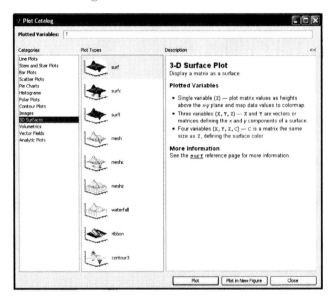

FIGURE 6.24:   MATLAB plot catalog window.

LABORATORY EXERCISES

1) Most fuels contain atoms of carbon (C), hydrogen (H), oxygen (O), nitrogen (N), and sulfur (S) with corresponding atomic weights:

   Carbon   12.011 g/g-mol
   Hydrogen   1.00794 g/g-mol
   Oxygen   15.9994 g/g-mol
   Nitrogen   14.00674 g/g-mol
   Sulfur   32.066 g/g-mol

   Develop a program in which the user can input the total number of each element present to compute the corresponding molecular weight of the fuel in g/g-mol.

2) A mixture of H_2S, SO_2, and COS is contained in a pressurized vessel. Create a MATLAB program that allows the user to encode the concentration of each compound then computes for the total sulfur of the gas mixture.

3) Develop a MATLAB program that will plot a NaOH concentration–boiling point diagram using data provided in Table 3.6.

4) Heavy hydrocarbon oil introduced at 100 lb_m/hr passes through a pipe of length 18 ft. Steam is used as the heating medium and is condensing outside the pipe wall. The inside temperature of the pipe is assumed constant at 370°F. Find the heat transfer coefficient (h) of the oil.

D = 1.5 in
N_{Re} = 300
N_{Pr} = 80
k = 0.090 Btu/h-ft°F
μ_{oil} = 12.52 lb_m/ft-h
μ_{water} = 0.046 lb_m/ft-h

$$\frac{hD}{k} = 1.86 \left[N_{Re} N_{Pr} \frac{D}{L} \right]^{\frac{1}{3}} \left[\frac{\mu_{oil}}{\mu_{water}} \right]^{0.14}$$

5) Using the data in Problem 4, determine the outlet temperature of the oil if the inlet temperature is 160°F. Use the following additional information to re-compute h:

Temp °F	μ_{oil} lb_m/ft-h	μ_{water} lb_m/ft-h
75	20.15	2.230
150	16.78	0.083
200	14.23	0.052
250	8.53	0.033
300	7.95	0.012
350	4.67	0.005

Cp = 0.65 Btu/lb_m

$$q = mCp \left[T_{outlet} - T_{inlet} \right]$$

$$q = hA\Delta T_m$$

where $\Delta T_m = \dfrac{\left[T_{wall} - T_{inlet} \right] + \left[T_{wall} - T_{outlet} \right]}{2}$

(Hint: Use looping statements to determine the final temperature T_{final}.)

Chapter

7

MATLAB FUNCTIONS FOR NUMERICAL COMPUTATION

Chapter 2 covered numerical methods that are quite useful in solving chemical engineering problems, which translate into C programs. MATLAB, however, is a mathematical programming language, so it already has numerous existing functions that can perform numerical computation. Some of these functions were developed based on the algorithm we learned when we developed functions in C.

In this chapter, we will further explore MATLAB existing functions including:

The Matrix Method for Linear Equations
    The `inv()` Function
    Matrix Left Division ($\backslash$)
Regression and Curve Fitting
    The `polyfit()` Function
    The `polyval()` Function
    Coefficient of Determination and Standard Error of Estimate
Finding Roots
    The `roots()` Function
    The `fzero()` Function

Interpolation
One-dimensional Interpolation
Two-dimensional Interpolation
Numerical Integration
Trapezoidal Integration
Quadrature
Ordinary Differential Equations (ODE)
The `ode23()` and `ode45()` Functions

 All program listings discussed in this chapter can be found on the CD-ROM.

```
\Program Listings\Chapter 7\PL7_1.m      Program Listing 7.1
\Program Listings\Chapter 7\PL7_2.m      Program Listing 7.2
\Program Listings\Chapter 7\coefdet.m    Program Listing 7.3
\Program Listings\Chapter 7\stderr.m     Program Listing 7.4
\Program Listings\Chapter 7\PL7_5.m      Program Listing 7.5
\Program Listings\Chapter 7\PL7_6.m      Program Listing 7.6
\Program Listings\Chapter 7\odeapp.m     Program Listing 7.7
\Program Listings\Chapter 7\PL7_8.m      Program Listing 7.8
```

THE MATRIX METHOD FOR LINEAR EQUATIONS

Let us consider a set of linear equations

$$a_{11} x_1 + a_{12} x_2 + \ldots a_{1n} x_n = b_1$$
$$a_{21} x_1 + a_{22} x_2 + \ldots a_{2n} x_n = b_2$$
$$\vdots$$
$$a_{n1} x_1 + a_{n2} x_2 + \ldots a_{nn} x_n = b_n$$

where:

$x_1, x_2, \ldots x_n$ are unknown variables
$a_{11}, a_{12} \ldots a_{nn}$ are known constant coefficients
$b_1, b_2 \ldots b_n$ are known constants
We can rewrite the equations using the following matrices,

representing the coefficients as A

$$A = \begin{bmatrix} a_{11} & a_{12} & a_{13} \\ a_{21} & a_{22} & a_{23} \\ a_{31} & a_{32} & a_{33} \end{bmatrix} \tag{7.1}$$

the unknown variable as x

$$x = \begin{bmatrix} x_1 \\ x_2 \\ x_3 \end{bmatrix}$$ (7.2)

and the right-hand values as b

$$B = \begin{bmatrix} b_1 \\ b_2 \\ b_3 \end{bmatrix}.$$ (7.3)

The `inv()` Function

With matrix multiplication, the system of equations can be represented as $Ax = B$. Through this equation, solving the unknown variables can be done using the inverse of A multiplied by B, which can be written as $x = A^{-1}B$. In MATLAB, it can be expressed as

```
>> x = inv(A)*B
```

Example Problem 7.1. Create a MATLAB program that will solve for simultaneous linear equations using inversion and multiplication.

Program Listing 7.1: Solving simultaneous linear equations using matrix inversion

```
1  clc
2  clear
3  dim = input ('Enter the number of unknowns: ');
4  for r = 1:dim
5  for c = 1:dim
6  fprintf ('Enter a(%d,%d): ',r,c);
7  A(r,c) = input ('');
8  end
9  fprintf('Enter b(%d): ',r);
10 B(r,1) = input('');
11 end
12 x = inv(A)*B
```

Line 2: The `clear` statement clears all existing arrays in the memory.
Lines 4 to 11: Utilize the `for` loop statements to collect all values needed to assign to A and B arrays.

Line 12: Performs matrix inversion and multiplication, and assigns the resulting values to vector x.

Utilize the simultaneous linear equations examples in Chapter 2.

$$3x_1 + x_2 + x_3 = 25$$
$$x_1 - 3x_2 + 2x_3 = 10$$
$$2x_1 + x_2 - x_3 = 6$$

Solving the values of x using Program Listing 7.1, we get

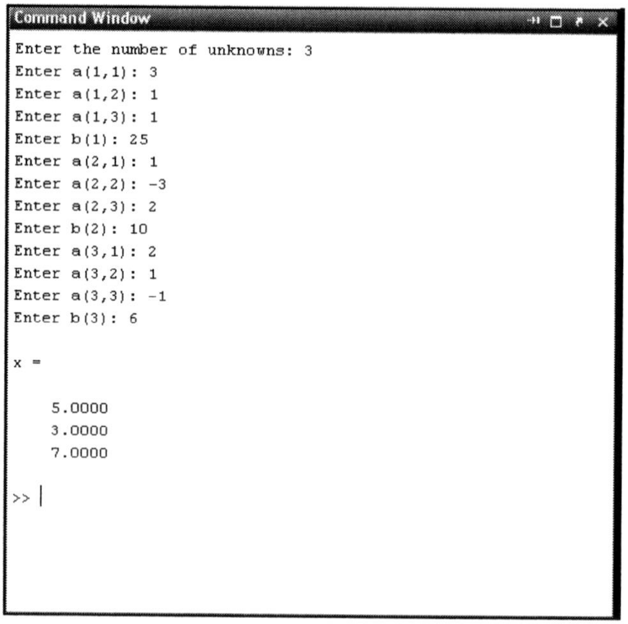

```
Command Window                                    ⁻¹ □ ⚓ ✕

Enter the number of unknowns: 3
Enter a(1,1): 3
Enter a(1,2): 1
Enter a(1,3): 1
Enter b(1): 25
Enter a(2,1): 1
Enter a(2,2): -3
Enter a(2,3): 2
Enter b(2): 10
Enter a(3,1): 2
Enter a(3,2): 1
Enter a(3,3): -1
Enter b(3): 6

x =

    5.0000
    3.0000
    7.0000

>> |
```

FIGURE 7.1:   Output of Program Listing 7.1.

Matrix Left Division (\)

Another alternative to solving systems of linear equations is through the use of matrix left division (\).

```
>> x = A\B
```

Substituting this expression in Line 12 of Program Listing 7.1 will get a similar answer. However, the difference is that this method utilizes Gaussian elimination and not matrix inversion. In this way the matrix left division operator is more efficient and accurate.

Program Listing 7.2: Solving simultaneous linear equations using matrix left division

```
1   clc
2   clear
3   dim = input ('Enter the number of unknowns: ');
4   for r = 1:dim
5     for c = 1:dim
6       fprintf ('Enter a(%d,%d): ',r,c);
7       A(r,c) = input ('');
8     end
9     fprintf('Enter b(%d): ',r);
10    B(r,1) = input('');
11  end
12  x = A\B
```

Line 12: Performs matrix left division, and assigns the resulting values to vector x.

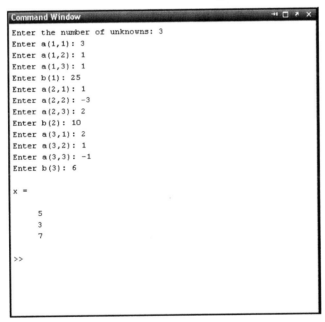

FIGURE 7.2: Output of Program Listing 7.2.

Note that the answers are presented as integers while the answers from Program Listing 7.1 are in float type.

The C code equivalent (from Program Listing 2.2) is as follows:

```
for(r=0;r<3;r++)
  {
    d=-1/a[r][r];
    for(c=0;c<3;c++)
    {
      if (c==r) (c++);
      a[r][c]=a[r][c]*d;
    }
    d=-d;
    for(i=0;i<3;i++)
    {
      if(i==r) (i++);
      e1=a[i][r];
      for(c=0;c<3;c++)
      {
        if(c==r) (a[i][r]=a[i][r]*d);
        else(a[i][c]=a[i][c]+a[r][c]*e1);
      }
    }
    a[r][r]=d;
  }
  for(i=0;i<3;i++)
  {
    for(c=0;c<3;c++)
    x[i]=x[i]+b[c]*a[i][c];
  }
```

can be reduced to MATLAB expression x = inv(a)*b or x = a\b.

REGRESSION AND CURVE FITTING

In Chapter 2, we covered linear and polynomial regression. The equations involving these types of regression have been studied and converted to C Programs. In this section we will no longer dwell on the topics already discussed, but instead, we will now focus more on MATLAB existing functions concerning regression and curve fitting, specifically polyfit() and polyval(), and how to apply them.

The polyfit() Function

The polyfit() function determines the least-square polynomial curve for given values of x and y pairs. The argument requires three input variables, mainly the vectors containing the x and y values, as well as the order number.

```
>> x = [0 .1 .2 .3 .4 .5 .6 .7 .8 .9 1.0];
>> y = [0 0.211 0.378 0.512 0.623 0.714 0.791 0.856 0.911
        0.959 1.0];
>> order = 5;
>> polyfit(x,y,order)


ans =
    0.4327   -1.6120    2.6895   -2.8803    2.3702    0.0001
```

Looking closely, we can observe that the output values are the same coefficients of the polynomial equation we derived in Chapter 3 for the Benzene-Toluene equilibrium curve, which was expressed as:

$$y = 0.432692x^5 - 1.612034x^4 + 2.689467x^3 - 2.880267x^2 + 2.370166x + 0.000068$$

The coefficients are presented in descending powers of x, of the *nth* order polynomial.

The `polyval()` Function

Another useful function is `polyval()`, which evaluates a polynomial at a set of data points and returns a vector of values that correspond to the x vector values. The function follows the format:

```
polyval(coeff,x)
```

where `coeff` is a vector containing the coefficients of a polynomial equation and x is a vector containing the x values. Note that we can use an expression of the `polyfit()` function to represent the `coeff` vector. For example:

```
>> x = [0 .1 .2 .3 .4 .5 .6 .7 .8 .9 1.0];
>> y = [0 0.211 0.378 0.512 0.623 0.714 0.791 0.856 0.911
        0.959 1.0];
>> order = 5;
>> polyval(polyfit(x,y,order),x)
ans =
  Columns 1 through 9
    0.0001    0.2108    0.3780    0.5125    0.6226    0.7140
    0.7909    0.8560    0.9113
  Columns 10 through 11
    0.9587    1.0001
```

As we learned in Chapter 3, it can be determined that the output of `polyval()` is basically the y_e values for 5th order in Table 3.3.

(from Table 3.3: Values of y_e using different polynomial orders.)

x	y (given)	y_e (5th order)
0.000	0.000	0.000068
0.100	0.211	0.210815
0.200	0.378	0.377965
0.300	0.512	0.512503
0.400	0.623	0.622580
0.500	0.714	0.714037
0.600	0.791	0.790923
0.700	0.856	0.856014
0.800	0.911	0.911332
0.900	0.959	0.958667
1.000	1.000	1.000092

Coefficient of Determination and Standard Error of Estimate

In Chapter 2, we learned some of the important statistical tools in determining which polynomial order number best fits the data points. We have identified that the coefficient of determination (r^2) and the standard error of estimate ($S_{y,x}$) help us measure the goodness of fit of a particular equation to the original data.

Let us now make a MATLAB function that will solve the coefficient of determination and standard error of estimate using the `polyfit()` and `polyval()` functions.

The coefficient of determination (r^2) function from Eqs 2.15 to 2.17:

$$S_t = \sum (y-\bar{y})^2 \qquad S_r = \sum (y-y_e)^2 \qquad r^2 = (S_t - S_r)/S_t$$

Program Listing 7.3: Coefficient of determination

```
1   % save as coefdet.m
2   function result=coefdet(x,y,order);
3   ye = polyval(polyfit(x,y,order),x);
4   yave= mean(y);
5   dify_yave = (y-yave).^2;
6   St = sum(dify_yave);
7   dify_ye = (y-ye).^2;
8   Sr = sum(dify_ye);
9   result = (St-Sr)/St;
```

Line 2: Declaration of `coefdet()` function where input information required are x and y values, and the order of the polynomial. The variable `result` returns the coefficient of determination.

Line 3: This line derives the different values of y_e by first generating the suitable polynomial equation using `polyfit()` and then calculates y_e using `polyval()`.

Lines 6, 8: Compute for S_t and S_r respectively.

Line 9: Finally, this line computes for the coefficient of determination and then assigns the value to `result`.

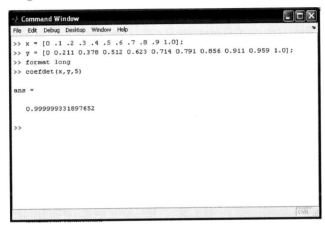

FIGURE 7.3: Output of Program Listing 7.3.

To develop the standard error of estimate $(S_{y,x})$ function, we consider Eqs 2.16 and 2.19:

$$S_r = \sum (y - y_e)^2 \qquad S_{y,x} = \sqrt{\frac{S_r}{\upsilon}} \quad \upsilon = n - (i+1) \text{ for } i^{\text{th}} \text{ order polynomial}$$

Program Listing 7.4: Standard error of estimate

```
1  % save as stderr.m
2  function result=stderr(x,y,order);
3  ye = polyval(polyfit(x,y,order),x);
4  dify_ye = (y-ye).^2;
5  Sr = sum(dify_ye);
6  df=length(x)-(order+1);
7  result = (Sr/df).^0.5;
```

Line 2: Similarly, this line declares the `stderr()` function where input information required are x and y values, and the order of the polynomial. The variable `result` returns the standard error of estimates.

Line 5: This computes for S_r.

Line 6: The degrees of freedom df is computed in this line.

Line 7: Finally, this line computes for the coefficient of determination and then assigns the value to `result`.

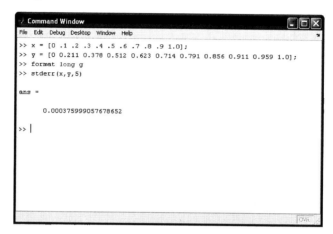

FIGURE 7.4:   Output of Program Listing 7.4.

FINDING ROOTS

The `roots()` Function

For values of polynomials equated to zero, finding the root using MATLAB is quite easy. The available function is `roots()`, which has a format `roots(coef)` where `coef` is a vector containing the coefficients of the polynomial presented in descending powers of x.

Let us apply the function with this example problem.

Example Problem 7.2. A cylinder tank is to be filled up with 170 g-mols of CO_2 to a pressure of 65 atm and a temperature of -70°C. Apply the van der Waals equation to determine the volume of the gas in the tank.

$$V^3 - \left(nb + \frac{nRT}{P} \right) V^2 + \frac{n^2 a}{P} V - \frac{n^3 ab}{P} = 0$$

The critical constant for carbon dioxide is:

$$a = 3.60 \times 10^6 \text{ atm} \left(\frac{cm^3}{g\text{-}mol} \right)^2 \quad b = 42.8 \left(\frac{cm^3}{g\text{-}mol} \right).$$

$R = 82.054$ cm^3-atm/g-mol K

```
>> format long g
>> coeff1=1;
>> coeff2=(170*42.8)+((170*82.054*(-70+273.15))/65);
>> coeff3=(170^2)*3600000/65;
>> coeff4=(170^3)*3600000*42.8/65;
>> coeff_V = [coeff1 -coeff2 coeff3 -coeff4];
>> roots(coeff_V)
ans =
            20589.6052668498 +      27884.0342689934i
            20589.6052668498 -      27884.0342689934i
            9693.34203553116
```

Based on the output, the volume of the gas tank is 9,693.342 cm^3 or 9.69 liters.

The `fzero()` Function

Another MATLAB function available to find the value of a variable in a one-dimensional equation equal to zero is `fzero()`. This function uses bisection and inverse quadratic equation interpolation to determine the unknown variable of the equation. The format of the function can be expressed as

```
var=fzero('equation',init_val)
```

where:

`var` = The approximated value of the unknown variable when `equation` is taken as zero.
`equation` = The equation containing the unknown `var`. Note that the equation must be a function, an inline function object, or a string expression.
`init_val` = The estimated initial value of the `var`. The nearer the value of the `init_val` to var, the faster MATLAB can derive the answer.

To illustrate the use of this function, let us consider this example problem.

Example Problem 7.3 Determine the temperature at which the vapor phase composition of benzene and toluene in an equilibrium mixture is 50%. The pressure of the closed system is at 1.013 bar.

The Antoine equation constants for benzene and toluene are:
Benzene (*a*) $A = 4.72583$, $B = 1660.652$, and $C = -1.461$
Toluene (*b*) $A = 4.08245$, $B = 1346.382$, and $C = -53.508$

Using the Antoine equation, we can determine p_a° and p_b°.

$$\log p_a^\circ = 4.72583 - \frac{1660.652}{T - 1.461} \quad \log p_b^\circ = 4.08245 - \frac{1346.382}{T - 53.508}$$

Getting the anti-log, the equation can be expressed as

$$p_a^\circ = 10^{4.72583 - \frac{1660.652}{T - 1.461}} \quad p_b^\circ = 10^{4.08245 - \frac{1346.382}{T - 53.508}}. \tag{7.4}$$

If x is taken as 0.5, we can solve for p_a and p_b.

$$p_a = 0.5 p_a^\circ \qquad p_b = (1\text{-}0.5) p_b^\circ \tag{7.5}$$

Also, the total pressure of the closed system (1.013) can be expressed as the sum of p_a and p_b.

$$1.013 = p_a + p_b.$$

When equated to zero, it can be presented as $0 = p_a + p_b - 1.013.$ $\tag{7.6}$

Eventually we can solve for y.

$$y = \frac{p_a}{p_a + p_b} \tag{7.7}$$

Program Listing 7.5: Using the `fzero()` function

```
1   clc
2   initval=300;
3   x=0.5;
4   temp0=inline('0.5.*(10.^(4.72583-(1660.652./(T-
    1.461))))+(1-0.5).*(10.^(4.08245-(1346.382/(T-
    53.508))))-1.013');
5   temp=fzero(temp0,initval);
6   y=x.*(10.^(4.72583-(1660.652./(temp-1.461))))/(
    x.*(10.^(4.72583-(1660.652./(temp-1.461))))+(1-
    x).*(10.^(4.08245-(1346.382/(temp-53.508)))));
7   fprintf('x \t\t\t y \t\t\t temp\n')
8   fprintf('%f \t %f \t %f\n',x,y,temp)
```

Line 1: Clears the Command Window.

Lines 2, 3: Initialize `initval` and x variables.

Line 4: Equates `temp0` with the inline function of Eq 7.6, wherein the relationships in Eqs 7.4 and 7.5 are substituted.

Line 5: Uses `fzero()` to determine T (expressed in the program as `temp`) from the inline function `temp0`.

Line 6: Once `temp` is determined, all y values are determined from the different given values of x using Eqs 7.4 and 7.5 substituted in Eq 7.7.

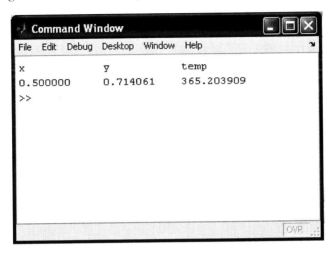

```
x                y                temp
0.500000         0.714061         365.203909
>>
```

FIGURE 7.5: Output of Program Listing 7.5.

INTERPOLATION

Another valuable numerical tool in estimating values between a given set of data points is interpolation. Although we have already discussed two types of interpolation in Chapter 2, linear and Lagrange polynomial interpolation, in this section we will cover other types, which are already available functions in MATLAB. Table 7.1 provides the different types of interpolation included in MATLAB.

Table 7.1: Types of interpolation included in MATLAB.

Type

`linear`	Assumes that the function between two data points can be estimated by a straight line.
`cubic`	Method used is based on a piecewise cubic Hermite interpolating polynomial.
`spline`	Treats the function between a pair of data points as a 3rd order polynomial and computes the function to ensure a smooth transition from the previous pair.
`nearest`	Nearest-neighbor method, provides the data point nearest to the given value. This is the least accurate among the four types.

MATLAB interpolation functions do not only consider one-dimensional values (`interp1()`), but also handle two (`interp2()`) dimensions using the following format:

```
interp1(x,y,x_val,type)
interp2(x,y,z,x_val,y_val,type)
```

where x, y, and z are vectors containing the data points. The `x_val` and `y_val` are vectors containing the values of x and y, in which the z values are being estimated through interpolation. Finally, `type` specifies the interpolation method to use.

Now let us consider these examples.

One-Dimensional Interpolation

Example Problem 7.4. Using the Benzene-Toluene Equilibrium data points, determine the interpolated value of y when x is 0.45. Compare results using the different types of interpolation methods.

```
>> x = 0:.1:1.0;
>> y = [0.0 0.211 0.378 0.512 0.623 0.714 0.791 0.856
       0.911 0.959 1.000];
>> interp1(x,y,0.45,'linear')
ans = 0.6685
>> interp1(x,y,0.45,'cubic')
ans = 0.6706
>> interp1(x,y,0.45,'spline')
ans = 0.6706
>> interp1(x,y,0.45,'nearest')
ans = 0.7140
```

To visualize the difference between each type, let us add the `plot()` function and consider another set of data points.

Program Listing 7.6: Using interpolation methods
```
1   x = 0:.1:.8;
2   y = [0 10 5 20 10 30 15 40 20];
3   x_val = 0:.01:.8;
4   y_val1 = interp1(x,y,x_val,'linear');
5   y_val2 = interp1(x,y,x_val,'spline');
6   y_val3 = interp1(x,y,x_val,'cubic');
```

    7 `plot(x_val,y_val1,'--',x_val,y_val2,'.',x_val,y_`
        `val3,x,y,'o'), title ('Linear, Cubic, and Spline`
        `Interpolation'),grid`

Line 1: Initializes values of x.
Line 2: Initializes values of y.
Line 3: Initializes values of x_val
Lines 4 to 6: Perform different methods of interpolation to determine values
        of y_val based on the given values of x_val using the x and y
        relationship.
Line 7: Plots x_val and the different values of y_val.

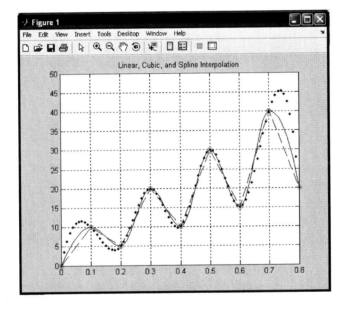

FIGURE 7.6: Output of Program Listing 7.6.

Two-Dimensional Interpolation

Example Problem 7.5. Revisiting the table for Absolute Humidity (H) and
Dew Point (DP) in Chapter 3, determine the approximate values of H and
DP when $RH = 60$ and temperature is 47.5°C using the different types of
interpolation. Compare the results from the actual output of the C Program
(Program Listing 3.3).

(from Table 3.5)

Relative Humidity (%RH)		10%	30%	50%	70%	90%
Dry Bulb [°C]						
50	H, g/m³	8.28	24.85	41.41	57.98	74.54
	DP, °C	10.11	27.54	36.78	42.88	48.12
45	H, g/m³	6.53	19.60	32.67	45.74	58.81
	DP, °C	6.38	23.35	32.16	38.30	42.85

Using MATLAB:

```
>> RH = 10:20:90;
>> T = 50:-5:45;
>> H = [8.28 24.85 41.41 57.98 74.54; 6.53 19.60 32.67
        45.74 58.81];
>> DP = [10.11 27.54 36.78 42.88 48.12; 6.38 23.35 32.16
        38.30 42.85];
```

Linear Interpolation:

```
>> actualH = interp2(RH,T,H,60.0, 47.5,'linear')
actualH = 44.4500
>> actualDP = interp2(RH,T,DP,60.0, 47.5,'linear')
actualDP = 37.5300
```

Cubic Hermite Interpolation:

```
>> actualH = interp2(RH,T,H,60.0, 47.5,'cubic')
actualH = 44.2070
>> actualDP = interp2(RH,T,DP,60.0, 47.5,'cubic')
actualDP = 37.8137
```

Cubic Spline Interpolation:

```
>> actualH = interp2(RH,T,H,60.0, 47.5,'spline')
actualH = 44.2067
>> actualDP = interp2(RH,T,DP,60.0, 47.5,'spline')
actualDP = 37.7599
```

Nearest-Neighbor Interpolation:

```
>> actualH = interp2(RH,T,H,60.0, 47.5,'nearest')
actualH = 45.7400
>> actualDP = interp2(RH,T,DP,60.0, 47.5,'nearest')
actualDP = 38.3000
```

Using the C Program (Program Listing 3.3), the results generated are $RH = 44.190876$ and $DP = 37.820486$. Therefore, based on the comparison, the cubic hermite interpolation is recommended.

Example Problem 7.6. Let's consider another example, this time involving three-dimensional plotting. We retrieve the flue temperature distribution data we established in Chapter 6, which was stored in array T.

200	200	200	200	200	200	200	200	200	200	200	200
200	227.2	254.4	279.5	289.8	293	293	289.8	279.5	254.4	227.2	200
200	254.4	312	369.6	387.9	392.2	392.2	387.9	369.6	312	254.4	200
200	279.5	369.6	500	500	500	500	500	500	369.6	279.5	200
200	289.8	387.9	500	500	500	500	500	500	387.9	289.8	200
200	293	392.2	500	500	500	500	500	500	392.2	293	200
200	293	392.2	500	500	500	500	500	500	392.2	293	200
200	289.8	387.9	500	500	500	500	500	500	387.9	289.8	200
200	279.5	369.6	500	500	500	500	500	500	369.6	279.5	200
200	254.4	312	369.6	387.9	392.2	392.2	387.9	369.6	312	254.4	200
200	227.2	254.4	279.5	289.8	293	293	289.8	279.5	254.4	227.2	200
200	200	200	200	200	200	200	200	200	200	200	200

Make a 3D plot of the data

```
>> [x,y]=meshgrid(1:12);
>> surf(x,y,T)
```

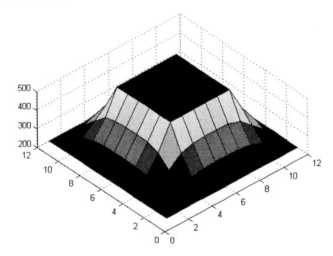

FIGURE 7.7: 3D representation of flue temperature distribution.

Generate a finer mesh for interpolation

```
>> [xi,yi]=meshgrid(1:.25:12);
```

Using nearest-neighbor interpolation:

```
>> zi1 = interp2(x,y,T,xi,yi,'nearest');
>> surf(xi,yi,zi1)
```

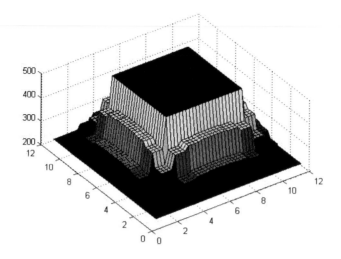

FIGURE 7.8: Using nearest -neighbor interpolation.

Using linear interpolation:

```
>> zi2 = interp2(x,y,T,xi,yi,'linear');
>> surf(xi,yi,zi2)
```

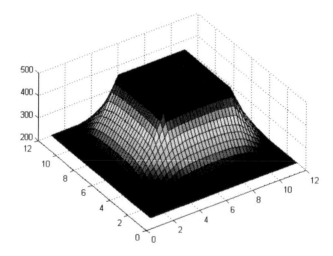

FIGURE 7.9: Using linear interpolation.

Using cubic interpolation:
```
>> zi3 = interp2(x,y,T,xi,yi,'cubic');
>> surf(xi,yi,zi3)
```

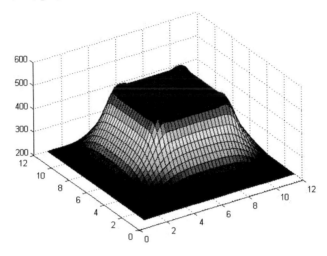

FIGURE 7.10:   Using cubic interpolation.

Using spline interpolation:
```
>> zi4 = interp2(x,y,T,xi,yi,'spline');
>> surf(xi,yi,zi4)
```

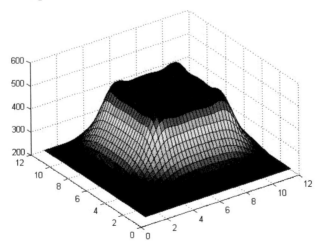

FIGURE 7.11:   Using spline interpolation.

By visualizing the result of the interpolation, it will be easier to evaluate which interpolation method is effective in providing the approximated value we want to get.

NUMERICAL INTEGRATION

The integral of a function $f(x)$ over the upper (a) and lower (b) limits of integration provides the area of the function. Sometimes the analytical method in computing the integral is very tedious, which is why the numerical method is employed to provide the approximate result.

Trapezoidal Integration

The common numerical method used to determine the area of the function is trapezoidal integration. This method uses the principle of the trapezoidal rule where the area of the function is approximated as an area of a trapezoid under a straight line from a to b.

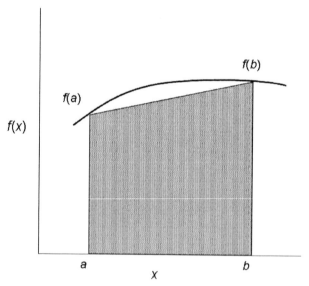

FIGURE 7.12: Approximating area of the function.

$$area = width \times height = (b-a)\frac{f(a)+f(b)}{2} \tag{7.8}$$

This area is approximately equal to the result of integration $\int_a^b f(x)\,dx$.

To further provide a higher accuracy, we divide the width $(b - a)$ into n smaller subsections, and then apply the trapezoidal rule on each of these trapezoids. These can be represented by the equation

$$\int_a^b f(x)dx \approx \frac{b-a}{2n}\left(f(x_0) + 2f(x_1) + 2f(x_2) + \ldots + 2f(x_{n-1}) + f(x_n)\right) \qquad (7.9)$$

where $x_0 = a$, $x_n = b$ and $x_i = $ the end points of each trapezoid.

MATLAB performs trapezoidal integration with the use of the `trapz()` function. These can be expressed as `trapz(x,y)`

where `y` contains the function values of `x`.

Let's take a look at some examples.

Example Problem 7.7. Determine the required heat (in cal) to raise the temperature of 1 g-mol of propane from 250°C to 750°C at a pressure of 1 atm.

$$Q = \int_{T_1}^{T_2} C_p dT \qquad (7.10)$$

where C_p is the molar heat capacity of propane taken as

$$C_p = 2.410 + 57.195 \times 10^{-3} T - 4.300 \times 10^{-6} T^2 . \qquad (7.11)$$

The heat Q therefore can be expressed as

$$Q = \int_{523}^{1023} 2.410 + 57.195 \times 10^{-3} T - 4.300 \times 10^{-6} T^2 dT . \qquad (7.12)$$

Using the `trapz()` function in MATLAB, we first try having the width of the subsections equal 100, meaning the difference between the temperature ($\Delta T = 1023 - 523 = 500$) is subdivided into five sections ($n = 5$) having a width equal to 100.

```
>> T = 523:(1023-523)/5:1023;
>> Cp = 2.410+ 57.195E-3.*T - 4.3E-6.*(T.^2);
>> trapz(T,Cp)

ans =
              21977.80515
```

However, once we increase the number of subsections and lessen the width of each section, let's say to 1 ($\Delta T = (1023 - 523)/500 = 1$), the new result will be

```
>> T = 523:(1023-523)/500:1023;
>> Cp = 2.410+ 57.195E-3.*T - 4.3E-6.*(T.^2);
>> trapz(T,Cp)

ans =
                21981.388125
```

Using the analytical method will yield an answer equal to 21,981.388 cal/g-mol as well.

When using the `trapz()` *function, try to increase the number of subsections or make the width of the trapezoid as thin as possible to get a more accurate answer.*

Quadrature

In trapezoidal integration, the subsection of the function's curve is represented by a line, which provides some difference in the actual result. However, there is another process of integrating $f(x)$, called quadrature, which represents the curve with a quadratic function for a higher accuracy.

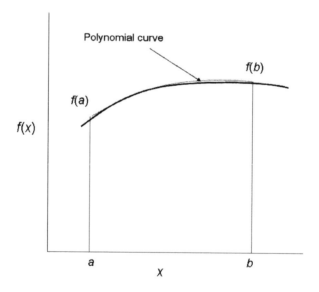

FIGURE 7.13:   Representing the given curve with a polynomial curve.

In MATLAB, several quadrature functions are available. Two of which will be discussed in this section: `quad()` and `quadl()`.

The quad() Function

This function employs the Simpson's rule, which works similarly to the Trapezoidal rule except that approximation is done using a polynomial instead of a line (see Figure 7.13). Simpson's rule can be derived by integrating a 3rd order Lagrange interpolating polynomial fit to the function at three equally spaced points.

This can be represented by the equation

$$\int_a^b f(x)dx \approx \frac{(b-a)}{6n}\left(f(x_0)+4f(x_1)+2f(x_2)+4f(x_3)+...+2f(x_{2n-2})+4f(x_{2n-1})+f(x_{2n})\right).$$

$$(7.13)$$

where x_0 = a, x_{2n} = b and x_i = end points of each section.

The MATLAB function can be expressed as

```
quad('function', a, b)
```

This will return the area of the function with limits set from a to b. Note that the function is a MATLAB function (not an equation).

To further understand, let us now apply the quad() to solve Example Problem 7.7. First, we develop the function but we are not going to use the Command Window, but instead will develop the program in the Editor.

```
% save as propane_Cp.m
function Cp = propane_Cp(T)
Cp = 2.410+ 57.195E-3.*T - 4.3E-6.*(T.^2);
```

Then from the Command Window we write

```
>> quad('propane_Cp',523,1023)

ans =
            21981.3884833333
```

To ensure M-file functions will work properly, make the file available in the current directory for easy access by MATLAB.

The quadl() Function

This function is based on a four-point Gauss-Lobatto formula and two successive Kronrod extensions established by Walter Gander and Walter Gautschi in their article "Adaptive Quadrature-Revisited" (1998).

For the interval from a to b, the Gauss-Lobatto formulas can be written as

$$\int_a^b f(x)dx \approx \frac{h}{6}\left[f(a) + 5f(m - \beta h) + 5f(m + \beta h) + f(b)\right] \tag{7.14}$$

and

$$\int_a^b f(x)dx \approx \frac{h}{1470}\left[77f(a) + 432f(m - \alpha h) + 625f(m - \beta h) + 627f(m) \tag{7.15}\right.$$

$$\left. + 625f(m - \beta h) + 432f(m + \alpha h) + 77f(b)\right]$$

where
$$h = (b-a)/2, \ m = (a+b)/2, \ \alpha = \sqrt{2/3}, \ \beta = 1/\sqrt{5}.$$

The Konrod extension is expressed as

$$\int_{-1}^1 f(x)dx = A\left[f(-1) + f(1)\right] + B\left[f(-x_1) + f(x_1)\right] + C\left[f(-\alpha) + f(\alpha)\right] + D\left[f(-x_2) + f(x_2)\right]$$

$$+ E\left[f(-\beta) + f(\beta)\right] + F\left[f(-x_3) + f(x_3)\right] + Gf(0) + R^{GLKK}(f) \tag{7.16}$$

where
$x_1 = 0.942882415695480$, $x_2 = 0.641853342345781$, $x_3 = 0.236383199662150$
$A = 0.0158271919734802$, $B = 0.094273840218850$, $C = 0.155071987336585$
$D = 0.188821573960182$, $E = 0.199773405226859$, $F = 0.224926465333340$
$G = 0.242611071901408$
$\alpha = \sqrt{2/3}$, $\beta = 1/\sqrt{5}$, the remainder $R^{GLKK}(P_{19}) = 0$.

The MATLAB function `quadl()`, which uses this formula, can be expressed similar to the `quad()` function.

```
quadl('function', a, b)
```

This function tries to approximate the integral of scalar value from *a* to *b* within a tolerable error of 1.e-6. Applying this to Example Problem 7.7, we get a similar result.

```
>> quadl('propane_Cp',523,1023)

ans =

    21981.3884833333
```

ORDINARY DIFFERENTIAL EQUATIONS (ODE)

MATLAB provides functions to solve a wide variety of problems involving differential equations. This section, however, will focus on MATLAB ODE functions to address initial value problems involving ordinary differential equations, which are common in chemical engineering studies.

The ode23() and ode45() Functions

These functions use the one-step Runge-Kutta (RK) method for 2nd to 5th order differential equations. The ode23() is used for problems where low accuracy results are acceptable, while ode45() provides results with moderate accuracy.

It must be remembered that in MATLAB, all ODEs must be in function form. The ODE must also be in 1st order form $dy/dx = f(y,x)$. To better illustrate its application, let us consider this example.

Example Problem 7.8. Solve the following set of differential equations assuming $x = 0$, $y_1 = 5$, $y_2 = 8$. Integrate to $x = 2$ with increment $h = 1.0$.

$$\frac{dy_1}{dx} = -0.6y_1 + 0.1x$$

$$\frac{dy_2}{dx} = 7.0 - 0.4y_2 - 0.2y_1 + 0.2x$$

Using the 4th order RK method,

$$y_{i+1} = y_i + \frac{1}{6}\left(k_{1,j} + 2k_{2,j} + 2k_{3,j} + k_{4,j}\right)h \tag{7.17}$$

$$k_{1,j} = f(x_i, y_i) \tag{7.18}$$

$$k_{2,j} = f\left(x_i + \tfrac{1}{2}h, y_i + \tfrac{1}{2}k_{i,1}h\right) \tag{7.19}$$

$$k_{3,j} = f\left(x_i + \tfrac{1}{2}h, y_i + \tfrac{1}{2}k_{i,2}h\right) \tag{7.20}$$

$$k_{4,j} = f\left(x_i + h, y_i + k_{i,3}h\right) \tag{7.21}$$

where $k_{n,j}$ is the nth value of k for the jth dependent variable:

$k_{1,1} = f(0,5,8) = -0.6(5) + 0.1(0) = -3$
$k_{1,2} = f(0,5,8) = 7 - 0.5(8) - 0.2(5) + 0.2(0) = 2$

and the midpoint values of y for each dependent variable

$y_{\text{mid } 1,1} = y_1 + \frac{1}{2} hk_{1,1} = 5 + \frac{1}{2}(1)(-3) = 3.5$
$y_{\text{mid } 1,2} = y_2 + \frac{1}{2} hk_{1,2} = 8 + \frac{1}{2}(1)(2) = 9$

$k_{2,1} = f(0.5,3.5,9) = -0.6(3.5) + 0.1(0.5) = -2.05$
$k_{2,2} = f(0.5,3.5,9) = 7 - 0.5(9) - 0.2(3.5) + 0.2(0.5) = 1.9$

$y_{\text{mid } 2,1} = y_1 + \frac{1}{2} hk_{2,1} = 5 + \frac{1}{2}(1)(-2.05) = 3.975$
$y_{\text{mid } 2,2} = y_2 + \frac{1}{2} hk_{2,2} = 8 + \frac{1}{2}(1)(1.9) = 8.95$

$k_{3,1} = f(0.5,3.975,8.95) = -0.6(3.975) + 0.1(0.5) = -2.335$
$k_{3,2} = f(0.5,3.975,8.95) = 7 - 0.5(8.95) - 0.2(3.975) + 0.2(0.5) = 1.83$

$y_{\text{mid } 3,1} = y_1 + hk_{3,1} = 5 + (1)(-2.335) = 2.665$
$y_{\text{mid } 3,2} = y_2 + hk_{3,2} = 8 + (1)(1.83) = 9.83$

$k_{4,1} = f(1,2.665,9.83) = -0.6(2.665) + 0.1(1) = -1.499$
$k_{4,2} = f(1,2.665,9.83) = 7 - 0.5(9.83) - 0.2(2.665) + 0.2(1) = 1.752$

when $x = 1$

$y_1 = 5\,[1/6\,(-3 + 2(-2.05) + 2(-2.335) - 1.499)](1) = 2.7885$
$y_2 = 8\,[1/6\,(2 + 2(1.9) + 2(1.83) + 1.752)](1) = 9.868667$

Repeating the process with a starting value for $x = 1$, $y_1 = 2.7885$, and $y_2 = 9.868667$:

$k_{1,1} = f(1,2.7885,9.868667) = -0.6(2.7885) + 0.1(1) = -1.5731$
$k_{1,2} = f(1,2.7885,9.868667) = 7 - 0.5(9.868667) - 0.2(2.7885) + 0.2(1) = 1.707967$

$y_{\text{mid } 1,1} = y_1 + \frac{1}{2} hk_{1,1} = 2.7885 + \frac{1}{2}(1)(-1.5731) = 2.00195$
$y_{\text{mid } 1,2} = y_2 + \frac{1}{2} hk_{1,2} = 9.868667 + \frac{1}{2}(1)(1.707967) = 10.72265$

$k_{2,1} = f(1.5,2.00195,10.72265) = -0.6(2.00195) + 0.1(1.5) = -1.05117$
$k_{2,2} = f(1.5,2.00195,10.72265) = 7 - 0.5(10.72265) - 0.2(2.00195) + 0.2(1.5) =$
1.538285

$y_{\text{mid 2,1}} = y_1 + \frac{1}{2} h k_{2,1} = 2.7885 + \frac{1}{2}(1)(-1.05117) = 2.262915$
$y_{\text{mid 2,2}} = y_2 + \frac{1}{2} h k_{2,2} = 9.868667 + \frac{1}{2}(1)(1.538285) = 10.63781$

$k_{3,1} = f(1.5, 2.262915, 10.63781) = -0.6(2.262915) + 0.1(1.5) = -1.20775$
$k_{3,2} = f(1.5, 2.262915, 10.63781) = 7 - 0.5(10.63781) - 0.2(2.262915) + 0.2(1.5)$
$= 1.528512$

$y_{\text{mid 3,1}} = y_1 + h k_{3,1} = 2.7885 + (1)(-1.20775) = 1.580751$
$y_{\text{mid 3,2}} = y_2 + h k_{3,2} = 9.868667 + (1)(1.528512) = 11.39718$

$k_{4,1} = f(2, 1.580751, 11.39718) = -0.6(1.580751) + 0.1(2) = -0.74845$
$k_{4,2} = f(2, 1.580751, 11.39718) = 7 - 0.5(11.39718) - 0.2(1.580751) + 0.2(2) =$
1.38526

when $x = 2$:

$y_1 = 2.7885 [1/6 (-1.5731 + 2(-1.05117) + 2(-1.20775) - 0.74845)](1) = 1.648602$
$y_2 = 9.868667 [1/6 (1.707967 + 2(1.538285) + 2(1.528512) + 1.38526)](1) = 11.40647$

Table 7.2 summarizes the results.

Table 7.2: Summary of x, y_1, and y_2 values

x	y_1	y_2
0	5	8
1	2.7885	9.868667
2	1.648602	11.40647

Solving a similar problem using MATLAB, let us consider the following program listing.

First, we write the `odeapp.m`, which contains the function `odeapp()`.

Program Listing 7.7: Ordinary differential equation function

```
% saved as odeapp.m
function dydx=odeapp(x,y)
dydx=[-0.6*y(1)+0.1*x;  7-0.5*y(2)-0.2*y(1)+0.2*x];
```

Then we make a separate m-file for the main program:

Program Listing 7.8: Main program

```
1   clc
2   clear
```

```
3  xspan=[0:1:2];
4  init_y=[5 8];
5  k1=-0.6*init_y(1)+0.1*xspan(1);
6  k2=7-0.5*init_y(2)-0.2*init_y(1)+0.2*xspan(1);
7  [x,y]=ode45('odeapp',xspan,init_y)
```

Line 1: Clears the Command Window.

Line 2: Clears the variables and functions from the memory.

Lines 3, 4: Initialize values of xspan and inti_y.

Lines 5, 6: Compute k1 and k2 (rate of reaction).

Line 7: Performs ordinary differential equations on function odeapp and returns the different values of x and y.

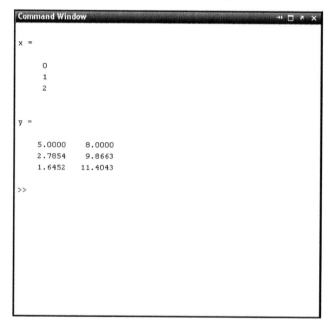

FIGURE 7.14:   Output of Program Listing 7.8.

LABORATORY EXERCISES

1) Develop a program that will solve the roots and plot the graph of any 4th order polynomial expression entered by the user.

2) How much heat is needed to raise 3.5 lb_m of a substance from 39.5°F to 157 °F? Given that

$$C_p = 9.678 - 43.964 \times 10^{-3} T + 7.696 \times 10^{-6} T^2 + 3.671 \times 10^{-9} T^3 \text{ Btu/lb}_m\text{-°F}$$

where T is in °F.

3) Modify Problem 2 and convert C_p to
a) KJ/kg-K
b) cal/g-°C
4) Develop a program that will interpolate the carbon dioxide (CO_2) gas volume from the set of tabulated data:

Temperature (°F)					
41	**42**	**43**	**44**		
3.17	3.12	3.06	3.01	**18**	
3.27	3.21	3.16	3.10	**19**	**Pressure**
3.37	3.31	3.25	3.20	**20**	**(psia)**
3.47	3.41	3.35	3.29	**21**	

Carbon dioxide gas volume is the unit of measurement to indicate the amount of CO_2 gas absorbed by a given amount of beverage. One "volume" is that volume of carbon dioxide (measured at 0 deg Centigrade or 273.16° Kelvin and at one atmosphere of pressure) that will dissolve in an equal volume of water at 15.6° Centigrade.

5) Based on the definition of gas volume in Problem 4, construct a new table for temperatures ranging from 10° to 40°C with an increment of 2 degrees, and pressure from 1.5 to 3.0 atm with an increment of 0.5. Use the Ideal Gas Law equation.

Chapter 8

APPLICATIONS USING MATLAB

I n this chapter we will apply what we have learned so far in MATLAB programming and develop programs that can handle more complicated chemical engineering problems.

These problems usually can be solved using functions for numerical computations discussed in the previous chapter, which covers:

Physical Properties Approximation

    Saturated Steam Pressure-Temperature Relationship

    Binary System – Equilibrium Curve

    Enthalpy-Concentration Relationship for Magnesium Sulfate System

    Phase Diagram – Magnesium Sulfate System

Chemical Engineering Sample Problems

    Cubic Equation of State

    Compressibility Factor

    Simple Differential Distillation

    Two-dimensional Conduction

    Reactor Tanks

    Reactions in Series

    Crystallization

All program listings discussed in this chapter can be found on the CD-ROM.

PHYSICAL PROPERTIES APPROXIMATION

Saturated Steam Pressure-Temperature Relationship

In Chapter 3, the saturated steam pressure-temperature relationship was estimated using polynomial regression. This time, interpolation will be used to obtain the saturated pressure given the saturated temperature. An evaluation will be made to determine which interpolation method is suitable.

From Eq 3.4, we obtained the following values of the saturated pressure generated from a saturated temperature ranging from 275 K to 645 K with a 5 degree increment. These can be declared in MATLAB as

```
t_sat = 275:5:645;
p_sat = [
698.4358601    991.7589281    1388.931054    1919.877195    2621.114915
3536.717587    4719.326831    6231.203640    8145.306525    10546.38439
13532.07139    17213.97117    21718.71798    27189.00309    33784.55558
41683.06781    51081.05720    62194.65721    75260.33206    90535.51115
108299.1406    128852.1507    152517.8399    179642.1761    210594.0186
245765.2635    285570.9185    330449.1096    380861.0286    437290.8252
500245.4516    570254.4663    647869.8037    733665.5172    828237.5035
932203.2148    1046201.367    1170891.653    1306954.465    1455090.638
1616021.224    1790487.301    1979249.829    2183089.567    2402807.053
```

```
2639222.675   2893176.829   3165530.193   3457164.132   3768981.251
4101906.126   4456886.241   4834893.154   5236923.951   5664003.019
6117184.197   6597553.393   7106231.723   7644379.306   8213199.841
8813946.135   9447926.819   10116514.55   10821156.12   11563385.02
12344837.38   13167272.39   14032599.33   14942914.30   15900552.42
16908166.25   17968853.73   19086393.24   20265769.53   21514941.10   ];
```

In this exercise we will disregard the nearest-neighbor method since it is obvious that this will not give an accurate result. Using the `interp1()` function of MATLAB we derived the following results:

Table 8.1: Summary of results using interpolation methods – linear, cubic, spline

Using IAWPS Equation		Temperature (K) Using Interpolation Method		
Pressure (Pa)	Temperature (K)	Linear	Cubic	Spline
612.101580	273.17	NAN	273.335196	273.239557
101417.993818	373.15	373.063136	373.150419	373.150295
1554939.222050	473.15	473.10225	473.150094	473.150053
8587867.486373	573.15	573.118352	573.150048	573.150015
22038358.010325	647.00	NAN	647.002794	647.007481

Both `cubic` and `spline` can also extrapolate data as shown in the first and last approximation. However, in this case `spline` provides a more accurate result, so we will use this method to create the `sat_p2t()` function.

Program Listing 8.1:

```
% save as sat_p2t.m
function result=sat_p2t(p_sati)
t_sat = 275:5:645;
p_sat = [698.4358601  991.7589281  1388.931054  1919.877195
2621.114915  3536.717587  4719.326831  6231.203640  8145.306525
10546.38439  13532.07139  17213.97117  21718.71798  27189.00309
33784.55558  41683.06781  51081.05720  62194.65721  75260.33206
90535.51115  108299.1406  128852.1507  152517.8399  179642.1761
210594.0186  245765.2635  285570.9185  330449.1096  380861.0286
437290.8252  500245.4516  570254.4663  647869.8037  733665.5172
828237.5035  932203.2148  1046201.367  1170891.653  1306954.465
1455090.638  1616021.224  1790487.301  1979249.829  2183089.567
2402807.053  2639222.675  2893176.829  3165530.193  3457164.132
3768981.251  4101906.126  4456886.241  4834893.154  5236923.951
5664003.019  6117184.197  6597553.393  7106231.723  7644379.306
```

```
8213199.841 8813946.135 9447926.819 10116514.55 10821156.12
11563385.02 12344837.38 13167272.39 14032599.33 14942914.30
15900552.42 16908166.25 17968853.73 19086393.24 20265769.53
21514941.10];
result = interp1(p_sat,t_sat,p_sati,'spline');
```

⚠**WARNING** *The* `sat_p2t.m` *is an m-file function and will not run on its own, but* `sat_p2t()` *can be used and called upon when we want to determine the saturated temperature of steam given the saturated pressure.*

Binary System – Equilibrium Curve

In Chapter 3, we determined the x and y values of the benzene-toluene equilibrium curve manually by doing trial-and-error to determine the correct temperature at which the sum of the total vapor pressures of both components equals the actual pressure provided. For this example, we will develop a MATLAB program that will do all the iterations to come up with the correct temperature and likewise provide the different values of x and y. Initial discussion of the use of `fzero()` was discussed in Chapter 7. In fact, Program Listing 8.2 is an expounded version of Program Listing 7.5.

Program Listing 8.2: Revised Program Listing 7.5

```
 1  clc
 2  initval=300;
 3  x0=0.0;
 4  x=x0;
 5  temp0=inline('0.0.*(10.^(4.72583-(1660.652./(T-
    1.461))))+(1-0.0).*(10.^(4.08245-(1346.382/(T-
    53.508))))-1.013');
 6  temp=fzero(temp0,initval);
 7  y0=x0.*(10.^(4.72583-(1660.652./(temp-1.461)))))/
    (x0.*(10.^(4.72583-(1660.652./(temp-1.461)))))+(1-
    x0).*(10.^(4.08245-(1346.382/(temp-53.508)))));
 8  y=y0;
 9  x1=0.1;
10  x=[x x1];
11  temp0=inline('0.1.*(10.^(4.72583-(1660.652./(T-
    1.461))))+(1-0.1).*(10.^(4.08245-(1346.382/(T-
    53.508))))-1.013');
12  temp=fzero(temp0,initval);
```

```
13  y1=x1.*(10.^(4.72583-(1660.652./(temp-1.461))))/
    (x1.*(10.^(4.72583-(1660.652./(temp-1.461))))+(1-
    x1).*(10.^(4.08245-(1346.382/(temp-53.508)))));
14  y=[y y1];
15  x2=0.2;
16  x=[x x2];
17  temp0=inline('0.2.*(10.^(4.72583-(1660.652./(T-
    1.461))))+(1-0.2).*(10.^(4.08245-(1346.382/(T-
    53.508))))-1.013');
18  temp=fzero(temp0,initval);
19  y2=x2.*(10.^(4.72583-(1660.652./(temp-1.461))))/
    (x2.*(10.^(4.72583-(1660.652./(temp-1.461))))+(1-
    x2).*(10.^(4.08245-(1346.382/(temp-53.508)))));
20  y=[y y2];
21  x3=0.3;
22  x=[x x3];
23  temp0=inline('0.3.*(10.^(4.72583-(1660.652./(T-
    1.461))))+(1-0.3).*(10.^(4.08245-(1346.382/(T-
    53.508))))-1.013');
24  temp=fzero(temp0,initval);
25  y3=x3.*(10.^(4.72583-(1660.652./(temp-1.461))))/
    (x3.*(10.^(4.72583-(1660.652./(temp-1.461))))+(1-
    x3).*(10.^(4.08245-(1346.382/(temp-53.508)))));
26  y=[y y3];
27  x4=0.4;
28  x=[x x4];
29  temp0=inline('0.4.*(10.^(4.72583-(1660.652./(T-
    1.461))))+(1-0.4).*(10.^(4.08245-(1346.382/(T-
    53.508))))-1.013');
30  temp=fzero(temp0,initval);
31  y4=x4.*(10.^(4.72583-(1660.652./(temp-1.461))))/
    (x4.*(10.^(4.72583-(1660.652./(temp-1.461))))+(1-
    x4).*(10.^(4.08245-(1346.382/(temp-53.508)))));
32  y=[y y4];
33  x5=0.5;
34  x=[x x5];
35  temp0=inline('0.5.*(10.^(4.72583-(1660.652./(T-
    1.461))))+(1-0.5).*(10.^(4.08245-(1346.382/(T-
    53.508))))-1.013');
36  temp=fzero(temp0,initval);
37  y5=x5.*(10.^(4.72583-(1660.652./(temp-1.461))))/
```

```
      (x5.*(10.^(4.72583-(1660.652./(temp-1.461)))))+(1-
      x5).*(10.^(4.08245-(1346.382/(temp-53.508)))))));
   38 y=[y y5];
   39 x6=0.6;
   40 x=[x x6];
   41 temp0=inline('0.6.*(10.^(4.72583-(1660.652./(T-
      1.461))))+(1-0.6).*(10.^(4.08245-(1346.382/(T-
      53.508))))-1.013');
   42 temp=fzero(temp0,initval);
   43 y6=x6.*(10.^(4.72583-(1660.652./(temp-1.461))))/
      (x6.*(10.^(4.72583-(1660.652./(temp-1.461))))+(1-
      x6).*(10.^(4.08245-(1346.382/(temp-53.508)))))));
   44 y=[y y6];
   45 x7=0.7;
   46 x=[x x7];
   47 temp0=inline('0.7.*(10.^(4.72583-(1660.652./(T-
      1.461))))+(1-0.7).*(10.^(4.08245-(1346.382/(T-
      53.508))))-1.013');
   48 temp=fzero(temp0,initval);
   49 y7=x7.*(10.^(4.72583-(1660.652./(temp-1.461))))/
      (x7.*(10.^(4.72583-(1660.652./(temp-1.461))))+(1-
      x7).*(10.^(4.08245-(1346.382/(temp-53.508)))))));
   50 y=[y y7];
   51 x8=0.8;
   52 x=[x x8];
   53 temp0=inline('0.8.*(10.^(4.72583-(1660.652./(T-
      1.461))))+(1-0.8).*(10.^(4.08245-(1346.382/(T-
      53.508))))-1.013');
   54 temp=fzero(temp0,initval);
   55 y8=x8.*(10.^(4.72583-(1660.652./(temp-1.461))))/
      (x8.*(10.^(4.72583-(1660.652./(temp-1.461))))+(1-
      x8).*(10.^(4.08245-(1346.382/(temp-53.508)))))));
   56 y=[y y8];
   57 x9=0.9;
   58 x=[x x9];
   59 temp0=inline('0.9.*(10.^(4.72583-(1660.652./(T-
      1.461))))+(1-0.9).*(10.^(4.08245-(1346.382/(T-
      53.508))))-1.013');
   60 temp=fzero(temp0,initval);
   61 y9=x9.*(10.^(4.72583-(1660.652./(temp-1.461))))/
      (x9.*(10.^(4.72583-(1660.652./(temp-1.461))))+(1-
      x9).*(10.^(4.08245-(1346.382/(temp-53.508)))))));
```

```
62 y=[y y9];
63 x10=1.0;
64 x=[x x10];
65 temp0=inline('1.0.*(10.^(4.72583-(1660.652./(T-
   1.461))))+(1-1.0).*(10.^(4.08245-(1346.382/(T-
   53.508)))))-1.013');
66 temp=fzero(temp0,initval);
67 y10=x10.*(10.^(4.72583-(1660.652./(temp-1.461))))/
   (x10.*(10.^(4.72583-(1660.652./(temp-1.461))))+(1-
   x10).*(10.^(4.08245-(1346.382/(temp-53.508)))));
68 y=[y y10];
69 fprintf('x \t\t\t y \t\t\t')
70 for i=1:11
71 fprintf('\n%f \t %f \t %f \n',x(i),y(i))
72 end
73 fprintf('\n');
74 diagonal =(0:1);
75 plot(x,y,diagonal,diagonal)
76 grid on
77 axis square
78 axis equal
79 xlabel('Benzene in Liquid (mole fraction)');
80 ylabel('Benzene in vapor (mole fraction)');
81 title('Benzene - Toluene Equilibrium System');
```

Line 1:   Clears the Command Window.

Lines 2 to 4: Initialize `initval`, `x0`, and `x` variables.

Line 5:   Equates `temp0` with the inline function using Eqs 7.4 and 7.5, and substitutes these in Eq 7.6.

Line 6:   Uses `fzero()` to determine T (expressed in the program as `temp`) from inline function `temp0`.

Lines 7, 8: Once `temp` is determined, all y values are determined from the different given values of x using Eqs 7.4 and 7.5 substituted in Eq 7.7.

Lines 9 to 68: The algorithm is repeated for each increment of x by 0.1 to obtain the corresponding results of y.

Lines 69 to 73: Generate a table of x and y values in the Command Window.

Line 74: Initializes coordinates for the diagonal line.

Line 75: Plots the x, y, and `diagonal` coordinates.

Line 76: Sets a grid line on the graph.

Lines 77, 78: Specify that the graph have equal length of intervals in both x and y axes.

Lines 79 to 81: Specify the label for the x and y axes, as well as the graph title.

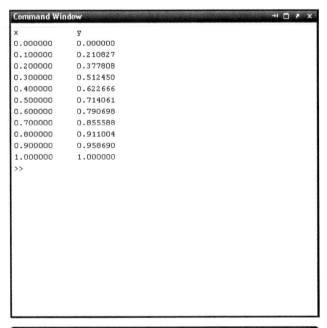

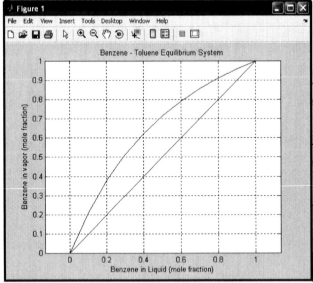

FIGURE 8.1:   Output of Program Listing 8.2.

For this lengthy code, an improved version of Program Listing 8.2 can be presented as follows:

Program Listing 8.2a: Revised Program Listing 8.2

```
clc
initval=300;
x = 0:0.1:1.0;
for i=1:length(x)
     temp0 = @(T)  x(i)  .*(10.^(4.72583-(1660.652./(T-
1.461))))+(1-x(i)).*(10.^(4.08245-(1346.382/(T-
53.508))))-1.013;
     temp=fzero(temp0,initval);
     y(i)=x(i).*(10.^(4.72583-(1660.652./(temp-1.461))
))/(x(i).*(10.^(4.72583-(1660.652./(temp-1.461)))))+(1-
x(i)).*(10.^(4.08245-(1346.382/(temp-53.508)))))));
end
fprintf('x \t\t\t y \t\t\t')
for i=1:11
fprintf('\n%f \t %f \t %f \n',x(i),y(i))
end
fprintf('\n');
diagonal =(0:1);
plot(x,y,diagonal,diagonal)
grid on
axis square
axis equal
xlabel('Benzene in Liquid (mole fraction)');
ylabel('Benzene in vapor (mole fraction)');
title('Benzene - Toluene Equilibrium System');
```

Notice that we have greatly reduced the length of the code by replacing the inline function with an anonymous function using the @ sign, a MATLAB operator that constructs a function handle. In this way, we can now use variables because the function is no longer in string format.

We can even make the code more flexible by assigning the Antoine equation constants as variables at the start of the code, and then using these variables in the anonymous function.

Program Listing 8.2b: Revised Program Listing 8.2 with Antoine equation constants

```
clc
initval=300;
A1 = 4.72583; B1 = 1660.652; C1 = -1.461; %Benzene
A2 = 4.08245; B2 = 1346.382; C2 = -53.508; %Toluene
```

```
x = 0:0.1:1.0;
for i=1:length(x)
        temp0 = @(T) x(i) .*(10.^(A1-(B1./(T+C1))))+(1-
x(i)).*(10.^(A2-(B2/(T+C2))))-1.013;
    temp=fzero(temp0,initval);
    y(i)=x(i).*(10.^(A1-(B1./(temp+C1)))))/(x(i).*(10.^(A1-
(B1./(temp+C1)))))+(1-x(i)).*(10.^(A2-(B2/(temp+C2)))));
end
fprintf('x \t\t\t y \t\t\t')
for i=1:11
fprintf('\n%f \t %f \t %f \n',x(i),y(i))
end
fprintf('\n');
diagonal =(0:1);
plot(x,y,diagonal,diagonal)
grid on
axis square
axis equal
xlabel('Benzene in Liquid (mole fraction)');
ylabel('Benzene in vapor (mole fraction)');
title('Benzene - Toluene Equilibrium System');
```

With this revised code, we can easily replace different substances with corresponding Antoine constants (refer to Appendix I).

Enthalpy-Concentration Relationship for Magnesium Sulfate System

Interpolation can also be applied to derive intermediate values in given tabulated data, wherein the algorithm and values can easily be incorporated into a MATLAB program.

Let us consider the enthalpy-concentration data for a $MgSO_4$-H_2O system based on 32°F.

Table 8.2: Magnesium sulfate enthalpy (Btu/lb) -concentration data.

°F	0.00	0.05	0.10	0.15	0.20	0.25	0.30	0.35	0.40	0.45	0.50
					Mass Fraction						
30	0	-14	-29	-43	-64	-111	-158	-205	-252	-299	-346
40	9	-7	-20	-35	-50	-67	-89	-111	-133	-155	-163
50	18	3	-11	-27	-43	-59	-81	-103	-125	-147	-158
60	29	12	-5	-20	-35	-51	-73	-95	-117	-139	-157
70	39	22	5	-10	-27	-44	-65	-89	-113	-137	-152
80	49	30	14	-3	-19	-35	-55	-81	-107	-133	-148

90	59	40	23	7	-11	-28	-47	-73	-99	-125 -145
100	69	50	32	15	-4	-21	-38	-64	-92	-120 -140
110	79	59	40	22	5	-14	-30	-55	-87	-119 -136
120	89	69	49	30	11	-6	-25	-45	-66	-87 -108
130	98	78	58	40	20	2	-17	-35	-59	-83 -107
140	109	87	70	47	28	9	-10	-28	-52	-76 -100
150	118	97	75	56	35	16	-3	-22	-45	-70 -95
160	129	107	85	65	45	24	5	-15	-26	-25 -24
170	139	116	94	73	52	31	11	-8	-24	-22 -20
180	149	125	102	80	59	38	17	-2	-20	-19 -18
190	159	134	111	90	67	46	25	5	-15	-16 -15
200	169	143	119	96	74	53	31	11	-10	-13 -12
210	179	152	128	104	83	60	38	16	-4	-10 -11
220	189	160	136	113	90	68	45	25	4	-6 -6
230	199	169	145	120	98	76	54	32	10	-5 -5

Data points were approximated and averaged from a magnesium sulfate enthalpy-concentration diagram of various sources. These are not results of a laboratory study and therefore may be inaccurate.

To translate into the MATLAB program, data in the table can be retrieved using linear interpolation. To achieve this, the `interp2()` will be used.

Program Listing 8.3: Enthalpy-concentration program for magnesium sulfate system

```
%save as enthal_MgSO4.m
function enthal=enthal_MgSO4(conc,temp)
x=[0:.05:0.5];
y=[30:10:230];
z = [0 -14 -29 -43 -64 -111 -158 -205 -252 -299 -346;
     9   -7 -20 -35 -50 -67  -89 -111 -133 -155 -163;
    18    3 -11 -27 -43 -59  -81 -103 -125 -147 -158;
    29   12  -5 -20 -35 -51  -73  -95 -117 -139 -157;
    39   22   5 -10 -27 -44  -65  -89 -113 -137 -152;
    49   30  14  -3 -19 -35  -55  -81 -107 -133 -148;
    59   40  23   7 -11 -28  -47  -73  -99 -125 -145;
    69   50  32  15  -4 -21  -38  -64  -92 -120 -140;
    79   59  40  22   5 -14  -30  -55  -87 -119 -136;
    89   69  49  30  11  -6  -25  -45  -66  -87 -108;
    98   78  58  40  20   2  -17  -35  -59  -83 -107;
   109  87  70  47  28   9  -10  -28  -52  -76 -100;
   118  97  75  56  35  16   -3  -22  -45  -70  -95;
```

```
129 107 85   65   45   24   5    -15 -26 -25 -24;
139 116 94   73   52   31   11   -8  -24 -22 -20;
149 125 102 80   59   38   17   -2  -20 -19 -18;
159 134 111 90   67   46   25   5   -15 -16 -15;
169 143 119 96   74   53   31   11  -10 -13 -12;
179 152 128 104 83   60   38   16  -4  -10 -11;
189 160 136 113 90   68   45   25  4   -6  -6;
199 169 145 120 98   76   54   32  10  -5  -5];
enthal=interp2(x,y,z,conc,temp,'linear');
```

From the function we developed we can now apply this to problems that require the resulting enthalpy given the concentration of magnesium sulfate.

Example Problem 8.1.
a) Determine the enthalpy of 27.5% concentration of magnesium sulfate at 110°F.
b) Determine the enthalpy of 27.5% concentration of magnesium sulfate at 50°F.

```
>> enthal_MgSO4(0.275,110)
ans =
   -22

>> enthal_MgSO4(0.275,50)
ans =
   -70
```

Solutions:
a) -22 Btu/lb
b) -70 Btu/lb

Phase Diagram – Magnesium Sulfate System
Similarly, the interpolation method can also be used to convert existing graphs or diagrams into numerical representations, which can be converted easily into the MATLAB program.

Let us consider the $MgSO_4$ phase diagram as an example.

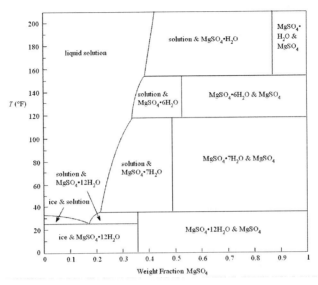

FIGURE 8.2: Magnesium sulfate–water system phase diagram (at 1 atm).

(Source: *Chemical Engineering Design and Analysis: An Introduction*, T. M. Duncan and J. A. Reimer, Cambridge University Press, 1998.)

Each line and curve in the diagram is represented into sets of coordinates, which are then used as a basis for interpolating the desired result. Program Listing 8.4 uses linear and cubic hermite interpolation to achieve this.

Program Listing 8.4: Phase diagram program for magnesium sulfate system

```
%save as phase.m
function MgSO4 = phase(tempsol)
if tempsol > 154
    temp_range=[154,210];
    fraction_sol=[0.385,0.425];
        solution=interp1(temp_range,fraction_sol,
tempsol,'linear');
    crystal=0.8698;
elseif tempsol <= 154 && tempsol >117
    temp_range=[117,144,154];
    fraction_sol=[0.34,0.36,0.385];
        solution=interp1(temp_range,fraction_sol,
tempsol,'cubic');
    crystal=0.5269;
elseif tempsol <= 117 && tempsol >34
```

```
temp_range=[34,79,117];
fraction_sol=[0.215,0.26,0.34];
   solution=interp1(temp_range,fraction_sol,
tempsol,'cubic');
  crystal=0.4884;
else
  disp('Temperature is out of range');
end
MgSO4=[solution crystal];
```

Example Problem 8.2. Determine the mass fraction of MgSO$_4$ in the mother liquor and in the magnesium sulfate crystal at 50°F.

```
>> phase(50)

ans =

0.2257    0.4884
```

Solution:

Mass fraction of MgSO$_4$ in mother liquor (50°F) = 0.2257.
Mass fraction of MgSO$_4$ in crystal slurry (50°F) = 0.4884.

The two resulting values provide the mass fraction of magnesium sulfate in the mother liquor (the first value provided) and in the slurry (the second value).

CHEMICAL ENGINEERING SAMPLE PROBLEMS

Equation of State – Cubic Volume

The equation of state that is cubic in volume can represent the behavior of both vapor and liquid. Two examples of this type are the equation proposed by van der Waals and the equation of Redlich and Kwong.

A Dutch physicist, Johannes van der Waals, established the equation of state,

$$\left(P + \frac{a}{V^2}\right)(V - b) = nRT \tag{8.1}$$

where a and b are constants, determined through fitting the van der Waals equation to experimental P-V-T data, especially the values at the critical points. Deriving for the cubic volume, the van der Waals equation of state can be presented as

$$V^3 - \left(nb + \frac{nRT}{P}\right)V^2 + \frac{n^2 a}{P}V - \frac{n^3 ab}{P} = 0 \tag{8.2}$$

where $a = \dfrac{27R^2 T_c^2}{64P_c}$ and $b = \dfrac{RT_c}{8P_c}$. The P_c and T_c are the critical pressure and temperature of the substance, respectively.

The Redlich-Kwong equation came about through the empirical modification of the van der Waals equation, which leads to the formulation of

$$\left[P + \left(\frac{a}{T^{1/2}V(V+b)}\right)\right](V - b) = nRT . \tag{8.3}$$

Like in the van der Waals equation, this equation has two constants, a and b, where

$$a = \frac{0.42748R^2 T_c^{2.5}}{P_c} \text{ and } b = \frac{0.08664RT_c}{P_c}.$$

Modifying the original Redlich-Kwong equation, the cubic volume can be obtained from the equation

$$V^3 - \frac{nRT}{P}V^2 + \left(n^2 b - \frac{n^2 RTb}{P} - \frac{n^2 a}{PT^{1/2}}\right)V - \frac{n^3 ab}{PT^{1/2}} = 0 . \tag{8.4}$$

Let us solve a problem using these equations.

Example Problem 8.3. Determine the specific volume of saturated steam at a temperature of 375°K. Use the van der Waals and Redlich-Kwong equations to determine the volume. Compare the results to the specific volume indicated in the saturated steam table in Appendix G.

van der Waals:

$$a = 5454293.072 \text{ atm}\left(\frac{cm^3}{gmol}\right)^2 \quad b = 30.42008933 \frac{cm^3}{gmol} \quad \text{basis: } n = 1gmol$$

$$V^3 - \left(30.42008933 + \frac{(82.06)(375)}{1.068829415}\right)V^2 + \left(\frac{(1^2)(5454293072)}{1.068829415}\right)V$$

$$- \left(\frac{(1^3)(5454293072)(30.42008933)}{1.068829415}\right) = 0$$

$$V^3 - 28821.2632V^2 + 5103052.924V - 155235325.8 = 0$$

$$V = 28643.29368 \frac{cm^3}{gmol} \times \frac{1gmol}{18g}$$

$$V = 1591.294093 \frac{cm^3}{g}$$

Redlich-Kwong:

$$a = 140623151.9 \; \frac{atm.K^{\frac{1}{2}}.cm^3}{gmol} \quad b = 21.08477231 \; \frac{cm^3}{gmol} \quad \text{assume: } n = 1gmol$$

$$V^3 - \left[\frac{(82.06)(375)}{1.068829415}\right]V^2 + \left[21.08477231 - \frac{(82.06)(375)(21.08477231)}{1.068829415}\right.$$

$$\left. - \frac{140623151.9}{(1.068829415)(\sqrt{375})}\right]V - \left(\frac{(140623151.9)(21.08477231)}{(1.068829415)(\sqrt{375})}\right) = 0$$

$$V = \left(29045.82204 \frac{cm^3}{gmol}\right) x \left(\frac{1gmol}{18g}\right)$$

$$V = 1613.656780 \frac{cm^3}{g}$$

Program Listing 8.5: Solving specific volume of saturated steam using equation of state

```
1   clc
2   Pc=input('Enter water vapor critical pressure (atm): ');
3   Tc=input('Enter water vapor critical temperature (K): ');
4   R=input('Enter gas constant R (cm3-atm/mol-K): ');
5   n=input('Enter number of moles: ');
6   T=input('Enter temperature (K): ');
7   P=input('Enter pressure (atm): ');
8   format long g;
9   a_waals = 27*R^2*Tc^2/(64*Pc);
10  b_waals = R*Tc/(8*Pc);
11  a_kwong = (0.42748*R^2*Tc^2.5)/Pc;
12  b_kwong = 0.08664*R*Tc/Pc;
13  eq_waals = [1 -(n*b_waals+(n*R*T/P)) ((n^2)*a_waals/P)
    -((n^3)*a_waals*b_waals/P)] ;
```

14 `V_waals = (roots(eq_waals))/18;`

15 `fprintf('Volume based on van der Waals equation: %f cu cm\n',V_waals(1));`

16 `eq_kwong = [1 -n*R*T/P ((n^2*b_kwong - (n^2)*R*T*b_kwong/P)-(n^2*a_kwong/(P*T^.5))) -(n^3*a_kwong*b_kwong/(P*T^.5))];`

17 `V_kwong = (roots(eq_kwong))/18;`

18 `fprintf('Volume based on Redlich-Kwong equation: %f cu cm\n',V_kwong(1));`

Line 1:   Clears the Command Window.

Lines 2 to 7: Request all necessary information such as the critical temperature and pressure, the gas constant, number of moles, temperature, and pressure.

Line 8:   Sets the display format to `long g` (shortest decimal places based on the significant number).

Lines 9, 10: Compute the constants *a* and *b* in the van der Waals equation (Eq 8.1).

Lines 11, 12: Compute the constants *a* and *b* in the Redlich-Kwong equation (Eq 8.3).

Line 13: Assigns the coefficients of the van der Waals equation (Eq 8.2) to `eq_waals`.

Line 14: Using the `roots()` function, this line computes for the volume in cubic cm per gram of water vapor. The molecular weight 18 g/gmol was used as a divisor to convert gmols to grams.

Line 15: Prints the result in the Command Window.

Line 16: Assigns the coefficients of the Redlich-Kwong equation (Eq 8.4) to `eq_kwong`.

Line 18: This line again prints the result in the Command Window.

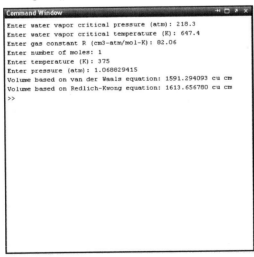

FIGURE 8.3:   Output of Program Listing 8.5.

Solutions:
Volume based on the van der Waals equation: 1591.294093 cu cm/g.
Volume based on the Redlich-Kwong equation: 1613.656780 cu cm/g.

Note: Based on the steam table (Appendix G), the volume is 1.57240 cu m/kg
or 1572.4 cu cm/g.

Compressibility Factor

The compressibility factor is a constant developed to tie together the concepts of
the law of corresponding states and the ideal gas law, which gives more accurate
results than a simple ideal gas law. The new equation of state is expressed as:

$$PV = znRT$$

$$z = PV\big/nRT \qquad\qquad\qquad\qquad (8.5)$$

where z is a function of pressure and temperature, $z = (P,T)$. The compressibility
factor z is a constant from Eq 8.5. For different gases we can obtain different
values of z by using the equation of state formulas and plotting it.

Example Problem 8.4. Determine the different compressibility factors using the
van der Waals equation for various P_r and T_r of carbon dioxide. The critical value
for temperature is 304.2°K and for pressure 72.9 atm. Plot the relationship.

To demonstrate manual computation we will consider a reduced temperature T_r
= 1.0 and a reduced pressure P_r = 0.5 as a basis.

$$T_r = 1.0 \qquad\qquad\qquad P_r = 0.5$$
$$T = T_r \times T_c \qquad\qquad\quad P = P_r \times P_c$$
$$T = 1.0 \times 304.2 \qquad\quad P = 0.5 \times 72.9$$
$$T = 304.2\,°\,K \qquad\qquad P = 36.45$$

$$a = \frac{27R^2T_c^2}{64P_c} \qquad\qquad b = \frac{RT_c}{8P_c}$$

$$a = \frac{27(82.06)^2(304.2)^2}{64(72.9)} \qquad b = \frac{(82.06)(304.2)}{8(72.9)}$$

$$a = 3606099.507 \qquad\qquad b = 42.8029$$

Using the van der Waals equation, we can solve for the specific volume.

$$V^3 - \left(nb + \frac{nRT}{P}\right)V^2 + \frac{n^2 a}{P}V - \frac{n^3 ab}{P} = 0$$

$$V^3 - \left((1)(42.8029) + \frac{(1)(82.06)(304.2)}{36.45}\right)V^2 + \frac{1^2(3606099.507)}{36.45}V$$

$$- \frac{1^3(3606099.507)(42.8029)}{36.45} = 0$$

V = 566.101484889818

Solving for the value of z,

$PV = zRT$

$$z = \frac{PV}{RT}$$

$$z = \frac{(36.45)(566.101484889818)}{(82.06)(304.2)}$$

$z = 0.826610855$

To make a more general MATLAB program to illustrate the relationship of the reduced pressure P_r and the compressibility factor z, we will use the same method discussed.

Program Listing 8.6: Compressibility factor program

```
1  clc
2  clear
3  Pc=input('Enter critical pressure (atm): ');
4  Tc=input('Enter critical temperature (K): ');
5  R=82.06;
6  Tr=[1.0:0.2:2.0];
7  Pr=[0.5:0.5:10];
8  a_waals = 27*R^2*Tc^2/(64*Pc);
9  b_waals = R*Tc/(8*Pc);
10 Ta=Tc.*Tr;
11 Pa=Pc.*Pr;
12 for i=1:6
13   T(1)=Ta(i);
```

```
14  for j=1:20
15    P(1)=Pa(j);
16     eq_waals = [1  -(b_waals+((R*T)/P))  (a_waals/P)  -
       ((a_waals*b_waals)/P)];
17     tempV_waals = roots(eq_waals);
18     for n=1:3
19      if isreal(tempV_waals(n))
20        nr=n;
21      end
22     end
23     V_waals(i,j)=abs(tempV_waals(nr));
24     z(i,j)=(P*V_waals(i,j)/(R*T));
25    end
26  end
27  Pr_new = 0.5:.05:10;
28  for i=1:6
29   z_new(i,:) = interp1(Pr,z(i,:),Pr_new,'cubic');
30  end
31  plot(Pr_new,z_new(1,:),Pr_new,z_new(2,:),Pr_new,z_
    new(3,:),Pr_new,z_new(4,:),Pr_new,z_new(5,:),Pr_new,z_
    new(6,:));
32  xlabel('Pr');
33  ylabel('Compressibility Factor z');
34  title('Compressibility Factor Diagram');
35  legend('Tr 1.0','Tr 1.2','Tr 1.4','Tr 1.6','Tr 1.8','Tr
    2.0');
```

Line 1: Clears the Command Window.
Line 2: Clears all array values in the memory.
Line 3: Requests the critical pressure (in atm) of the gas being considered.
Line 4: Requests the critical temperature (in K) of the gas being considered.
Line 5: Initializes the value for gas constant R.
Line 6: Initializes the value of Tr as 1.0 to 2.0 with a 0.2 increment.
Line 7: Initializes the value of Pr as 0.5 to 10 with a 0.5 increment.
Line 16: Assigns the coefficients of the van der Waals equation (Eq 8.2) to eq_waals for varying values of P and T as determined by the for loop statement.
Line 17: Using the roots() function, this line computes for the volume in cubic cm per mole of the gas.
Lines 12 to 23: This loop statement determines the real value from the array of values generated by the roots() function, which is then assigned to the variable V_waals().

Line 24: Computes for the compressibility factor from the derived volume, using Eq 8.5.

Lines 27 to 30: Generate more data points for `Pr` and `z` using cubic interpolation.

Lines 31 to 35: Plot the newly generated data points, where the x-axis is the reduced pressure `Pr` and the y-axis is the compressibility factor `z`.

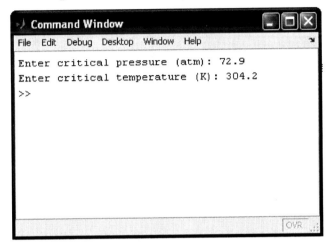

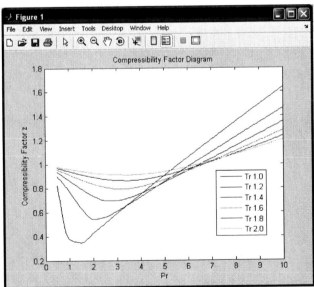

FIGURES 8.4: Output of Program Listing 8.6.

Simple Differential Distillation

In simple differential distillation, the liquid is heated slowly to boiling and the vapor is collected and condensed. At the start, the vapor condensed contains the richest fraction of the more volatile component. As the vaporization continues, its concentration decreases.

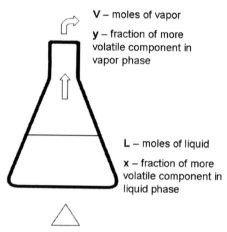

V – moles of vapor

y – fraction of more volatile component in vapor phase

L – moles of liquid

x – fraction of more volatile component in liquid phase

FIGURE 8.5: Simple differential distillation.

The concentration x changes through time, therefore we can formulate the equation using material balance based on the more volatile component.

initial amount = amount in liquid + amount in vapor

Let dx = change of concentration of the more volatile component in liquid.
dL = small amount of liquid evaporated.

$$xL = (x - dx)(L - dL) + ydL$$

$$xL = xL - Ldx - xdL + dxdL + ydL$$

Since the value of $dxdL$ is too small it becomes negligible.

$$xL = xL - Ldx - xdL + ydL$$

$$xdL - ydL = xL - xL - xdL$$

$$dL(x - y) = -Ldx$$

$$\frac{-dL}{L} = \frac{dx}{x - y}$$

Integrating both sides we get

$$-\int_{L_1}^{L_2} \frac{dL}{L} = \int_{x_1}^{x_2} \frac{dx}{x-y}$$

$$-\ln \frac{L_2}{L_1} = \int_{x_1}^{x_2} \frac{dx}{x-y}$$

(8.6)

where L_1 = initial moles of liquid
        L_2 = moles of liquid after evaporation takes place.

Example Problem 8.5. A solution containing 60 moles of benzene and 50 moles of toluene is distilled using the simple differential distillation process at 1.013 bar until such time only 70 moles of liquid is left. Determine the composition of the liquid after the distillation process.

In solving this problem, we first need to determine the initial concentration of benzene in the solution.

Initial conc. of benzene $= \frac{60}{(60+50)} = 0.5454$.

Likewise. solving for $-\ln \frac{L_2}{L_1} = -\ln\left(\frac{70}{60+50}\right) = 0.452$.

Then we derive the x and y values using the Antoine equation. In this case we can use Program Listing 8.2 to generate these values. Table 8.3 provides the mass fraction of benzene in the liquid and vapor phases.

Table 8.3: Mass fraction of benzene in thebenzene-toluene solution.

x	y
0.000000	0.000000
0.100000	0.210827
0.200000	0.377808
0.300000	0.512450
0.400000	0.622666
0.500000	0.714061
0.600000	0.790698
0.700000	0.855588
0.800000	0.911004
0.900000	0.958690
1.000000	1.000000

Once the sample points of the benzene-toluene equilibrium curve are obtained, a MATLAB program is developed to solve the problem.

Program Listing 8.7: Simple differential distillation program.

```
 1  clc
 2  clear
 3  x=[0:0.1:1.0];
 4  y=[0  0.210827  0.377808  0.512450  0.622666  0.714061
    0.790698 0.855588 0.911004 0.958690 1.0];
 5  eq_y=polyfit(x,y,5);
 6  eq_x=[0 0 0 0 1 0];
 7  eq=eq_x-eq_y;
 8  x1=.5454;
 9  exp_area=.452;
10 for x2=0.54:-0.001:0
11  x=x1:-0.0001:x2;
12  area=trapz(x,1./(polyval(eq,x)));
13  delta=area-exp_area;
14  if delta <0.001 & delta >-0.001
15    correct_x2=x2;
16  end
17 end
18 fprintf('The composition of the 70 moles liquid after
   distillation\n\n');
19 fprintf('Benzene: %.2f %% \n',correct_x2*100);
20 fprintf('Toluene: %.2f %% \n\n',(1-correct_x2)*100);
```

Lines 3, 4: Initialize x and y variables with values from Table 8.3.

Line 5: Derives the polynomial equation (5th order) that best fits the x and y coordinates and assigns the coefficients of the equation to vector eq_y.

Line 6: Assigns x to eq_x.

Line 7: Assigns the equation (x-y) to eq.

Line 8: Equates x1 with the initial concentration of benzene in the solution.

Line 9: Declares the value 0.452 (expected area under the curve) to exp_area.

Lines 10 to 17: Perform iterations to determine the values of x2, in which during the iteration, numerical integration using trapz() is implemented.

Lines 18 to 20: Print the result in the Command Window.

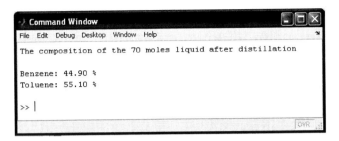

FIGURE 8.6: Output of Program Listing 8.7.

Two-Dimensional Conduction

Example Problem 8.6. Consider a flat square plate shown in Figure 8.7. The sides of the plate are held at a constant temperature with values of 200°C and 500°C as shown in figure 8.7. Find the temperature at each node.

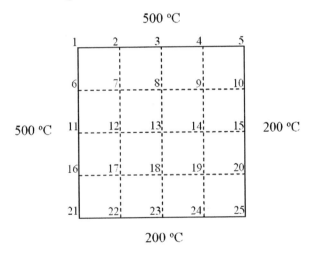

FIGURE 8.7: Flat square metal plate.

A heat balance on the volume on a sample node (7) gives

$$q_7 = \frac{kA}{\Delta x}(T_6 - T_7) + \frac{kA}{\Delta x}(T_8 - T_7) + \frac{kA}{\Delta y}(T_{12} - T_7) + \frac{kA}{\Delta y}(T_2 - T_7).$$

Using a square net, Δx is equal to Δy, so by rearranging kA, the equation becomes

$$\frac{q_7}{kA} = T_8 + T_{12} - 4T_7 + 1000.$$

When the steady state is reached, the imaginary sink (q_7/kA) is equal to zero. By completing the equation at various nodes, we yield

$$T_1 = T_2 = T_3 = T_4 = T_6 = T_{11} = T_{16} = 500°C$$

$$T_{10} = T_{15} = T_{20} = T_{25} = T_{24} = T_{23} = T_{22} = 200°C$$

$$T_5 = \frac{500 + 200}{2} = 350°C$$

$$T_{21} = \frac{500 + 200}{2} = 350°C$$

At node 7: $-4T_7 + T_8 + T_{12} = -1000$ At node 14: $-4T_{14} + T_{13} + T_{19} + T_9 = -200$

At node 8: $-4T_8 + T_7 + T_{13} + T_9 = -500$ At node 17: $-4T_{17} + T_{18} + T_{12} = -700$

At node 9: $-4T_9 + T_8 + T_{14} = -700$ At node 18: $-4T_{18} + T_{17} + T_{19} + T_{13} = -200$

At node 12: $-4T_{12} + T_{13} + T_{17} + T_7 = -500$ At node 19: $-4T_{19} + T_{18} + T_{14} = -400$

At node 13: $-4T_{13} + T_{12} + T_{14} + T_{18} + T_8 = 0$

$$A = \begin{vmatrix} -4 & 1 & 0 & 1 & 0 & 0 & 0 & 0 & 0 \\ 1 & -4 & 1 & 0 & 1 & 0 & 0 & 0 & 0 \\ 0 & 1 & -4 & 0 & 0 & 1 & 0 & 0 & 0 \\ 1 & 0 & 0 & -4 & 1 & 0 & 1 & 0 & 0 \\ 0 & 1 & 0 & 1 & -4 & 1 & 0 & 1 & 0 \\ 0 & 0 & 1 & 0 & 1 & -4 & 0 & 0 & 1 \\ 0 & 0 & 0 & 1 & 0 & 0 & -4 & 1 & 0 \\ 0 & 0 & 0 & 0 & 1 & 0 & 1 & -4 & 1 \\ 0 & 0 & 0 & 0 & 0 & 1 & 0 & 1 & -4 \end{vmatrix} \begin{Vmatrix} T_7 \\ T_8 \\ T_9 \\ T_{12} \\ T_{13} \\ T_{14} \\ T_{17} \\ T_{18} \\ T_{19} \end{Vmatrix} = \begin{vmatrix} -1000 \\ -500 \\ -700 \\ -500 \\ 0 \\ -200 \\ -700 \\ -200 \\ -400 \end{vmatrix}$$

In developing a MATLAB program, we will generate a similar matrix using array manipulation, which we studied in Chapter 6.

Program Listing 8.8: Two-dimensional conduction program

```
1   clc
2   clear
3   node=input('Enter number of nodes in the square plate: ');
4   up_temp=input('Enter temperature at the upper portion: ');
5   right_temp=input('Enter temperature at the right
    portion: ');
6   low_temp=input('Enter temperature at the lower portion: ');
7   left_temp=input('Enter temperature at the left portion: ');
8   for i=2:(sqrt(node)-1)
9     n_temp(i)=up_temp;
10    n_temp(i*sqrt(node))=right_temp;
11    n_temp(node+2-i)=low_temp;
12    n_temp(1+sqrt(node)+(i-2)*sqrt(node))=left_temp;
13  end
14  n_temp(1)=(left_temp+up_temp)/2;
15  n_temp(sqrt(node))=(up_temp+right_temp)/2;
16  n_temp(node+1-sqrt(node))=(left_temp+low_temp)/2;
17  n_temp(node)=(right_temp+low_temp)/2;
18  dimn=(sqrt(node)-2)^2;
19  x1=-4*eye(dimn);
20  x2=diag(linspace(1,1,dimn-3),-3);
21  x3=diag(linspace(1,1,dimn-3),3);
22  temp=linspace(1,1,dimn-1);
23  for i=1:floor((dimn-1)/3)
24    temp(1,i*3)=temp(1,i*3)-1;
25  end
26  x4=diag(temp,-1);
27  x5=diag(temp,1);
28  A=x1+x2+x3+x4+x5;
29  for i=1:dimn
30    B(1,i)=0;
31  end
32  for i=1:sqrt(dimn)
33    B(i,1)=-up_temp;
34    B(i*sqrt(dimn),1)=-right_temp;
35    B(dimn+1-i,1)=-low_temp;
36    B(1+(i-1)*sqrt(dimn),1)=-left_temp;
```

```
37 end
38 B(1,1)=-up_temp-left_temp;
39 B(sqrt(dimn),1)=-up_temp-right_temp;
40 B(dimn+1-sqrt(dimn),1)=-low_temp-left_temp;
41 B(dimn,1)=-low_temp-right_temp;
42 T=A\B;
43 counter=1;
44 for i=sqrt(node)+1:sqrt(node):node-2*(sqrt(node))+1
45   for j=1:sqrt(dimn)
46     n_temp(i+j)=T(counter);
47     counter=counter+1;
48   end
49 end
50 for i=1:node
51   fprintf('Temperature   at   node   %d   =   %f\n',i,n_
   temp(i));
52 end
```

Lines 3 to 7: Request the temperature at each side of the square plate.

Lines 8 to 13: Assign the given temperature to each side node at its corresponding portion (example nodes 1 to 5 have a temperature of 500°K).

Lines 14 to 17: Assign the average temperature of each side to the corner nodes.

Line 18: Computes for the total number of nodes excluding side nodes. This will be the dimension of the matrix. In this example, the dimension of the matrix is 9×9.

Line 19: Assigns a -4 value on all elements in the main diagonal.

Lines 20 to 28: Set up other values of the elements in the matrix.

Lines 29 to 41: Generate the values for the right-hand side.

Line 42: Performs matrix left division.

Lines 44 to 52: Print the generated result in the Command Window.

Using the developed **MATLAB** program for solving for simultaneous linear

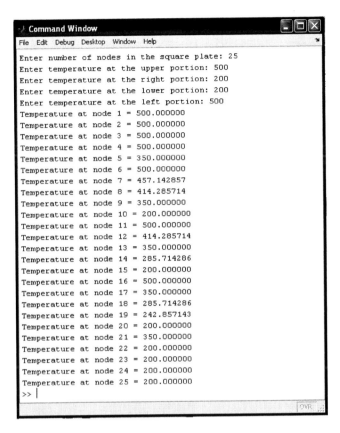

```
Command Window                                                    _ □ ✕
File  Edit  Debug  Desktop  Window  Help                                  ⅴ
Enter number of nodes in the square plate: 25
Enter temperature at the upper portion: 500
Enter temperature at the right portion: 200
Enter temperature at the lower portion: 200
Enter temperature at the left portion: 500
Temperature at node  1 = 500.000000
Temperature at node  2 = 500.000000
Temperature at node  3 = 500.000000
Temperature at node  4 = 500.000000
Temperature at node  5 = 350.000000
Temperature at node  6 = 500.000000
Temperature at node  7 = 457.142857
Temperature at node  8 = 414.285714
Temperature at node  9 = 350.000000
Temperature at node 10 = 200.000000
Temperature at node 11 = 500.000000
Temperature at node 12 = 414.285714
Temperature at node 13 = 350.000000
Temperature at node 14 = 285.714286
Temperature at node 15 = 200.000000
Temperature at node 16 = 500.000000
Temperature at node 17 = 350.000000
Temperature at node 18 = 285.714286
Temperature at node 19 = 242.857143
Temperature at node 20 = 200.000000
Temperature at node 21 = 350.000000
Temperature at node 22 = 200.000000
Temperature at node 23 = 200.000000
Temperature at node 24 = 200.000000
Temperature at node 25 = 200.000000
>> |
                                                               OVR
```

FIGURE 8.8:   Output of Program Listing 8.8.

equations the temperature of each node is determined, and the results are presented below:

Reactor Tanks

Multiple reactor tanks do not necessarily start at a steady state condition. In Chapter 4, we dealt with reactor tanks considered to have achieved a steady state condition and solved for the concentration at each tank. In this section we will try to determine the condition of the tank as it progresses toward the steady state condition. We will derive the ordinary differential equation that will determine the concentration at each tank, from the start until it reaches a steady state.

Material balance

$$\text{Input} = \text{Output} + \text{Accumulation}$$
$$QC_{\text{input}} = QC_{\text{output}} + V \, dC/dt \tag{8.7}$$

Rearranging the equation, we can present it as the rate of change of concentration C with respect to time t.

$$dC/dt = (QC_{input} - QC_{output})/V \qquad (8.8)$$

Example Problem 8.7 (Based on Example Problem 4.1). Four reactor tanks are connected by pipes where directions of flow are depicted by means of arrows. The transfer rate of fluid in the tanks is taken as the product of the flow rate (Q, liters per second) and the concentration (C, grams per liter). The volume of each tank is taken as 500 liters, and the flow rates are

$Q_{13} = 75$ liters/sec $Q_{24} = 20$ liters/sec $Q_{33} = 60$ liters/sec
$Q_{21} = 25$ liters/sec $Q_{32} = 45$ liters/sec $Q_{43} = 30$ liters/sec

At time $t_{init} = 0$, all concentration in the reaction tanks are zero. Determine how the concentration changes through time until it reaches a near steady state (t = 340 sec).

Reactor 1

$$75C_1 - 25C_2 = 350$$
$$dC_1/dt = (-75C_1 + 25C_2 + 350)/500$$
$$dC_1/dt = -0.15C_1 + 0.05C_2 + 0.7$$

Reactor 2

$$45C_3 - 45C_2 = 0$$
$$dC_2/dt = (45C_3 - 45C_2)/500$$
$$dC_2/dt = 0.09C_3 - 0.09C_2$$

Reactor 3

$$75C_1 + 30C_4 - 105C_3 = 0$$
$$dC_3/dt = (75C_1 + 30C_4 - 105C_3)/500$$
$$dC_3/dt = 0.15C_1 + 0.06C_4 - 0.21C_3$$

Reactor 4

$$30C_4 - 20C_2 = 150$$
$$dC_4/dt = (-30C_4 + 20C_2 + 150)/500$$
$$dC_4/dt = -0.06C_4 + 0.04C_2 + 0.3$$

Summarizing

$$dC_1/dt = -0.15C_1 + 0.05C_2 + 0.7$$
$$dC_2/dt = 0.09C_3 - 0.09C_2$$

$$dC_3/dt = 0.15C_1 + 0.06C_4 - 0.21C_3$$
$$dC_4/dt = -0.06C_4 + 0.04C_2 + 0.3$$

After solving manually, we can develop the MATLAB program to solve this example problem:

Program Listing 8.9: Reactor m-file function

```
%saved as reactor.m
function dCdt=reactor(t,C)
dCdt=[-0.15*C(1)+0.05*C(2)+0.7;0.09*C(3)-0.09*C(2);0.15*C(1
)+0.06*C(4)-0.21*C(3);-0.06*C(4)+0.04*C(2)+0.3];
```

Program Listing 8.10: Reactor tank program

```
1   clc
2   clear
3   timespan=[0:20:340];
4   init_C=[0 0 0 0];
5   [t,C]=ode45('reactor',timespan,init_C);
6   fprintf('\t\t\t\t\t Concentration(g/l)\n');
7   fprintf('Time(sec) Tank 1 \t Tank 2 \t Tank 3 \t Tank
    4\n');
8   for i=1:(340/20)+1
9   fprintf('%3d  \t%f\t%f\t%f\t%f\n',t(i),C(i,1),C(i,2),
    C(i,3),C(i,4));
10  end
11  plot(t,C(:,1),'*-',t,C(:,2),'o-',t,C(:,3),'+-
    ',t,C(:,4),'x-');
12  xlabel('time (sec)');
13  ylabel('Concentration (g per liter)');
14  title('Multiple Reactors');
15  legend('Tank 1','Tank 2','Tank 3','Tank 4');
```

Line 3:   Initializes `timespan` with values from 0 to 340 with an increment of 20 seconds.

Line 4:   Declares that the initial concentration at each tank is zero.

Line 5:   Performs an ordinary differential equation on the `reactor` function declared in the `reactor.m` file.

Lines 6 to 10:   Generate a table presenting the concentration in each tank at each time interval.

Lines 11 to 15: Plot the result.

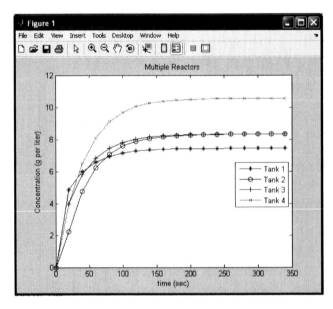

```
Command Window                                        _ □ X
File  Edit  Debug  Desktop  Window  Help                     ↘
                         Concentration(g/1)
Time(sec)   Tank 1      Tank 2      Tank 3      Tank 4
    0       0.000000    0.000000    0.000000    0.000000
   20       4.870886    2.227789    3.982370    3.987199
   40       5.988705    4.757951    5.818338    6.486537
   60       6.591399    6.224659    6.838249    8.102722
   80       6.939959    7.081883    7.444477    9.090048
  100       7.144602    7.589631    7.802948    9.682651
  120       7.265447    7.892369    8.016525    10.035436
  140       7.336824    8.067109    8.153602    10.246299
  160       7.386112    8.172440    8.221529    10.373256
  180       7.407093    8.239874    8.267054    10.446111
  200       7.422524    8.277704    8.293431    10.490508
  220       7.427831    8.308676    8.300339    10.514756
  240       7.436905    8.309670    8.325927    10.533160
  260       7.439708    8.321706    8.324953    10.541809
  280       7.441836    8.325710    8.329264    10.547530
  300       7.443301    8.329858    8.328349    10.550722
  320       7.441468    8.333202    8.330654    10.551845
  340       7.443859    8.331637    8.332696    10.553869
>>
```

FIGURES 8.9:   Output of Program Listing 8.10.

The concentrations of the tanks at a steady state, determined by material balance and solving for the simultaneous linear equation, are presented below:

Tank 1: 7.44444 g/l

Tank 2: 8.33333 g/l

Tank 3: 8.33333 g/l
Tank 4: 10.55556 g/l

Reaction in Series

Consider the reactions

$$X \xrightarrow{k_1} Y \xrightarrow{k_2} Z.$$

In a suitable condition, the rate equation of the reactions can be presented as

$$\frac{dC_X}{dt} = -k_1 C_X \tag{8.9}$$

$$\frac{dC_Y}{dt} = k_1 C_X - k_2 C_Y \tag{8.10}$$

$$\frac{dC_Z}{dt} = k_2 C_Y \tag{8.11}$$

Initially, only material X is present with concentration C_{X0}. As reaction takes place, part of the material X is converted to Y; likewise some material Y is converted to Z. If Y is highly reactive, the concentration of Y, taken as C_Y, will be minimal, as it will be converted immediately to Z.

By doing some integration on Eqs 8.9 to 8.11, concentration of X, Y, and Z in a given holding time t and reaction rates k_1 and k_2 can be expressed as

$$C_X = C_{X0} e^{-k_1 t} \tag{8.12}$$

$$C_Y = C_{X0} \frac{k_1}{k_2 - k_1} \left(e^{-k_1 t} - e^{-k_2 t} \right) \tag{8.13}$$

$$C_S = C_{X0} - C_X - C_Y \tag{8.14}$$

If the intermediate material Y is the desired reaction product, which will be the basis for designing and selecting a reactor system, then the amount of Y at which concentrations at maximum in a plug-flow or batch reactor, can then be computed

$$C_{Y\max} = C_{X0} \left(\frac{k_1}{k_2} \right)^{k_2 /(k_2 - k_1)} . \tag{8.15}$$

Likewise, C_{Ymax} is obtained at holding time

$$t_{\max} = \frac{\ln(k_2 / k_1)}{k_2 - k_1}. \tag{8.16}$$

The volume of the reactor can be derived by the equation

$$V = vt \qquad\qquad (8.17)$$

where v is the volumetric flow rate and t is the holding time.

Example Problem 8.8.

Under a certain favorable condition X decomposes as follows:

$$X \xrightarrow{\ k_1\ } Y \xrightarrow{\ k_2\ } Z$$

The desired material is Y. Determine the maximum Y concentration, the holding time t at which $C_{Y_{max}}$ was achieved, and the volume of the batch reactor at which concentration of Y is at maximum given the flow rate of X entering is 25 liters/min. The concentration of X is 2.5 moles/l while Y and Z initially have zero concentration. The reaction rates are taken as $k_1 = 0.2$ and $k_2 = 0.1$ /min.

Program Listing 8.11:

```
%saved as series.m
function dCdt=series(t,C)
global k1 k2
dCdt=[-k1*C(1);k1*C(1)-k2*C(2);k2*C(2)];
```

Program Listing 8.12:

```
 1  clc
 2  clear
 3  format short g
 4  global k1 k2
 5  disp('Ensure all units are consistent');
 6  Cx0=input('Enter initial concentration of X (moles/l): ');
 7  Cy0=input('Enter initial concentration of Y (moles/l): ');
 8  Cz0=input('Enter initial concetration of Z (moles/l): ');
 9  v_rate=input('Enter inlet volumetric flow rate (liters/
    min): ');
10  k1=input('Enter k1: ');
11  k2=input('Enter k2: ');
12  t_limit=input('Enter maximum holding time to consider
    (min): ');
13  timespan=[0:0.0005:t_limit];
14  init_C=[Cx0 Cy0 Cz0];
15  [t,C]=ode45('series',timespan,init_C);
16  plot(t,C(:,1),'-',t,C(:,2),':',t,C(:,3),'--');
```

```
17 xlabel('time (min)');
18 ylabel('Concentration (moles per liter)');
19 title('Reaction in Series');
20 legend('Conc X','Conc Y','Conc Z');
21 C_max=max(C(:,2))
22 t_max=t(find(C(:,2)==(max(C(:,2)))))
23 Vol=v_rate*t_max
```

Lines 4: Declares `k1` and `k2` variables as global. These variables are used in `series.m`.

Lines 5 to 12: Request all the necessary information from the user.

Line 13: Initializes `timespan` with values from 0 to `t_limit` with an increment of 0.0005. Note that the lower the increment, the more accurate the graph is.

Line 14: Declares the initial concentration based on the inputted data.

Line 15: Performs an ordinary differential equation on the `reactor` function declared in the `series.m` file.

Lines 16 to 20: Generate a table presenting the concentration in each tank at each time interval.

Lines 21 to 23: Calculate and display in the Command Window the maximum Y concentration, the holding time t at which $C_{Y_{max}}$ was achieved, and the volume of the batch reactor at which the concentration of Y is at maximum.

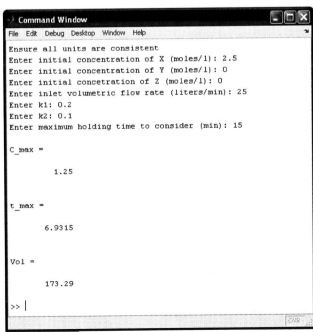

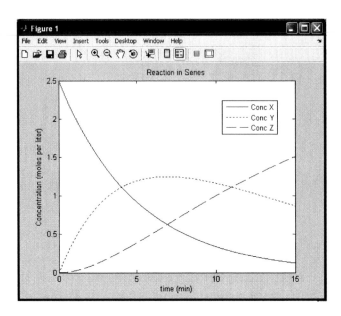

FIGURES 8.10:   Output of Program Listing 8.12.

Manual computation of the problem yields the following answers:
C_{Ymax} (using eq. 8.15) = 1.25 moles per liter
t_{max} (using eq. 8.16) = 6.931472 min
V (using eq. 8.17) = 173.2868 liters

Crystallization

Crystallization is a process of concentrating a solution to a super-saturated state so as to cause the formation of solid particles within a homogeneous phase. This may occur in the freezing of water to form ice, solidification of liquid melts, vapors, and crystallization from a liquid solution.

It is considered one of the most important industrial processes because many materials are marketed in crystalline form. The specific crystallization process widely used in industries is the crystallization of substances from liquid solutions. An example of such material is magnesium sulfate ($MgSO_4$).

Example Problem 8.9. A solution of 27.5% $MgSO_4$ at 110°F is cooled, without appreciable evaporation, to 50°F in a batch water-cooled crystallizer. Determine the heat that must be removed from the solution per ton of slurry (crystals).

Using Program Listing 8.3 we can directly determine the enthalpy to be -22.0 Btu/lb given that the mass fraction is at 0.275 and the temperature is 110°F.

Computing the same way on the second value for the enthalpy at mass fraction of 0.275 and at 50°F is -70.0 Btu/lb.

Per 1 lb of original solution, the ΔH of the solution is computed as

$$1(-22+70)=48 \text{ Btu.}$$

At 50°F, fig. 8.2 Magnesium sulfate–water system phase diagram, it can be determined that the mass fraction of $MgSO_4$ in the mother liquor is 0.22575, and in the crystal slurry is 0.4884.

Computing for the weight of the slurry per lb of original solution, we obtain:

$$1\left(\frac{0.275-0.2257}{0.4884-0.2257}\right)=0.1877 \text{ lb slurry}$$

So the heat evolved per ton of slurry can be determined to be

$$\left(\frac{48}{0.1877}\right)=255.727 \text{ Btu/lb}$$

255.727 (2000 lb/ton) = 511454.45 Btu/ton

In developing a MATLAB program for solving this type of problem, we can reuse existing functions `enthal_MgSO4()` (Program Listing 8.3) and the `phase()` function (Program Listing 8.4).

Program Listing 8.13:

```
1  clc
2  clear
3  conc_init=input('Enter initial mass fraction of MgSO4
   in the solution: ');
4  init_temp=input('Enter initial temperature of solution
   in deg F: ');
5  final_temp=input('Enter final temperature in the solution
   in deg F: ');
6  enth1=enthal_MgSO4(conc_init,init_temp);
7  enth2=enthal_MgSO4(conc_init,final_temp);
8  deltaH = enth1-enth2;
9  conc=phase(final_temp);
10 conc_sol=conc(1);
11 conc_crystal=conc(2);
12 wt_slurry=(conc_init-conc_sol)/(conc_crystal-conc_sol);
13 heat = (deltaH/wt_slurry)*2000;
14 fprintf('Heat to be removed is %f Btu/ton of slurry.\
   n',heat);
```

Lines 3 to 5: Request the necessary information from the user.

Lines 6 and 7: Given the initial concentration of magnesium sulfate, and the final temperature, theses lines derive the enthalpy using the `enthal_MgSO4()` function from Program Listing 8.3.

Line 8:   Computes for the delta H.

Line 7:   Determines the concentration of both the solution and the crystal slurry using the `phase()` function from Program Listing 8.4.

Lines 9 to 11: Equate `x1` with the initial concentration of benzene in the solution.

Line 12: Computes for the weight of the slurry.

Line 13: Computes the heat needed for the crystallization process.

Line 14: Prints the result in the Command Window.

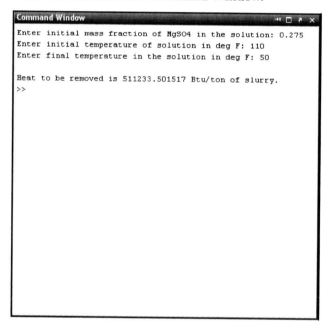

FIGURE 8.11:   Output of Program Listing 8.13.

Note: The slight difference in the result is due to the rounding off of numbers during manual computation.

LABORATORY EXERCISES

1)   Sugar factories classify their refined sugar according to its color:

Refined Sugar	**Color, ICUMSA Unit (IU)**
Bottler's Grade	≤ 35
Premium Grade	≤ 50
Standard Grade	≤ 100

Using a spectrophotometer, a chemical engineer has established a relationship between the absorbance reading of a sugar solution and its color. He obtained the following data:

Absorbance	0.0	0.01	0.02	0.035	0.067	0.093	0.105
Color	0	10	20	40	70	100	120

Make a program that will compute for the color of the sugar solution given its absorbance reading, and then classify the type of sugar produced.

2) A solution of $MgSO_4$ crystals is cooled to 30°F. From the total water present in the system, 7% evaporated during cooling. How many kilograms of crystals were formed if the 2000 kg original solution is comprised of 35% $MgSO_4$ and 65% H_2O?

3) Five-hundred moles of an equimolar mixture of benzene and toluene is subjected to a differential distillation at 2 atm (2.026 bars). The composition of benzene in the residue should be 20%. What percentage of the feed was distilled? Modify Program Listing 8.2b for the x and y values.

4) A cylinder containing 32.4 grams of butane is left in a room. A pressure gauge connected to the cylinder reads 710 psig. It was found out later that the temperature inside the cylinder is 652 R.

   a) Assuming butane follows the van der Waals equation of state, find the volume of butane in the cylinder.

   b) Calculate the volume of butane if it behaves like a Redlich-Kwong gas.

   For butane (C_4H_{10}) $T_c = 425.17$ K, $P_c = 37.47$ atm

   Atomic weight: $C = 12$ g/gmol $H = 1.008$ g/gmol

5) Equal amounts of reactants A and B are fed into a Continuous Stirred Tank Reactor (CSTR) at a rate of 25 L/min. The volume of the first reactor is $85m^3$ and the overall rate of reaction is 0.013 L/gmol. What should be the volume of the second reactor if 95.2% conversion is desired?

Chapter 9

INTERFACING MATLAB® WITH C

Having appreciated the power of C/C++ and the convenience of MATLAB, this chapter will teach us how to take advantage of these programming languages. Based on the fact that MATLAB was developed using C, and C++, interfacing between the languages is quite feasible.

This chapter will teach us how to call C routines from MATLAB and use MATLAB's existing libraries from the C program through the following:

Introduction to MEX-files
    Components of MEX-files
    MATLAB Supported C Compilers
Handling Scalars
    One Scalar Input and One Scalar Output
    Two Scalar Inputs and One Scalar Output
Handling Vectors
    One Vector Input and One Scalar Output
Handling Matrices
    Multiple Matrix Inputs and One Vector Output
Previous Evaporation Problems Revisited
    Single-Effect Evaporator Problem
    Triple-Effect Evaporator Problem

Introduction to MATLAB Engine
  Components of MATLAB Engine
  Example on Plotting
  Compiling Engine Programs
Modified Distillation Program

 All program listings discussed in this chapter can be found on the CD-ROM.

INTRODUCTION TO MEX-FILES

Throughout the book we have been discussing the benefits and features of C, C++, and MATLAB. Each of the languages has its own pros and cons. One way of combining the best of both C and MATLAB is through the development of MEX-files. MEX-files are C program routines that can be called from MATLAB. This means we can use existing C routines and incorporate them into our MATLAB program, thereby saving time for encoding or translating to MATLAB. Also, there are routines that C can perform faster, and if combined with MATLAB, the developed program becomes faster compared to a program done purely with MATLAB. Note, however, that it is only advisable for MEX-files to substitute built-in MATLAB operations where C routines are known to be efficient.

Components of MEX-files

The basic components of a MEX-file are:
1) The header file "`mex.h`." The C program to be developed must include the `#include mex.h` directive to be able to support the MX, and MEX functions.
2) The `mexFunction` gateway. The `mexFunction` is the entry point to the MEX-file and can be written using the format:
   ```
   void mexFunction( int nlhs, mxArray *plhs[], int nrhs,
   const mxArray *prhs[])
   ```
   where:

`nlhs` – is an integer containing the number of expected mxArrays (*number at left-hand* side).

`plhs` – is an array of pointers to the expected output. Initially this contains the null pointer; the programmer creating any output array will assign the array pointers to plhs (*pointer at left-hand* side).

`nrhs` – is an integer containing the input arguments (*number at right-hand* side).

`prhs` – is an array of pointers to the input data(*pointer at right-hand* side). Note that the input data is read-only and should not be altered within the program.

3) The `mxArray` data. All data in MATLAB are arrays, so all input and output arguments to MEX-files are `mxArray` data. This is the C representation of MATLAB arrays, which contains information of the variable names, dimensions, data type, and indicates whether it is a real or complex number.

⚠**WARNING** *In C, data are stored in an array from left to right, while MATLAB stores data in an array from top to bottom.*

4) The Application Program Interface (API) function. This refers to all C functions incorporated into the MEX-file.

MATLAB Supported C/C++ Compilers

MATLAB supports a number of C/C++ compilers. Table 9.1 provides a list of C/C++ compilers compatible with MATLAB 7.4 under Windows (32-bit) format.

Table 9.1: Compatible C/C++ compilers.

Compiler	Version	Home Page
Borland C++ Builder 6	5.6	www.borland.com
Borland C++ Builder 5	5.5	
Borland C++ Compiler	5.5	
Intel C++	7.1 and 9.1	www.intel.com/software/products/compilers
Lcc - Win32	2.4.1	www.cs.virginia.edu/~lcc-win32
Microsoft Visual C++ 2005	8.0	msdn.microsoft.com/visualc
Microsoft Visual C++	7.1	
Microsoft Visual C/C++	6.0	
Open Watcom	1.3	www.openwatcom.org

To select an existing C/C++ compiler for interface with MATLAB use the command

```
>> mex -setup
```

MATLAB will then ask if it can locate installed compilers on your computer.

```
Please choose your compiler for building external interface
(MEX) files:
Would you like mex to locate installed compilers [y]/n? y
```

By choosing yes (y), MATLAB will then search based on the known path directory.

```
Select a compiler:
[1] Lcc-win32 C 2.4.1 in C:\PROGRA~1\MATLAB\R2007a\sys\lcc
[2] Microsoft Visual C++ 6.0 in C:\Program Files\Microsoft
Visual Studio

[0] None
```

In this case, there are only two available C/C++ compilers in the system (Lcc-Win32 and Microsoft Visual C++). By choosing Lcc (option 1), MATLAB then sets Lcc as the default C/C++ compiler for running C MEX-files.

```
Compiler: 1

Please verify your choices:

Compiler: Lcc-win32 C 2.4.1
Location: C:\PROGRA~1\MATLAB\R2007a\sys\lcc

Are these correct?([y]/n): y

Trying to update options file: C:\Documents and Settings\
Owner\Application Data\MathWorks\MATLAB\R2007a\mexopts.bat
From template:          C:\PROGRA~1\MATLAB\R2007a\bin\
win32\mexopts\lccopts.bat
 Done . . .
```

The MATLAB package already includes a C compiler Lcc-win32 for creating C MEX-files. This is the C compiler used by default.

HANDLING SCALARS

One Scalar Input and One Scalar Output

Let's consider a simple C function and the corresponding MEX-file. This first example handles a single scalar value, processes it, and returns a single-valued output.

Example Problem 9.1. Write a C Program and the equivalent C MEX-file that will convert a temperature from °C to °F.

The C program is presented as follow:

Program Listing 9.1: °C to °F temperature conversion (C function)

```
int CtoF(double tempC)
    {
     double ans;
     ans = tempC*1.8 + 32.0;
     return(ans);
    }
```

The MEX-file equivalent can be written as

Program Listing 9.2: °C to °F temperature conversion (m-file)

```
 1  /* save as CtoF.c */
 2  #include "mex.h"
 3  void mexFunction( int nlhs, mxArray *plhs[], int nrhs,
    const mxArray *prhs[])
 4  {
 5   double tempC,*ans;
 6   tempC=mxGetScalar(prhs[0]);
 7   plhs[0]=mxCreateDoubleMatrix(1,1,mxREAL);
 8   ans=mxGetPr(plhs[0]);
 9   *ans = tempC*1.8 + 32.0;
10  }
```

Line 2: The `#include` statement tells the C compiler to include the `mex.h` header file during compiling. This header file provides support to MEX functions.

Line 3: The entry point to the program. The `mexFunction()` function is required to interface with the MATLAB environment. The variable `nlhs` contains the number of expected `mxArrays`, the `*plhs[]` is the pointer to the output, `nrhs` contains the input, and `*prhs[]` is the pointer to the input.

Line 5: Declares `empC`, and `*ans` as a double.

Line 6: Gets the input scalar value of the temperature and assigns it to `tempC`.

Lines 7, 8: Create a scalar (1 × 1 matrix) for storing the pointer of the output.

Line 9: Converts the input temperature from degrees C to degrees F and the result is assigned to `ans`.

Example Problem 9.1. Using `CtoF()`, convert 100°C to °F.

Running the program in MATLAB Command Window we get the following answer…

FIGURE 9.1:   Output of Program Listing 9.2.

The program converts 100°C, and returns a value of 212.

Two Scalar Inputs and One Scalar Output

Let us try another sample problem, this time passing on two scalar values.

Example Problem 9.2. Write a C MEX-file that will calculate the dew point (dp), given the % relative humidity (%RH) and the temperature (T) using the equations from Jensen et al (1990).

$$p_{sat} = 0.611e^{(17.27T/(T+237.3))} \qquad \text{in kPa} \qquad \textbf{(9.1)}$$
$$p_{act} = p_{sat} \times (\%RH/100) \qquad \text{in kPa} \qquad \textbf{(9.2)}$$
$$dp = (116.9+237.3\times\ln(p_{act}))/(16.78-\ln(p_{act})) \qquad \text{in °C} \qquad \textbf{(9.3)}$$

Test the function by determining the dew point of air having a relative humidity of 50% and a temperature of 30°C.

Program Listing 9.3: Dew point function

```
1   /* save as dewpoint.c (dewpoint(RH,temp)) */
2   #include <math.h>
3   #include "mex.h"
4   void mexFunction( int nlhs, mxArray *plhs[], int nrhs,
    const mxArray *prhs[])
5   {
6     double RH,temp,pact,psat,*ans;
```

```
7   RH=mxGetScalar(prhs[0]);
8   temp=mxGetScalar(prhs[1]);
9   plhs[0]=mxCreateDoubleMatrix(1,1,mxREAL);
10  ans=mxGetPr(plhs[0]);
11  psat = 0.611*exp(17.27*temp/(temp+237.2));
12  pact = psat*(RH/100.0);
13  *ans = (116.9+237.3*log(pact))/(16.78-log(pact));
14  }
```

Lines 2, 3: The `#include` statement tells the C compiler to include both the `math.h` and `mex.h` header files during compiling.

Line 4: The `mexFunction()` function is required to interface with the MATLAB environment. The variable `nlhs` contains the number of expected `mxArrays`, the `*plhs[]` is the pointer to the output, `nrhs` contains the input, and `*prhs[]` is the pointer to the input.

Line 6: Declares `RH`, `temp`, `pact`, `psat`, and `*ans` as doubles.

Line 7: Gets the first input scalar value and assigns it to `RH`.

Line 8: Gets the second input scalar value and assigns it to `temp`.

Lines 9 to 10: Create a scalar (1 × 1 matrix) for storing the pointer of the output.

Line 11: This line computes for the saturated pressure based on Eq 9.1 and assigns the result to `psat`.

Line 12: Computes for the actual pressure based on Eq 9.2 and assigns the result to `pact`.

Line 13: Computes for the dew point based on Eq 9.3 and assigns the result to `ans`.

Testing the `dewpoint()` function.

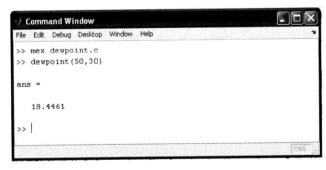

FIGURE 9.2: Output of Program Listing 9.3.

HANDLING VECTORS

One Vector Input and One Scalar Output

So far we have only dealt with variables having one value. However, it is also possible to develop a MEX-file with vector inputs. To illustrate, let us consider a program that solves for the root of a 3rd order polynomial equation using the Müller method.

Partial code has already been provided in Chapter 5's Laboratory Exercise2, which is then completed as a MEX-file.

Program Listing 9.4: Müller function

```c
/* save as muller.c */
#include <math.h>
#include "mex.h"
void mexFunction( int nlhs, mxArray *plhs[], int nrhs,
const mxArray *prhs[])
{
 double xr,h[3],d[3],a,b,c,fx[4],x[4],rad2,rad,dxr,den;
 double *ax,*ans;
 int i,j,n;
 ax=mxGetPr(prhs[0]);
 plhs[0]=mxCreateDoubleMatrix(1,1,mxREAL);
 ans=mxGetPr(plhs[0]);
 xr=1.0;
 x[2]=xr;
 x[1]=xr+(0.1*xr);
 x[0]=xr-(0.1*xr);
 do{
  fx[2]=0;fx[1]=0;fx[0]=0;
  for(i=0;i<3;i++)
  {
   for(j=3;j>=0;j--) fx[i]=fx[i]+(ax[3-j]*pow(x[i],j));
  }
  h[0]=x[1]-x[0];
  h[1]=x[2]-x[1];
  d[0]=(fx[1]-fx[0])/h[0];
  d[1]=(fx[2]-fx[1])/h[1];
  a=(d[1]-d[0])/(h[1]+h[0]);
  b=a*h[1]+d[1];
  c=fx[2];
```

```
    rad2=(b*b)-(4*a*c);
    rad2=fabs(rad2);
    rad=sqrt(rad2);
    if(fabs(b+rad)>fabs(b-rad)) den=b+rad;
    else den=b-rad;
    dxr=-2*c/den;
    xr=x[2]+dxr;
    x[0]=x[1];x[1]=x[2];x[2]=xr;
    }while(fabs(dxr)>0.0001);
    *ans=xr;
}
```

Example Problem 9.3 (Based on Example Problem 7.2). One hundred seventy gmol of carbon dioxide gas is filled up in a cylinder to a pressure of 65 atm and temperature of -70°C. Apply the van der Waals equation to determine the volume of the gas in the tank.

$$V^3 - \left(nb + \frac{nRT}{P} \right) V^2 + \frac{n^2 a}{P} V - \frac{n^3 ab}{P} = 0. \tag{9.4}$$

The critical constant for carbon dioxide:

$$a = 3.60 \times 10^6 \text{ atm} \left(\frac{cm^3}{g \text{ - mol}} \right)^2 \qquad b = 42.8 \left(\frac{cm^3}{g \text{ - mol}} \right)$$

$R = 82.054$ cm³-atm/gmol K

To solve for the coefficients of Eq 9.4, we substitute all the given information.

First coefficient: 1
Second coefficient: -[(170×42.8) + ((170×82.054×(-70+273.15))/65)]
Third coefficient: (170^2)×3600000/65
Fourth coefficient: -[(170^3)×3600000×42.8/65]

These values will comprise the input vector for `muller()`.

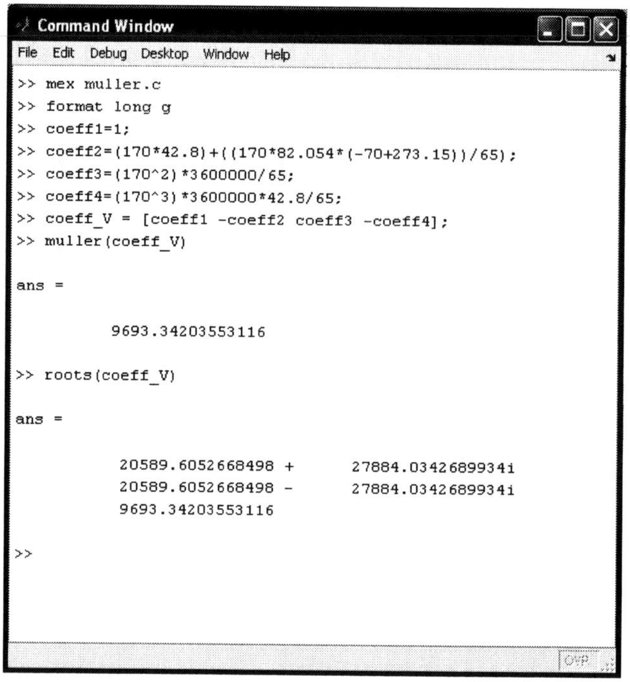

```
Command Window

File  Edit  Debug  Desktop  Window  Help

>> mex muller.c
>> format long g
>> coeff1=1;
>> coeff2=(170*42.8)+((170*82.054*(-70+273.15))/65);
>> coeff3=(170^2)*3600000/65;
>> coeff4=(170^3)*3600000*42.8/65;
>> coeff_V = [coeff1 -coeff2 coeff3 -coeff4];
>> muller(coeff_V)

ans =

             9693.34203553116

>> roots(coeff_V)

ans =

             20589.6052668498 +    27884.0342689934i
             20589.6052668498 -    27884.0342689934i
             9693.34203553116

>>
```

FIGURE 9.3:   Output of Program Listing 9.4.

Note that the answer from the developed `muller()` was compared to the pre-existing MATLAB `roots()` function.

HANDLING MATRICES

Multiple Matrix Inputs and One Vector Output

A more complicated MEX-file handles matrices. Program Listing 9.5 demonstrates how to develop MEX-files that require numerous data inputs as well as numerous data outputs. Also, in this case, we are now required to create a matrix output.

Program Listing 9.5: Simplex mex-file

```
/* save as simplex.c */
#include <math.h>
#include "mex.h"
void mexFunction( int nlhs, mxArray *plhs[], int nrhs,
const mxArray *prhs[])
```

```
{
 int cntr2,cf1,ne,nc,nopt,pivot1,pivot2,xerr=0,i,j;
 double cf2,simtab[20][20],flag,aux,xmax;
 double *obj,*cons,*ans;
 cf1=mxGetScalar(prhs[0]);
 obj=mxGetPr(prhs[1]);
 cons=mxGetPr(prhs[2]);
 ne=mxGetN(prhs[1]);
 nc=mxGetM(prhs[2]);
 plhs[0]=mxCreateDoubleMatrix(ne+1,1,mxREAL);
 ans=mxGetPr(plhs[0]);
 for (j = 1; j<=ne; j++)
 {
  cf2=obj[j-1];
  simtab[1][j+1] = cf2 * cf1;
 }
 cf2=0;
 simtab[1][1] = cf2 * cf1;
 cntr2=0;
 for (i=1; i<=nc; i++)
 {
  for (j=1; j<=ne; j++)
   {
   cf2=cons[cntr2];
   simtab[i + 1][j + 1] = -cf2;
   cntr2++;
   }
  simtab[i+1][1]=cons[cntr2];
  cntr2++;
 }
 for(j=1; j<=ne; j++)  simtab[0][j+1] = j;
 for(i=ne+1; i<=ne+nc; i++)  simtab[i-ne+1][0] = i;
 do{
  xmax = 0.0;
  for(j=2; j<=ne+1; j++)
  {
   if (simtab[1][j] > 0.0 && simtab[1][j] > xmax)
   {
    xmax = simtab[1][j];
    pivot2 = j;
   }
```

```
  }
  flag = 1000000.0;
  for (i=2; i<=nc+1; i++)
  {
   if (simtab[i][pivot2] < 0.0)
   {
    aux = fabs(simtab[i][1] / simtab[i][pivot2]);
    if (aux < flag)
    {
     flag = aux;
     pivot1 = i;
    }
   }
  }
  aux = simtab[0][pivot2];
  simtab[0][pivot2] = simtab[pivot1][0];
  simtab[pivot1][0] = aux;
  for (i=1; i<=nc+1; i++)
  {
   if (i != pivot1)
   {
    for (j=1; j<=ne+1; j++)
    {
    if (j != pivot2) simtab[i][j] -= simtab[pivot1][j] *
    simtab[i][pivot2] / simtab[pivot1][pivot2];
    }
   }
  }
  simtab[pivot1][pivot2] = 1.0 / simtab[pivot1][pivot2];
  for (j=1; j<=ne+1; j++)
  {
   if (j != pivot2) simtab[pivot1][j] *= fabs(simtab[piv
ot1][pivot2]);
  }
  for (i=1; i<=nc+1; i++)
  {
     if (i != pivot1) simtab[i][pivot2] *= simtab
[pivot1][pivot2];
  }
  for (i=2; i<=nc+1; i++)
  if (simtab[i][1] < 0.0)  xerr = 1;
```

```
  nopt = 0;
  if (xerr != 1)
  {
   for (j=2; j<=ne+1; j++)
   if (simtab[1][j] > 0.0)   nopt = 1;
  }
 }while(nopt==1);
 if (xerr == 0)
 {
  cntr2=0;
  for (i=1; i<=ne;  i++)
  for (j=2; j<=nc+1; j++)
  {
   if (simtab[j][0]  == 1.0*i)
   {
    ans[cntr2]=simtab[j][1];
    cntr2++;
   }
  }
  ans[cntr2]=simtab[1][1];
 }
 else mexPrintf("No Feasible Solution.\n");
}
```

Example Problem 9.4 (based on Example Problem 2.14).

Maximize: $z = 80x_1 + 100x_2$ (Objective Function)
Constraints:
$0.5x_1 + 0.5x_2 \leq 25$
$0.2x_1 + 0.6x_2 \leq 10$
$0.8x_1 + 0.4x_2 \leq 14$
$x_1 \geq 0, x_2 \geq 0$

Using the `simplex()` function:

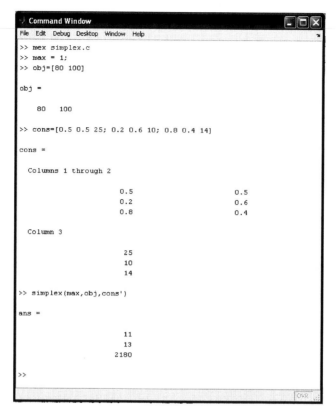

FIGURE 9.4: Output of Program Listing 9.5 based on Example Problem 9.4.

We need to get the transpose (`'`) matrix of constraint variable `cons` for it to function correctly. The expression `simplex(max,obj,cons)` is used to solve the given linear programming problem. The variable `max` has a value of 1 to signify that the objective function is to be maximized (-1 is provided to minimize objective function).

Solution: $x_1 = 11$, $x_2 = 13$, $z = 2180$
Let's try a process optimization problem.

Example Problem 9.5. A plastic company produces crates for soft drinks and supplies them to the nearby beverage company. The demand is so high that all of the plastic crates produced are sold immediately. Currently, the company produces two different qualities of crates:

   Standard Quality – which uses 50% recycled plastic, and 50% virgin material
   Premium Quality – which uses 30% recycled plastic, and 70% virgin material

The plastic materials are supplied as follows:
Recycled: 15 tons per week
Virgin: 19 tons per week

Profit from the two types of crates are given as:
Standard Quality: \$400/ton of crates
Premium Quality: \$525/ton of crates

About 500 pieces of plastic crates weigh about one ton.
Maximum production capacity of the plant is 30 tons per week. Determine how much of each quality of crates (in tons) should be manufactured to maximize profit.

Write the objective function and the constraints:
Let x_1 – tons of standard quality crates
         x_2 – tons of premium quality crates

Maximize profit $z = 400x_1 + 525x_2$
Constraints $x_1 + x_2 \le 30$ (production capacity constraints)
                     $0.5x_1 + 0.3x_2 < 15$ (supply constraints on recycled material)
                     $0.5x_1 + 0.7x_2 < 19$ (supply constraints on virgin material)

Now, let us use again the `simplex()` function to solve this problem.

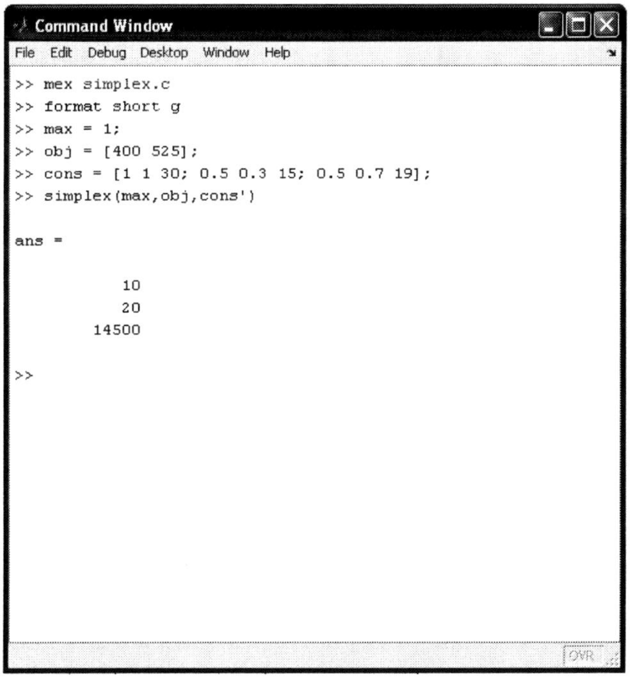

```
Command Window
File  Edit  Debug  Desktop  Window  Help

>> mex simplex.c
>> format short g
>> max = 1;
>> obj = [400 525];
>> cons = [1 1 30; 0.5 0.3 15; 0.5 0.7 19];
>> simplex(max,obj,cons')

ans =

            10
            20
         14500

>>
```

FIGURE 9.5: Output of Program Listing 9.5 based on Example Problem 9.5.

Solution: To maximize profit, 10 tons of standard crates and 20 tons of premium crates must be produced per week. This will yield a total profit of $14,500.00 per week.

PREVIOUS EVAPORATION PROBLEMS REVISITED

The following functions are needed to generate all the necessary information in developing a program to solve any evaporation problem involving:
- forward-feed type of evaporators
- evaporating solutions with no boiling point elevations, and
- those liquid being evaporated that has similar characteristics as water.

In summary, the functions needed include:
sat_p2t() – M-file function to convert saturated pressure (in Pa) to temperature (in K)
enthalpy_l() – Mex-file function that provides the enthalpy for liquid (in J/kg)
enthalpy_v() – Mex-file function that provides the enthalpy for vapor (in J/kg)

Program Listing 9.8 is a MATLAB program solving evaporation problems where the algorithm used is similar to that of the C version (Program Listing 4.4).

sat_p2t.m (Program Listing 8.1)

```
% save as sat_p2t.m
function result=sat_p2t(p_sati)
t_sat = 275:5:645;
p_sat  = [698.4358601  991.7589281  1388.931054  1919.877195
2621.114915 3536.717587 4719.326831 6231.203640 8145.306525
10546.38439 13532.07139 17213.97117 21718.71798 27189.00309
33784.55558 41683.06781 51081.05720 62194.65721 75260.33206
90535.51115 108299.1406 128852.1507 152517.8399 179642.1761
210594.0186 245765.2635 285570.9185 330449.1096 380861.0286
437290.8252 500245.4516 570254.4663 647869.8037 733665.5172
828237.5035 932203.2148 1046201.367 1170891.653 1306954.465
1455090.638 1616021.224 1790487.301 1979249.829 2183089.567
2402807.053 2639222.675 2893176.829 3165530.193 3457164.132
3768981.251 4101906.126 4456886.241 4834893.154 5236923.951
5664003.019 6117184.197 6597553.393 7106231.723 7644379.306
8213199.841 8813946.135 9447926.819 10116514.55 10821156.12
```

```
11563385.02 12344837.38 13167272.39 14032599.33 14942914.30
15900552.42 16908166.25 17968853.73 19086393.24 20265769.53
21514941.10];
result = interp1(p_sat,t_sat,p_sati,'spline');
```

enthalpy_l.c (Extracted from `steam.h` in Chapter 3)

Program Listing 9.6:

```c
#include <math.h>
#include "mex.h"
void mexFunction( int nlhs, mxArray *plhs[], int nrhs,
const mxArray *prhs[])
{
 double temp, ans, *val;
 double temp_var1,temp_var2,temp_var3, temp_var4, theta,
tau,  pres,ln_tau0;
 double ln_tau,ln_tau1, ln_tau2, dpdt, alpha ,rho_l;
  double  a[6]={-7.85951783,  1.84408259,-11.7866497,
22.6807411,-15.9618719, 1.80122502};
  double  b[6]={1.99274064,  1.09965342,-0.510839303,  -
1.75493479,-45.5170352,  -674694.45};
  double d[5]={-0.0000000565134998,2690.66631, 127.287297,-
135.003439, 0.981825814};
 temp=mxGetScalar(prhs[0]);
 plhs[0]=mxCreateDoubleMatrix(1,1,mxREAL);
 val=mxGetPr(plhs[0]);
 theta = temp/647.096;
 tau = 1 - theta;
 temp_var1 =(647.096/temp)*(a[0]*tau + a[1]*pow(tau,1.5)
+ a[2]*pow(tau,3) + a[3]*pow(tau,3.5) + a[4]*pow(tau,4)
+ a[5]*pow(tau,7.5));
 pres = 22064000*exp(temp_var1);
 ln_tau0 = 6.5*log(tau);
 ln_tau1 = 2.5*log(tau);
 ln_tau2 = 0.5*log(tau);
 temp_var2 = (7.5*a[5]*exp(ln_tau0)) +(4*a[4]*pow(tau,3))
+ (3.5*a[3]*exp(ln_tau1)) + (3*a[2]*pow(tau,2)) +
(1.5*a[1]*exp(ln_tau2)) + a[0]+ (log(pres/22064000));
 dpdt = -1*(pres/temp)*temp_var2;
 ln_tau=log(tau);
 temp_var3 = 1 + b[0]*exp(ln_tau/3.0) + b[1]*exp(2.0*ln_
```

```
tau/3.0)  +  b[2]*exp(5.0*ln_tau/3.0)  +  b[3]*exp(16.0*ln_
tau/3.0)  +  b[4]*exp(43.0*ln_tau/3.0)  +  b[5]*exp(110.0*ln_
tau/3.0);
    rho_l = 322.0* (temp_var3*1.0);
    temp_var4  =  -1135.905627715  +  d[0]*pow(theta,-19)  +
  d[1]*pow(theta,1) + d[2]*pow(theta,4.5) + d[3]*pow(theta,5)
  +  d[4]*pow(theta,54.5);
  alpha = 1000*temp_var4;
  ans= alpha+((temp/rho_l)*dpdt);
  *val=ans;
  }
```

enthalpy_v.c (Extracted from `steam.h` in Chapter 3)
Program Listing 9.7:

```
#include <math.h>
#include "mex.h"
void mexFunction( int nlhs, mxArray *plhs[], int nrhs,
const mxArray *prhs[])
{
 double temp, ans, *val;
 double temp_var1,temp_var2,temp_var3, temp_var4, theta,
tau, pres, ln_tau0;
 double ln_tau,ln_tau1, ln_tau2, dpdt, alpha ,rho_v;
  double  a[6]={-7.85951783,  1.84408259,-11.7866497,
22.6807411,-15.9618719, 1.80122502};
  double  c[6]={-2.03150240,  -2.68302940,-5.38626492,  -
17.2991605,-44.7586581,  -63.9201063};
 double d[5]={-0.0000000565134998,2690.66631, 127.287297,-
135.003439, 0.981825814};
 temp=mxGetScalar(prhs[0]);
 plhs[0]=mxCreateDoubleMatrix(1,1,mxREAL);
 val=mxGetPr(plhs[0]);
 theta = temp/647.096;
 tau = 1 - theta;
 temp_var1 = (647.096/temp)*(a[0]*tau + a[1]*pow(tau,1.5)
+ a[2]*pow(tau,3) + a[3]*pow(tau,3.5) + a[4]*pow(tau,4)
+ a[5]*pow(tau,7.5));
 pres = 22064000*exp(temp_var1);
 ln_tau0 = 6.5*log(tau);
 ln_tau1 = 2.5*log(tau);
 ln_tau2 = 0.5*log(tau);
```

```
 temp_var2 = (7.5*a[5]*exp(ln_tau0)) +(4*a[4]*pow(tau,3))
+ (3.5*a[3]*exp(ln_tau1)) + (3*a[2]*pow(tau,2)) +
(1.5*a[1]*exp(ln_tau2)) + a[0]+(log(pres/22064000));
 dpdt = -1*(pres/temp)*temp_var2;
 ln_tau=log(tau);
 temp_var3 = c[0]*exp(ln_tau/3.0) + c[1]*exp(2.0*ln_
tau/3.0) + c[2]*exp(4.0*ln_tau/3.0) + c[3]*exp(3.0*ln_
tau) + c[4]*exp(37.0*ln_tau/6.0) + c[5]*exp(71.0*ln_
tau/6.0);
 rho_v = 322.0* exp(temp_var3*1.0);
 temp_var4 = -1135.905627715 + d[0]*pow(theta,-19) +
d[1]*pow(theta,1) + d[2]*pow(theta,4.5) + d[3]*pow(theta,5)
+ d[4]*pow(theta,54.5);
 alpha = 1000*temp_var4;
 ans= alpha+((temp/rho_v)*dpdt);
 *val=ans;
}
```

Program Listing 9.8:Evaporation Program (MATLAB Version).

```
1  clc
2  clear
3  k = input('Enter number of evaporators: ');
4  conc(1) = input('Enter feed concentration (mass
   fraction): ');
5  tl(1) = input('Enter feed temperature (F): ');
6  conc(k+1) = input('Enter product concentration (mass
   fraction): ');
7  f = input('Enter amount of feed in lb/hr: ');
8  ps = input ('Enter pressure of steam (psi): ');
9  pk = input ('Enter pressure of last evaporator (psi): ');
10 for j = 2:k+1
11  fprintf('Enter OHT Coefficient (Btu/sqft-hr-f) for
    effect no. %d: ',j-1);
12 u(j)= input('');
13 end
14 pres =ps*6894.757;
15 temp= sat_p2t(pres);
16 tv(1) = ((temp-273.15)*1.8)+32;
17 pres = pk*6894.757;
18 temp = sat_p2t(pres);
19 tv(k+1) = ((temp-273.15)*1.8)+32;
20 ts = tv(1);
```

```
21 p = f*conc(1)/conc(k+1);
22 vt = (f-p)/k;
23 for j = 2:k
24   conc(j)= f*conc(1)/(f-(vt*(j-1)));
25 end
26 ut = 0;
27 for j = 2:k+1
28   ut = ut+(1/u(j));
29 end
30 dta = tv(1)-tv(k+1);
31 dtn = 0;
32 for j = 2:k
33   dt(j)=dta*((1/u(j))/ut);
34   dtn=dtn+dt(j);
35 end
36 dt(k+1)=dta-dtn;
37 flag='n';
38 while(flag=='n')
39   for i=(k+1):-1:2
40     tl(i) = tv(i);
41     tv(i-1)=tl(i)+dt(i);
42   end
43   tv(1) = ts;
44   clc;
45   for j=1:k+1
46     fprintf('temperature of vapor[%d] in deg F = %f\n',j-
      1,tv(j));
47     fprintf('temperature of liquid[%d] in deg F = %f\
      n',j-1,tl(j));
48   end
49   fprintf('product in lb/hr = %f \n',p);
50   for j=1:k+1
51     temp_t=((tv(j)-32)/1.8)+273.15;
52     hv(j)=enthalpy_v(temp_t);
53     hv(j)=hv(j)/(1055.056*2.2046226);
54     temp_t=((tl(j)-32)/1.8)+273.15;
55     hl(j)=enthalpy_l(temp_t);
56     hl(j)=hl(j)/(1055.056*2.2046226);
57   end
58   for j=1:k
59     temp_t=((tv(j)-32)/1.8)+273.15;
```

```
60    hd(j+1)=enthalpy_l(temp_t);
61    hd(j+1)=hd(j+1)/(1055.056*2.2046226);
62  end
63  xmatrx = zeros(k);
64  constant(1,1)=f*(hl(1)-hl(2));
65  for j=2:k-1
66    constant(j,1)=f*(hl(j)-hl(j+1));
67  end
68  constant(k,1)=f*(hl(k)-hv(k+1)+(conc(1)*(hv(k+1)-
    hl(k+1))/conc(k+1)));
69  xmatrx(k,1)=0;
70  for j=2:k-1
71    xmatrx(k,j)=hl(k)-hv(k+1);
72  end
73  xmatrx(k,k)=hd(k+1)+hl(k)-hv(k)-hv(k+1);
74  xmatrx(1,1)=hd(2)-hv(1);
75  for j=1:k-1
76    xmatrx(j,j+1)=hv(j+1)-hl(j+1);
77  end
78  for j=2:k-1
79    xmatrx(j,j)=hd(j+1)+hl(j)-hv(j)-hl(j+1);
80  end
81  for j=3:k-1
82    for w=2:k-2
83      xmatrx(j,w)=hl(j)+hl(j+1);
84    end
85  end
86  vapor=xmatrx\constant;
87  totv=0;
88  for j=2:k
89    totv=totv+vapor(j,1);
90  end
91  vapor(k+1,1)= f-p-totv;
92  for j=1:k+1
93    fprintf('vapor[%d]  in  lb/hr  =  %f\n',j-
    1,vapor(j,1));
94  end
95  for j=1:k
96    heat = vapor(j,1)*(hv(j)-hd(j+1));
97    fprintf('Rate of heat transfer at effect %d is %f
    Btu/hr.\n',j,heat);
```

```
98   end
99   stemcon=(totv+vapor(k+1))/vapor(1);
100  fprintf('Steam economy = %f\n',stemcon);
101  for j=2:k+1
102    area(j)=(vapor(j-1)*(hv(j-1)-hd(j)))/(u(j)*(tv(j-
       1)-tl(j)));
103    fprintf('Area for evaporator no. %d = %f sq ft\n',j-
       1,area(j));
104  end
105  if k==1
106    flag='y';
107  else
108    flag=input('Are the areas OK? (y/n)','s');
109    flag=lower(flag);
110    if flag=='n'
111      tarea=0;
112      for j=2:k+1
113        tarea=tarea+(dt(j)*area(j));
114      end
115      aream=tarea/dta;
116      for j=2:k+1
117        dt(j)=dt(j)*area(j)/aream;
118      end
119    end
120  end
121  end
```

Line 1: Clears the Command Window.

Line 2: Clears all array values in the memory.

Lines 3 to 13: Request all the needed values using the input() and the looping statement for loop.

Lines 14 to 20: Convert the given pressure of steam to a saturation temperature using sat_p2t() and then conducts the necessary conversion of units.

Line 21: Solves for the mass flow of products based on given mass flow of feed and concentrations.

Lines 22 to 36: Solve for the estimated ΔT in each effect.

Lines 39 to 43: Solve for the estimated temperature of liquid and vapor in each effect.

Lines 44 to 49: Clear the screen, then print the temperature of liquid and vapor, including the product mass flow rate.

Lines 50 to 62: Approximate the enthalpies of liquid and vapor based on the derived temperature using the `enthalpy_v()` and `enthalpy_l()` functions.

Lines 63 to 85: Fill up the matrix Eq 4.9 (energy balance).

Line 86: Performs matrix left division to solve for the simultaneous linear equation and determines mass flow rate of steam and vapor in each effect.

Lines 87 to 121: Compute for the heat transfer area and steam economy. The user will be asked if the areas provided are acceptable; if the areas are not acceptable, then the program will redo all the computation until the areas of all effects are almost equal.

Revisited Example Problem 4.7

A single vertical-tube evaporator is used to concentrate 29,000 lb/hr of organic colloid (60°F) from 25% to 60% solid. The solution is considered to have a negligible boiling point elevation, and its properties are similar to that of water at all concentrations. Saturated steam is introduced at 25 psia and the pressure in the condenser is 1.69 psia. The overall heat transfer coefficient is 300 Btu/ft²-h°F. Determine the heat transfer area in ft², and the steam consumption in lb per hr.

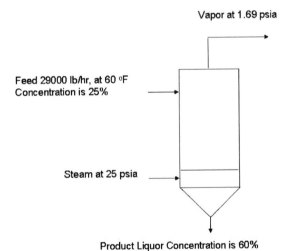

FIGURE 9.6: Single-effect evaporator.

Prepare all MEX-files
```
>> mex enthalpy_l.c
>> mex enthalpy_v.c
```

Note: The `sat_p2t.m` file must also be in the current directory.

Run `PL9_8.m` as viewed in the Command Window and provide all the necessary data.

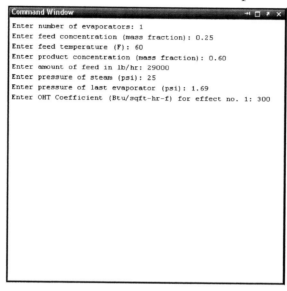

FIGURE 9.7: Output of Program Listing 9.8 based on Example Problem 4.7.

Figure 9.8 shows the computed output.

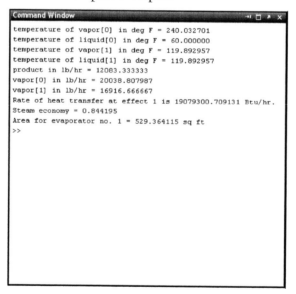

FIGURE 9.8: Output of Program Listing 9.8 based on Example Problem 4.7.

Revisited Example Problem 4.8

An organic colloidal solution with a negligible boiling point rise is concentrated from 13% to 55% solid in a triple-effect forward-feed evaporator. Steam is available at 29.82 psia and the third effect is 1.27 psia. Feed enters at a rate of 45000 lb/hr at a temperature of 60°F. The specific heat of the solution at all concentrations is the same as water. The overall coefficients are $U_1 = 530$, $U_2 = 350$, $U_3 = 195$ Btu/ft²-hr-°F. The area of each effect has to be the same. Calculate the area of the heating surface (A), the steam consumption (m_{v0}), and the steam economy.

By providing all the necessary data we obtain the following results:

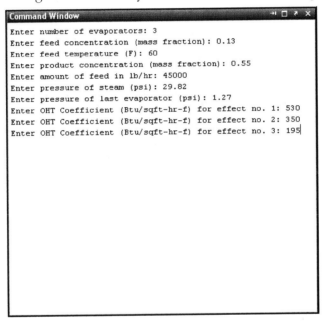

```
Command Window                                    ⇥ □ ↗ ✕
Enter number of evaporators: 3
Enter feed concentration (mass fraction): 0.13
Enter feed temperature (F): 60
Enter product concentration (mass fraction): 0.55
Enter amount of feed in lb/hr: 45000
Enter pressure of steam (psi): 29.82
Enter pressure of last evaporator (psi): 1.27
Enter OHT Coefficient (Btu/sqft-hr-f) for effect no. 1: 530
Enter OHT Coefficient (Btu/sqft-hr-f) for effect no. 2: 350
Enter OHT Coefficient (Btu/sqft-hr-f) for effect no. 3: 195
```

FIGURE 9.9: Output of Program Listing 9.8 based on Example Problem 4.8.

The results of the first iteration are:

```
Command Window                                    ┤ ☐ ⤢ ✕
temperature of vapor[0] in deg F = 249.954979
temperature of liquid[0] in deg F = 60.000000
temperature of vapor[1] in deg F = 223.171670
temperature of liquid[1] in deg F = 223.171670
temperature of vapor[2] in deg F = 182.614088
temperature of liquid[2] in deg F = 182.614088
temperature of vapor[3] in deg F = 109.818428
temperature of liquid[3] in deg F = 109.818428
product in lb/hr = 10636.363636
vapor[0] in lb/hr = 18252.092920
vapor[1] in lb/hr = 10284.857118
vapor[2] in lb/hr = 11455.596865
vapor[3] in lb/hr = 12623.182381
Rate of heat transfer at effect 1 is 17257836.977573 Btu/hr.
Rate of heat transfer at effect 2 is 9905000.463463 Btu/hr.
Rate of heat transfer at effect 3 is 11321708.217316 Btu/hr.
Steam economy = 1.882723
Area for evaporator no. 1 = 1215.755555 sq ft
Area for evaporator no. 2 = 697.773386 sq ft
Area for evaporator no. 3 = 797.575598 sq ft
Are the areas OK? (y/n)|
```

FIGURE 9.10:   First iteration

Figure 9.11 provides the result after the 6th iteration.

```
Command Window                                    ┤ ☐ ⤢ ✕
temperature of vapor[0] in deg F = 249.954979
temperature of liquid[0] in deg F = 60.000000
temperature of vapor[1] in deg F = 212.216867
temperature of liquid[1] in deg F = 212.216867
temperature of vapor[2] in deg F = 178.136605
temperature of liquid[2] in deg F = 178.136605
temperature of vapor[3] in deg F = 109.818428
temperature of liquid[3] in deg F = 109.818428
product in lb/hr = 10636.363636
vapor[0] in lb/hr = 17963.591428
vapor[1] in lb/hr = 10442.349046
vapor[2] in lb/hr = 11415.583359
vapor[3] in lb/hr = 12505.703958
Rate of heat transfer at effect 1 is 16985051.180681 Btu/hr.
Rate of heat transfer at effect 2 is 10129345.886651 Btu/hr.
Rate of heat transfer at effect 3 is 11313084.181524 Btu/hr.
Steam economy = 1.912960
Area for evaporator no. 1 = 849.201642 sq ft
Area for evaporator no. 2 = 849.200868 sq ft
Area for evaporator no. 3 = 849.200297 sq ft
Are the areas OK? (y/n) y|
```

FIGURE 9.11:   Sixth iteration

INTRODUCTION TO THE MATLAB ENGINE

In developing MEX-files we are able to call C subroutines from MATLAB. In this section we will learn to call MATLAB as part of the C program through the MATLAB Engine. It allows transferring of data from C to MATLAB, performing MATLAB commands while running the C program, and transferring back the output of MATLAB operations to C before shutting down.

Components of the MATLAB Engine

The MATLAB Engine programs are C programs (with the addition of basic components) that can interface with MATLAB and therefore can access all its features.

The basic components of the MATLAB Engine are the following:
1) The header file "engine.h." The C program must include engine.h to be able to support MX, MEX, and the MATLAB Engine functions.
2) The mxArray data. Just as when developing MEX-files, mxArray needs to be created for storing information.
3) The engOpen function. It is needed to start a MATLAB Engine session.
4) The mxDestroyArray function. It frees up the allocated memory for mxArray prior to shutdown.
5) The engClose function. It is used to properly terminate the MALAB Engine session.
6) Other MATLAB Engine functions such as engPutVariable() and engEvalString(). Descriptions of these functions and other important Engine functions are presented in Appendix D.

Plotting Example

Let us consider a basic example from which we utilize the plot function of MATLAB and call it from a C program.

Program Listing 9.9: MATLAB Engine program utilizing plot function

```
1   /* save PL9_9.c */
2   #include <stdlib.h>
3   #include <stdio.h>
4   #include "engine.h"
5   int main()
6   {
7     Engine *ep;
8     mxArray *X = NULL, *Y = NULL;
```

```
 9   double x[11] = { 0.0, 0.1, 0.2, 0.3, 0.4, 0.5, 0.6,
     0.7, 0.8, 0.9, 1.0 };
10   double y[11] = { 0.0, 0.211, 0.378, 0.512, 0.623,
     0.714, 0.791, 0.856,  0.911, 0.959, 1.0 };
11   if (!(ep = engOpen("\0")))
12   {
13     fprintf(stderr, "\nCan't access MATLAB!\n");
14     return 0;
15   }
16   X = mxCreateDoubleMatrix(1, 11, mxREAL);
17   Y = mxCreateDoubleMatrix(1, 11, mxREAL);
18   memcpy((void *)mxGetPr(X), (void *)x, sizeof(x));
19   memcpy((void *)mxGetPr(Y), (void *)y, sizeof(y));
20   engPutVariable(ep, «X», X);
21   engPutVariable(ep, "Y", Y);
22   engEvalString(ep, "diagonal =(0:1);");
23   engEvalString(ep, "plot(X,Y,diagonal,diagonal);");
24   engEvalString(ep, "grid on");
25   engEvalString(ep, "title('Benzene - Toluene Equilibrium
     System');");
26   engEvalString(ep, "xlabel('Benzene in Vapor(mole
     fraction)');");
27   engEvalString(ep, "ylabel('Benzene in Liquid (mole
     fraction)');");
28   system("pause");
29   mxDestroyArray(X);
30   mxDestroyArray(Y);
31   engClose(ep);
32   return 0;
33 }
```

Line 4: The `#include` statement tells the C compiler to include the `engine.`
`h` header file during compiling. This headerfile provides support to MX,
MEX, and MATLAB Engine functions.

Lines 7,11 to15: These lines open an engine session `ep` with an error trap routine
to check if it is possible to do so.

Lines 8, 16 to 21: Create an mxArray `X,Y`, and transfer all data from `x[]`,`y[]`
(declared in lines 9 and 10) to these mxArrays using `memcpy()` and
`engPutVariable()` functions.

Lines 22 to 27: Plot the `X` and `Y` using MATLAB syntax through
`engEvalString()`.

Lines 29 to 31: Destroy the mxArrays and close the MATLAB session.
Line 32: The main program returns a null value before terminating.

This is an example of calling the `plot()` function of MATLAB from the C program. The next thing to do is compile it to generate an executable (.exe) file.

Compiling Engine Programs

Compiling an Engine program is not as simple as compiling a MEX-file. We need to determine first what C compiler to use. Presented in Table 9.1 is the list of C/C++ compilers compatible with MATLAB. In this case, we are using the LCC Compiler that is provided by MathWorks as part of the MATLAB 2007's installation package.

Once the C compiler has been identified, we need to locate the batch file (`.bat`) for that particular compiler in order to properly specify it during compiling. To compile Program Listing 9.9, the following command should be typed in the MATLAB Command Window.

```
>> mex  -f  c:\Progra~1\MATLAB\R2007a\bin\win32\mexopts\
lccengmatopts.bat pl9_9.c
```

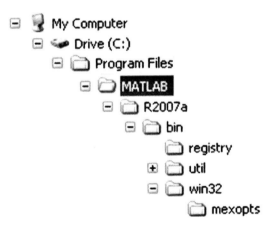

FIGURE 9.12: Location of the batch file.

Other Batch files available are:

Table 9.1: C compilers compatible with MATLAB.

Compiler	Version	Batch Filename
Borland C++ Builder 6	5.6	`bccengmatopts.bat`
Borland C++ Compiler	5.5	`bcc55freematops.bat`
Intel C++	7.1	`intelc71engmatopts. bat`
Intel C++	9.1	`intelc91engmatopts. bat`
Microsoft Visual C++ 2005	8.0	`msvc60engmatopts.bat`
Microsoft Visual C++	7.1	`msvc71engmatopts.bat`
Microsoft Visual C/C++	6.0	`msvc60engmatopts.bat`
Open Watcom	1.3	`openwatc13engmatopts.bat`

After compiling `p19_.c`, an executable `p19_9.exe` file will be generated. Executing this file will produce the following output:

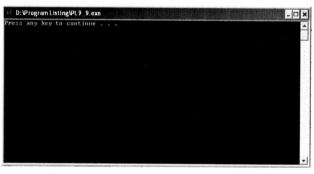

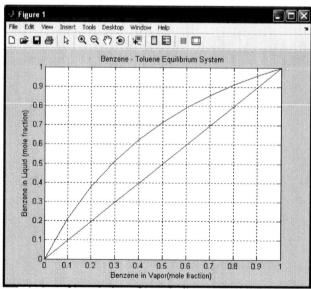

FIGURES 9.13:   Output of Program Listing 9.9.

MODIFIED DISTILLATION PROGRAM

To further illustrate the benefits of utilizing the MATLAB Engine, let's try to improve an existing C program. Let us consider the C program previously developed, which solves the distillation problem (Program Listing 4.5), and incorporate the plot features of MATLAB.

By adding a graph, the user can further appreciate the program by visually verifying the result.

Program Listing 9.10:

```
1   #include <stdio.h>
2   #include <stdlib.h>
3   #include "engine.h"
4   #include "benz_tol.h"
5   int main()
6   {
7   Engine *ep;
8   mxArray *X=NULL,*Y=NULL,*XMIN=NULL,*YMIN=NULL,*XACT=
    NULL,*YACT =NULL;
9   double x[11] = { 0.0, 0.1, 0.2, 0.3, 0.4, 0.5, 0.6,
    0.7, 0.8, 0.9, 1.0 };
10  double y[11] = { 0.0, 0.211, 0.378, 0.512, 0.623,
    0.714, 0.791, 0.856, 0.911, 0.959, 1.0 };
11  double pts[3];
12  double xd=0,xf=0,xb=0,xcoor[100],ycoor[100],q,ycoorf
    ,xcoorr, xact[100],yact[100],xmin[100],ymin[100];
13  double ycoorr,mr,mb,rmin,bmin,bact,dcoorr,ref,delm,d
    elb;
14  int j,n=0,nmin=0,fplate;
15  printf("\nThis program will solve problems on
    Distillation of Benzene-Toluene\n");
16  printf("\nEnter fraction of benzene in distillate : ");
17  scanf("%lf",&xd);
18  printf("Enter fraction of benzene in feed : ");
19  scanf("%lf",&xf);
20  printf("Enter fraction of benzene in bottoms : ");
21  scanf("%lf",&xb);
22  printf("Enter q value : ");
23  scanf("%lf",&q);
24  j=0;ycoor[0]=xd;
25  printf("Minimum Number of Plates\n");
```

```
26  do{
27   j++;
28   xcoor[j]=benzenetoluene(ycoor[j-1],1);
29   ycoor[j]=xcoor[j];
30   nmin=j;
31  }while(xcoor[j]>xb);
32  xmin[0]=xd;
33  ymin[0]=xd;
34  for(j=1;j<nmin+1;j++)
35  {
36   printf("fraction    x=%lf        y=%lf          plate
    no.=%d\n\r",xcoor[j],ycoor[j-1],j);
37   xmin[(2*j)-1]=xcoor[j];xmin[2*j]=xcoor[j];
38   ymin[(2*j)-1]=ycoor[j-1];ymin[2*j]=ycoor[j];
39  }
40  /* for Minimum Reflux Ratio */
41  if (q==1)
42  {
43   xcoorr=xf;
44   ycoorr=benzenetoluene(xcoorr,2);
45   }
46  else if (q==0)
47  {
48   ycoorr=xf;
49   xcoorr=benzenetoluene(ycoorr,1);
50  }
51  else
52  {
53   if (q>1) xcoorr= xf;
54   else if (q < 0)xcoorr= benzenetoluene(xb,1);
55   else xcoorr= benzenetoluene(xf,1);
56   do
57   {
58    xcoorr=xcoorr+0.00000005;
59    ycoorf=(((-1*q)/(1-q))*xcoorr)+(xf/(1-q));
60    ycoorr= benzenetoluene(xcoorr,2);
61    dcoorr=ycoorf-ycoorr;
62   }while(dcoorr > 0.000001 || dcoorr< -0.000001);
63  }
64  mr=(xd-ycoorr)/(xd-xcoorr);
65  bmin=ycoorr-(mr*xcoorr);
```

```
66  rmin=(xd/bmin)-1;
67  printf("Minimum reflux is = %lf\r\n",rmin);
68  printf("Enter reflux ratio : ");
69  scanf("%lf",&ref);
70  /* for Actual number of Plates */
71  if(q==1)(xcoorr=xf);
72  else
73  {
74   delm=(xd/(ref+1))-(xf/(1-q));
75   delb=((-1*q)/(1-q))-(ref/(ref+1));
76   xcoorr=delm/delb;
77  }
78  ycoorr=((ref/(ref+1))*xcoorr)+(xd/(ref+1));
79  mb=(ycoorr-xb)/(xcoorr-xb);
80  bact=ycoorr-(mb*xcoorr);
81  j=0;ycoor[0]=xd;
82  do
83  {
84   j++;
85   xcoor[j]=benzenetoluene(ycoor[j-1],1);
86   if(xcoor[j] >= xcoorr)
87   {
88    ycoor[j]=((ref/(ref+1))*xcoor[j])+(xd/(ref+1));
89    fplate=j+1;
90   }
91   if(xcoor[j] < xcoorr)(ycoor[j]=(mb*xcoor[j])+bact);
92   n=j;
93  }while(xcoor[j]>xb);
94  printf("Actual Number of Plates\r\n");
95  xact[0]=xd;
96  yact[0]=xd;
97  for(j=1;j<n+1;j++)
98  {
99   printf("fraction     x=%lf        y=%lf         plate
    no.=%d\n\r",xcoor[j],ycoor[j-1],j);
100  xact[(2*j)-1]=xcoor[j]; xact[2*j]=xcoor[j];
101  yact[(2*j)-1]=ycoor[j-1];yact[2*j]=ycoor[j];
102 }
103 printf("\nFeed is introduced in plate number %d\r\
    n",fplate);
104 if (!(ep = engOpen("\0"))) {
```

```
       fprintf(stderr, "\nCan't access MATLAB!\n");
       return EXIT_FAILURE;
105  }
106  X = mxCreateDoubleMatrix(1, 11, mxREAL);
107  Y = mxCreateDoubleMatrix(1, 11, mxREAL);
108  XACT = mxCreateDoubleMatrix(1, (2*n+1), mxREAL);
109  YACT = mxCreateDoubleMatrix(1, (2*n+1), mxREAL);
110  XMIN = mxCreateDoubleMatrix(1, (2*nmin+1), mxREAL);
111  YMIN = mxCreateDoubleMatrix(1, (2*nmin+1), mxREAL);
112  memcpy((void *)mxGetPr(X), (void *)x, sizeof(x));
113  memcpy((void *)mxGetPr(Y), (void *)y, sizeof(y));
114  memcpy((void *)mxGetPr(XACT), (void *)xact,
       sizeof(xact));
115  memcpy((void *)mxGetPr(YACT), (void *)yact,
       sizeof(yact));
116  memcpy((void *)mxGetPr(XMIN), (void *)xmin,
       sizeof(xmin));
117  memcpy((void *)mxGetPr(YMIN), (void *)ymin,
       sizeof(ymin));
118  engPutVariable(ep, "X", X);
119  engPutVariable(ep, "Y", Y);
120  engPutVariable(ep, "XACT", XACT);
121  engPutVariable(ep, "YACT", YACT);
122  engPutVariable(ep, "XMIN", XMIN);
123  engPutVariable(ep, "YMIN", YMIN);
124  engEvalString(ep, "diagonal =(0:1);");
125  engEvalString(ep, "plot(X,Y,diagonal,diagonal,'--
       ');");
126  engEvalString(ep, "grid on");
127  engEvalString(ep, "title('Benzene - Toluene Equilibrium
       System');");
128  engEvalString(ep, "xlabel('Benzene in Vapor(mole
       fraction)');");
129  engEvalString(ep, "ylabel('Benzene in Liquid (mole
       fraction)');");
130  engEvalString(ep, "set(line(XACT,YACT),'Color','k','
       LineWidth',2);");
131  engEvalString(ep, "set(line(XMIN,YMIN),'LineStyle','-
       -','Color','r', 'LineWidth',2);");
132  system("pause");
133  mxDestroyArray(X);
```

```
134 mxDestroyArray(Y);
135 mxDestroyArray(XACT);
136 mxDestroyArray(YACT);
137 mxDestroyArray(XMIN);
138 mxDestroyArray(YMIN);
139 engClose(ep);
140 return 0;
141 }
```

Line 3: The `#include` statement tells the compiler to include the `engine.h` header file during compiling. This headerfile provides support to MX, MEX, and MATLAB Engine functions.

Lines 7, 104 to 105: These lines open an engine session with an error trap routine to check if it is possible to do so.

Lines 8, 106 to123: Create an mxArray `X`, `Y`, `XACT`, `YACT`, `XMIN`, `YMIN`, and transfer all data from `x[]`, `y[]`, `xmin[]`, `ymin[]`, `xact[]`, `yact[]` to these mxArrays using `memcpy()` and `engPutVariable()` functions.

Lines 12 to 103: These lines were extracted from Program Listing 4.5 (Distillation C Program)

Lines 124 to 131: Plot the `X`, `Y`, `XACT`, `YACT`, `XMIN`, `YMIN` using MATLAB syntax through `engEvalString()`.

Lines 133 to 139: Destroy the mxArrays and close the MATLAB session.

Line 140: The main program returns a null value before terminating.

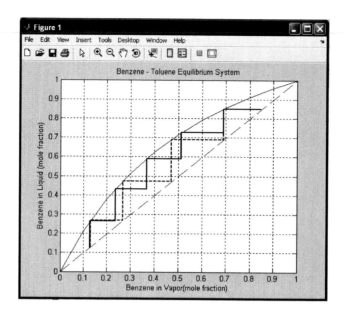

FIGURES 9.14:   Output of Program Listing 9.10.

LABORATORY EXERCISES

1) Write a C MEX-file that will do the following conversions:
   a) joule to erg (1 J = 10^7 erg)
   b) cP to lb_f-s/ft^2 (1 cP = 2.0886×10^{-5} lb_f-s/ft^2)
   c) atm to mmHg (1 atm = 760 mmHg)
   d) hp to kW (1 hp= 0.74570 kW)
2) Fifteen thousand kg/h of a 5°Bx sugar solution at 50°C is concentrated to 15°C Bx in a forward-feed quadruple-effect evaporator. Saturated steam is utilized at 125°C. The temperature at the last effect is maintained at 105°C so as not to caramelize the sugar crystals. The overall heat transfer coefficients (OHTC) at each effect are 50,100,150, and 200 W/m^2-K. Determine the heating surface required for the evaporators assuming the solution, due to low concentration, has a negligible boiling point elevation and the specific heat is equal to water.
   1°Bx = 1g sugar per 100 g of solution
   Revise Program Listing 9.10 to solve this problem.
3) Modify Program Listing 3.6 (the Friction Factor Program) and develop a MATLAB Engine to include plotting the diagram for a friction factor for flow in smooth pipe. Use the `loglog(x,y)` function of MATLAB to create a log-log scale plot.

4) Develop a MEX-File version of Program Listing 3.5 (the Enthalpy-Concentration Program for NaOH Solution) to create a function that will give the boiling point of NaOH given the corresponding boiling point of water and the concentration of NaOH in the solution.

5) In an experiment, an engineer has established a relationship between the temperature and pressure of a certain gas:

$$P = 5.324 \times 10^{-7} T^3 + 2.696 \times 10^{-3} T^2 - 5.3T + 2242.05 \ (P \text{ in atm}, T \text{ in K})$$

Assuming a molar volume of about 13000L/mol, at what temperature and pressure will the gas behave ideally?

Develop a MATLAB Engine that utilizes the MATLAB quad() function to solve this problem.

CONTINUING ON

The best way to end this book is actually not to end it. There is still much work to be done, more problems to solve, and more programs to develop. Let this book be a catalyst for you to cultivate that interest in making more application programs for chemical engineers.

Below are some suggested areas of study that you might want to explore:
   Transportation Problem – Optimization
   Steady State Heat Exchanger – Heat Transfer
   Calculation of Corrected Log Mean Temperature Difference – Heat Transfer
   Multistage Leaching – Mass Transfer
   Filtration (Solid-Liquid Separation) – Mass Transfer
   Measurement of Flow of Fluids – Flow of Fluids
   Drying – Heat and Mass Transfer
   Absorption – Mass Transfer

Finally, I wish you good luck on your future programming endeavors and happy computing.

Appendix A — C/C++ COMPILERS

The Open Watcom C/C++ compiler was used by the author in developing the programs in this book. A copy of the compiler is included on the CD-ROM included with this book. In this section, we will briefly discuss this and other compilers that can be used.

OPEN WATCOM C/C++

Open Watcom C/C++ is an Open Source successor to WATCOM, a commercial compiler previously marketed by Sybase and Powersoft. The current version of Open Watcom C/C++ (v. 1.6) is a professional, optimizing, multi-platform C and C++ compiler with a comprehensive suit of development tools for developing and debugging both 16-bit and 32-bit applications for DOS, extended DOS, Novell NLMs, 16-bit OS/2, 32-bit OS/2, Window 3.x, Windows 95/98/Me, Windows NT/2000/XP, and 32-bit OS/2.

Included in the package is an Integrated Development Environment (IDE), a text editor, C++ class libraries, sample programs, and much more.

Integrated Development Environment (IDE)

The IDE runs under Windows 3.x, Windows 95/98/Me, Windows NT/2000/XP, and 32-bit OS/2. This easily allows the user to edit, link, compile, debug, and build applications for 16-bit and 32-bit systems. Projects can include multiple files even EXEs and DLLs. The environment even produces "makefiles" for the project that can be viewed and edited in its accompanying text editor.

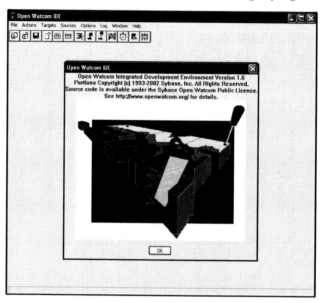

FIGURE A.1:   Open Watcom IDE.

The copy of the Sybase Open Watcom Public License version 1.0 is included in Appendix L.

Text Editor

The Open Watcom Text Editor can be called from the IDE or can be run independently. The editor contains a toolbar and menu items for easy access of commands. This also allows users to highlight different parts of the syntax through different colors and/or fonts, therefore making it easy to find parts of code. The Editor even uses common C formatting conventions such as smart indenting with braces.

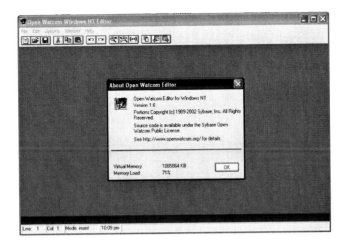

FIGURE A.2: Open Watcom Editor.

Other Free C/C++ Compilers and IDE

LCC Win 32

The operating system for LCC Win 32 must be at least **Windows 2000**. Even though the compiler and command line utilities will work with older OS like Windows 95 without any difficulty, the Integrated Development Environment (IDE) uses features that are not available in older versions.

Developed by Logiciels/Informatique in 1995, this compiler is one of the C/C++ compilers that is compatible with MATLAB.

The LCC Win 32 site can be accessed at *http://www.cs.virginia.edu/~lcc-win32/*

Digital Mars C/C++

Digital Mars C and C++ compilers for Win32, Win16, DOS32, and DOS can be considered to have the fastest compile/link times, uses powerful optimization technology, Design by Contract, and with a complete library source, HTML browseable documentation, disassembler, librarian, resource compiler, etc. It also includes command line and GUI versions, tutorials, sample code, online updates, Standard Template Library, and much more. Visit the Digital Mars web site for more information.

The compiler can be downloaded from *http://www.digitalmars.com/*

Code::Blocks

Code::Blocks is a free C/C++ IDE (Integrated Development Environment) built to meet the most demanding needs of its users. It is designed to be very extensible and fully configurable.

Built around a custom-made plug-in framework, Code::Blocks can be extended with plug-ins.

For more information visit *http://www.codeblocks.org/*

Appendix B MATLAB®

MATLAB, short for MATrix LABoratory, not only performs as a programming language, but also provides an environment for users to do numerical computation. The original MATLAB program was created by Cleve Moler in 1970 and was rewritten using C language in 1984, the same year MathWorks was founded.

Today, MATLAB has been used by more than one million people both from academia and industries. Recognized to be a valuable tool in the engineering field of study, it finds its way into chemical engineering and is being used to teach students computer programming.

The MathWorks Site provides support to MATLAB users. For more information, visit *http://www.mathworks.com/products/matlab/*

The site also provides links to their other products, services, and the MATLAB user community.

Appendix C

C/C++ FUNCTIONS AND STATEMENTS

H ere are some of C/C++ available functions and statements.

C STANDARD INPUT/OUTPUT FUNCTIONS

Included in the `<stdio.h>` **header file:**

`printf(string)` – writes formatted output to the screen

`scanf(string &address)` – scans and formats input from the stream

`getchar(void)` – gets character

`putchar(char)` – output character

`puts(string)` – output a string to the stream

C Console Input/Output Functions

Included in the `<conio.h>` **header file:**

`clreol(void)` – clears to end of line in text window

`clrscr(void)` – clears the text mode window

`cprintf(string)` – writes formatted output to the screen

`cscanf(string)` – scans and formats input from the console

`getch(void)` – gets characters from keyboard but not print to screen

`getche(void)` – gets characters from keyboard and echoes to screen

`gotoxy(int x, int y)` – position cursor in windows

`kbhit(void)` – checks for currently available keystrokes

`wherex(void)` – gives horizontal position within window

`wherey(void)` – gives vertical cursor position within window

C Basic Mathematical Functions

Included in the `<math.h>` **header file:**

`acos(double x)` – returns the arc cosine value of x

`asin(double x)` – returns the arc sine value of x

`atan(double x)` – returns the arc tangent value of x

`cos(double x)` – returns the cosine value of x

`exp(double x)` – returns the value of the natural logarithm e raise to the x power

`fabs(double x)` – returns the absolute value of x

`fmod(double x, double y)` – returns the remainder of x/y

`log(double x)` – returns the natural logarithm of x

`pow(double x, double y)` – returns the base (x) raised to the exp (y)

`sin(double x)` – returns the sine value of x

`sqrt(double x)` – returns the square root value of x

`tan(double x)` – returns the tangent value of x

C Character Classification and Conversion Functions

Included in the `<ctype.h>` **header file:**

`tolower(char)` – converts characters to lowercase

`toupper(char)` – converts characters to uppercase

C Standard Library Functions

Included in the `<stdlib.h>` **header file:**

`abs(int)` – returns the absolute value of an integer

`atof(string)` – converts a string to a float

`atoi(string)` – converts a string to an integer

`rand(void)` – returns a random positive integer

`system(string *command)` – issues a command to the operating system

C String Manipulation Functions

Included in the `<string.h>` **header file:**

`strcat(string ,string)` – appends one string to another

`strchr(string ,char)` – scans a string for the first occurrence of a given
character
`strcmp(string,string)` – compares one string to another
`strcpy(string variable, string)` – copies one string into another
`strlen(string)` – determines the length of the string
`strrev(string)` – reverses a string
`strstr(string, substring)` – scans a string for the occurrence of a given
substring

C++ Input/Output Stream Objects

Included in the `<iostream.h>` **library:**
`cout` – displays data to the standard output device, equivalent to `printf()` in C
`cin` – reads data from the standard input device, equivalent to `scanf()` in C
`cerr` – writes unbuffered output to the standard error device
`clog` – like `cerr`, but uses buffered output

C/C++ Keywords

Some of these keywords were used in the book. Note that this is not a complete
list of all keywords available in C and C++.

Common to C and C++	Exclusive C++ Keywords
break	bool
case	class
char	delete
const	false
default	private
do	public
double	throw
else	true
float	using
for	virtual
if	
int	
return	
switch	
void	
while	

Appendix D

MATLAB Functions and Commands

H ere are some of MATLAB's functions and commands.

MATLAB INPUT/OUTPUT FUNCTIONS AND COMMANDS

`clc` – clears the Command Window
`disp(string)` – displays an array or string
`fprintf(string)` – writes formatted output to the screen
`input(string)` – formats input from the steam
`;` – prevents screen printing
`format` – numeric display format, expressed together with `short`, `long`, `e`, `bank`, or `+`

MATLAB Mathematical Functions

`abs(x)` – returns the absolute value of x
`acos(x)` – returns the arc cosine value of x
`asin(x)` – returns the arc sine value of x
`atan(x)` – returns the arc tangent value of x

ceil(x) – returns the rounded value nearest integer toward positive infinity

cos(x) – returns the cosine value of x

exp(x) – exponential, returns the value of e raise to the power of x. The constant e (equals 2.71828182845904) is the base of the natural logarithm.

floor(x) – returns the rounded value of the nearest integer toward negative infinity

log(x) – returns the natural logarithm of x

log10(x) – returns the common (base 10) logarithm of x

round(x) – returns the rounded value of the nearest integer

sin(x) – returns the sine value of x

sqrt(x) – returns the square root value of x

tan(x) – returns the tangent value of x

Polynomial Functions

eig() – returns the computed eigenvalues of a matrix

polyfit() – returns the coefficients of the best fit polynomial equation

polyval() – returns the value of the evaluated polynomial

roots() – returns computed polynomial roots

Optimization Functions

fminbnd() – returns the minimum of a function of one variable

fminsearch() – returns the minimum of a function with multiple variables

fzero() – returns the value of the variable when the function is equal to zero

Interpolation Functions

interp1() – performs interpolation on a function with one variable

interp2() – performs interpolation on a function with one variable

Numerical Integration and ODE Functions

quad() – performs numerical integration using the adaptive Simpson's rule

quadl() – performs interpolation on a function with one variable

ode23() – low-order solver

ode45() – medium-order solver

MATLAB Plot Functions and Commands

axis – command to set the axis limits

grid – command to set the grid on or off

loglog – generates a log-log plot

mesh – generates a 3D mesh surface plot

plot() – function to plot an xy graph

surf – generates a shaded 3D mesh surface plot

title() – provides text title at top of the graph

xlabel() – provides text label to the x-axis

ylabel() – provides text label to the y-axis

MX and MEX Functions

mexFunction – an entry point to the MEX-file

mexGetVariable – gets a copy of the specified variable and return mxArray

mexGetVariablePtr – gets a read-only pointer to specified variable

mexPutVariable – copies the mxArray, at pointer, from the MEX-function

mxGetScalar – gets the value of the first real element of the mxArray

mxCreateDoubleMatrix – create a two-dimensional, double precision, floating-point mxArray

mxCreateDoubleScalar – creates a double precision scalar

mxGetPr – determines the starting address of the real data in the mxArray

mxDestroyArray – free dynamic memory allocated by mxCreate

MATLAB Engine Functions

engPutVariable – puts a variable in the MATLAB Engine workspace

engEvalString – evaluates an expression in a string in the MATLAB Engine session

engGetVariable – copies a variable from the MATLAB Engine workspace

engOpen – allows the Engine start a MATLAB process

engClose – sends a quit command to the MATLAB Engine session

MATLAB Graphical User Interface (GUI) Development Overview

Appendix **E**

The graphical user interface (GUI) allows the user to perform tasks through components of a graphical display. This permits interaction between computer and user when performing a certain task. In MATLAB, there are GUI components available that can be used during the development of a graphical front-end. Listed are some of the components commonly used.

Push Button – referred to also as command button, the push button is a rectangular figure that contains a text label usually indicating the action the computer will perform once it is pressed or clicked.

Slider – more commonly known as a scroll bar, this is composed of a rectangular trough, a slider, and arrow buttons. This slider, or scroll bar, is mostly used to allow the user to select a value within a range of values by moving the slider along the trough by dragging it with a mouse or clicking on the arrow button.

Radio Button –a button with a label on the side. The label indicates the option to which this button represents. An unselected button will have a hallow portion, once selected or clicked on, the hollow portion will be filled with color. This is used to select one or more options which are exclusive of each other.

Check Box – its difference with the radio button is its shape. The radio button is usually circular, while the check box is square, and the selected box contains a check mark.

Edit Text – this is a square space for in which the user can add text or numbers. It may contain default text, but can be edited or changed by the user.

Static Text – this is used to display text such as titles, labels, or instructions which can not be edited or changed by the user.

Pop-up Menu – also known as drop-down menu, it is used in representing available options and presents them in a list. Once an option is selected and the menu closes, the chosen option is the only visible label.

List Box –a box with multiple lines of text which allows the user to select one or more options from the list. Unlike the pop-up menu, the list box can not be closed and the list is always visible.

Toggle Button – a button that represents two conditions, on (when the button is down), or off (when the button is up) and each condition corresponds to a certain action as designed by the programmer.

Axes – allows display of graphs or images on the selected area of the GUI.

Panel – basically a shaded rectangular area bounded by a line. This is used to group other components.

Button Group –similar to panel except that it only allows exclusive selection of radio buttons and toggle buttons. When a user clicks on a button, that button is selected and all other buttons in the group are deselected.

These components are linked to a written instruction or set of programmed routines called callbacks. The callback is performed once the user interacts with the graphical interface by either clicking on a button, or selecting an item from a menu, for example.

To minimize programming, MATLAB provides an environment called GUIDE (GUI Development Environment) where one can create and lay-out an interface by just clicking and dragging the available components. To initiate GUIDE we click on the START button of the MATLAB Main Desktop window (see Figure E.1).

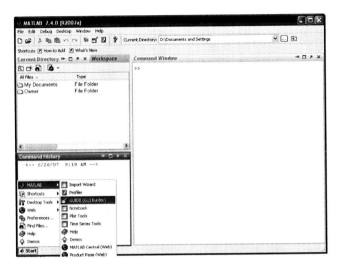

FIGURE E.1: START Button.

By selecting GUIDE (GUI Builder), the GUIDE Quick Start window will pop-up and wait for an action or item to be selected.

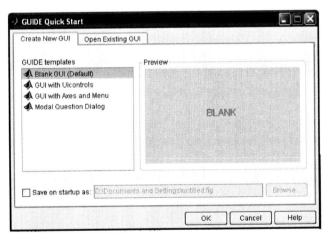

FIGURE E.2: GUIDE Quick Start.

We can thereby choose whether to create a new GUI (selecting the default option) or open an existing one (those with the .fig filename extension). Let us try to open a new GUI by selecting the Blank GUI template. This is a pre-built GUI which allows modifying and applying the design layout we want.

After choosing the template, the Layout Editor will appear with a blank layout area. We can fill up the area with the different components available from the left-side portion of the Layout Editor. We can add components by dragging the components in the area where we want it to be.

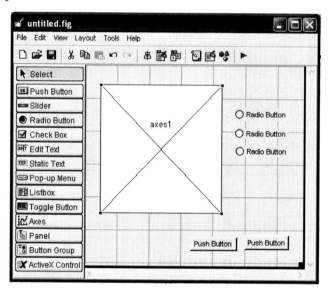

FIGURE E.3: MATLAB Layout Editor.

We can then save the layout to generate both the Fig-file and the M-file. The Fig-file contains the figure layout of the GUI, while the M-file contains the GUI callbacks. It is still necessary to do some programming on the generated M-file for it to perform the way we want it to.

Let us consider another example and open an existing program developed by Claudio Gelmi. It is posted in the File Exchange section of the MATLAB Central web site: http://www.mathworks.com/matlabcentral/fileexchange/loadFile. do?objectId=10387&objectType=file

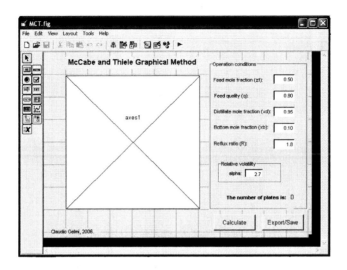

FIGURE E.4: `MCT.fig` **Layout.**

Also provided is the corresponding M-file to run the program together with the interface.

```
% MCT.m
function varargout = MCT(varargin)
% MCT M-file for MCT.fig
% MCT, by itself, creates a new MCT or raises the existing
singleton*.
%
%   H = MCT returns the handle to a new MCT or the handle to
%   the existing singleton*.
%
%   MCT('Property','Value',...) creates a new MCT using
the
%   given property value pairs. Unrecognized properties
are passed via
%   varargin to MCT_OpeningFcn.   This calling syntax
produces a
%   warning when there is an existing singleton*.
%
%   MCT('CALLBACK') and MCT('CALLBACK',hObject,...) call
the
%   local function named CALLBACK in MCT.M with the given
input
%   arguments.
```

```
%
%    *See GUI Options on GUIDE's Tools menu.  Choose "GUI
allows only one
%    instance to run (singleton)".
%
% See also: GUIDE, GUIDATA, GUIHANDLES
% Edit the above text to modify the response to help
MCT
% Last Modified by GUIDE v2.5 11-Feb-2006 15:04:42
% Begin initialization code - DO NOT EDIT
gui_Singleton = 1;
gui_State = struct('gui_Name',        mfilename, ...
    'gui_Singleton',  gui_Singleton, ...
    'gui_OpeningFcn', @MCT_OpeningFcn, ...
    'gui_OutputFcn',  @MCT_OutputFcn, ...
    'gui_LayoutFcn',  [], ...
    'gui_Callback',   []);
if nargin && ischar(varargin{1})
    gui_State.gui_Callback = str2func(varargin{1});
end

if nargout
    [varargout{1:nargout}] = gui_mainfcn(gui_State,
varargin{:});
else
    gui_mainfcn(gui_State, varargin{:});
end
% End initialization code - DO NOT EDIT

% --- Executes just before MCT is made visible.
function MCT_OpeningFcn(hObject, eventdata, handles,
varargin)
% This function has no output args, see OutputFcn.
% hObject    handle to figure
% eventdata  reserved - to be defined in a future version
of MATLAB
% handles    structure with handles and user data (see
GUIDATA)
% varargin   unrecognized PropertyName/PropertyValue pairs
from the
%            command line (see VARARGIN)
```

```
% Choose default command line output for MCT
handles.output = hObject;

% Update handles structure
guidata(hObject, handles);

% UIWAIT makes MCT wait for user response (see
UIRESUME)
% uiwait(handles.figure1);

% --- Outputs from this function are returned to the
command line.
function varargout = MCT_OutputFcn(hObject, eventdata,
handles)
% varargout   cell array for returning output args (see
VARARGOUT);
% hObject      handle to figure
% eventdata    reserved - to be defined in a future version
of MATLAB
% handles      structure with handles and user data (see
GUIDATA)

% Get default command line output from handles structure
varargout{1} = handles.output;

function zf_Callback(hObject, eventdata, handles)
% hObject      handle to zf (see GCBO)
% eventdata    reserved - to be defined in a future version
of MATLAB
% handles      structure with handles and user data (see
GUIDATA)

% Hints: get(hObject,'String') returns contents of zf as
text
%             str2double(get(hObject,'String')) returns
contents of zf as a double

% --- Executes during object creation, after setting all
properties.
function zf_CreateFcn(hObject, eventdata, handles)
% hObject      handle to zf (see GCBO)
```

```
% eventdata  reserved - to be defined in a future version
of MATLAB
% handles    empty - handles not created until after all
CreateFcns called

% Hint: edit controls usually have a white background on
Windows.
%         See ISPC and COMPUTER.
if ispc && isequal(get(hObject,'BackgroundColor'), get(0
,'defaultUicontrolBackgroundColor'))
    set(hObject,'BackgroundColor','white');
end


function q_Callback(hObject, eventdata, handles)
% hObject    handle to q (see GCBO)
% eventdata  reserved - to be defined in a future version
of MATLAB
% handles    structure with handles and user data (see
GUIDATA)

% Hints: get(hObject,'String') returns contents of q as
text
%             str2double(get(hObject,'String')) returns
contents of q as a double

% --- Executes during object creation, after setting all
properties.
function q_CreateFcn(hObject, eventdata, handles)
% hObject    handle to q (see GCBO)
% eventdata  reserved - to be defined in a future version
of MATLAB
% handles    empty - handles not created until after all
CreateFcns called

% Hint: edit controls usually have a white background on
Windows.
%         See ISPC and COMPUTER.
if ispc && isequal(get(hObject,'BackgroundColor'), get(0
,'defaultUicontrolBackgroundColor'))
    set(hObject,'BackgroundColor','white');
end
```

```
function xd_Callback(hObject, eventdata, handles)
% hObject     handle to xd (see GCBO)
% eventdata   reserved - to be defined in a future version
of MATLAB
% handles     structure with handles and user data (see
GUIDATA)

% Hints: get(hObject,'String') returns contents of xd as
text
%              str2double(get(hObject,'String')) returns
contents of xd as a double


% --- Executes during object creation, after setting all
properties.
function xd_CreateFcn(hObject, eventdata, handles)
% hObject     handle to xd (see GCBO)
% eventdata   reserved - to be defined in a future version
of MATLAB
% handles     empty - handles not created until after all
CreateFcns called

% Hint: edit controls usually have a white background on
Windows.
%        See ISPC and COMPUTER.
if ispc && isequal(get(hObject,'BackgroundColor'), get(0
,'defaultUicontrolBackgroundColor'))
    set(hObject,'BackgroundColor','white');
end


function xb_Callback(hObject, eventdata, handles)
% hObject     handle to xb (see GCBO)
% eventdata   reserved - to be defined in a future version
of MATLAB
% handles     structure with handles and user data (see
GUIDATA)
% Hints: get(hObject,'String') returns contents of xb as
text
%              str2double(get(hObject,'String')) returns
contents of xb as a double
% --- Executes during object creation, after setting all
properties.
```

```
function xb_CreateFcn(hObject, eventdata, handles)
% hObject     handle to xb (see GCBO)
% eventdata   reserved - to be defined in a future version
of MATLAB
% handles     empty - handles not created until after all
CreateFcns called

% Hint: edit controls usually have a white background on
Windows.
%        See ISPC and COMPUTER.
if ispc && isequal(get(hObject,'BackgroundColor'), get(0
,'defaultUicontrolBackgroundColor'))
    set(hObject,'BackgroundColor','white');
end

function R_Callback(hObject, eventdata, handles)
% hObject     handle to R (see GCBO)
% eventdata   reserved - to be defined in a future version
of MATLAB
% handles     structure with handles and user data (see
GUIDATA)

% Hints: get(hObject,'String') returns contents of R as
text
%             str2double(get(hObject,'String')) returns
contents of R as a double

% --- Executes during object creation, after setting all
properties.
function R_CreateFcn(hObject, eventdata, handles)
% hObject     handle to R (see GCBO)
% eventdata   reserved - to be defined in a future version
of MATLAB
% handles     empty - handles not created until after all
CreateFcns called

% Hint: edit controls usually have a white background on
Windows.
%        See ISPC and COMPUTER.
if ispc && isequal(get(hObject,'BackgroundColor'), get(0
,'defaultUicontrolBackgroundColor'))
```

```matlab
    set(hObject,'BackgroundColor','white');
end

function alpha_Callback(hObject, eventdata, handles)
% hObject    handle to alpha (see GCBO)
% eventdata  reserved - to be defined in a future version
of MATLAB
% handles    structure with handles and user data (see
GUIDATA)

% Hints: get(hObject,'String') returns contents of alpha
as text
%          str2double(get(hObject,'String')) returns
contents of alpha as a double

% --- Executes during object creation, after setting all
properties.
function alpha_CreateFcn(hObject, eventdata, handles)
% hObject    handle to alpha (see GCBO)
% eventdata  reserved - to be defined in a future version
of MATLAB
% handles    empty - handles not created until after all
CreateFcns called

% Hint: edit controls usually have a white background on
Windows.
%       See ISPC and COMPUTER.
if ispc && isequal(get(hObject,'BackgroundColor'), get(0
,'defaultUicontrolBackgroundColor'))
    set(hObject,'BackgroundColor','white');
end

% --- Executes on button press in calculate.
function calculate_Callback(hObject, eventdata, handles)
% hObject    handle to calculate (see GCBO)
% eventdata  reserved - to be defined in a future version
of MATLAB
% handles    structure with handles and user data (see
GUIDATA)

% Get data from GUI
```

```
alpha = str2double(get(handles.alpha,'String'));
R = str2double(get(handles.R,'String'));
q = str2double(get(handles.q,'String'));
zf = str2double(get(handles.zf,'String'));
xb = str2double(get(handles.xb,'String'));
xd = str2double(get(handles.xd,'String'));

% Error checks
if any([zf xb xd] >= 1) || any([zf xb xd] <= 0)
    errordlg('The molar fractions (zf,xb,xd) must be
between 0 and 1!')
    return
end

if q > 1 || q < 0
    errordlg('The feed quality must be between 0 and 1 (0
<= q <= 1)!')
    return
end

if alpha < 1
    errordlg('Alpha must be greater than 1!')
    return
end

% Computation of equilibrium curve
ye = 0:0.01:1;
xe = equilib(ye,alpha);

% Computing the intersection of feed line and operating
lines
xi = (-(q-1)*(1-R/(R+1))*xd-zf)/((q-1)*R/(R+1)-q);
yi = (zf+xd*q/R)/(1+q/R);

% plotting operating feed lines and equilibrium curve
axes(handles.axes1)
cla
hold on
plot(xe,ye,'r','LineWidth',1)
xlabel('X','FontWeight','bold'), ylabel('Y','FontWeight
','bold')
```

```
axis([0 1 0 1])
set(line([xd  xi,zf  xi,xb  xi],[xd  yi,zf  yi,xb
yi]),'Color','b')
set(line([0 1],[0 1]),'Color','k')

% Stepping off stages
% Rectifying section
i = 1;
xp(1) = xd;
yp(1) = xd;
y = xd;
while xp(i) > xi
    xp(i+1)= equilib(y,alpha);
    yp(i+1)= R/(R+1)*xp(i+1)+xd/(R+1);
    y = yp(i+1);
    set(line([xp(i) xp(i+1)],[yp(i) yp(i)]),'Color','m')
    text(xp(i+1),yp(i),num2str(i))
    if xp(i+1) > xi
            set(line([xp(i+1)  xp(i+1)],[yp(i)
yp(i+1)]),'Color','m')
    end
    i = i+1;
    if i > 20
        errordlg('The distillation is not possible! Try
a different operation condition')
        return
    end
end

% Stripping section
ss = (yi-xb)/(xi-xb);
yp(i) = ss*(xp(i)-xb)+xb;
y = yp(i);
set(line([xp(i) xp(i)],[yp(i-1) yp(i)]),'Color','m')

while xp(i) > xb
    xp(i+1) = equilib(y,alpha);
    yp(i+1) = ss*(xp(i+1)-xb)+xb;
    y = yp(i+1);
        set(line([xp(i)  xp(i+1)],[yp(i)
yp(i)]),'Color','m');
```

```
      text(xp(i+1),yp(i),num2str(i))
      if xp(i+1) > xb
            set(line([xp(i+1)  xp(i+1)],[yp(i)
yp(i+1)]),'Color','m')
      end
      i = i+1;
end
hold off


% Write on the GUI the final number of plates
set(handles.numplates,'String',i-1);


%%%%%%%%%%%%%%%%%%%%%%%%%%%%%%
function x = equilib(y,alpha)
% Constant relative volatility model
x = y./(alpha-y*(alpha-1));


% --- Executes on button press in pushbutton9.
function pushbutton9_Callback(hObject, eventdata,
handles)
% hObject      handle to pushbutton9 (see GCBO)
% eventdata    reserved - to be defined in a future version
of MATLAB
% handles      structure with handles and user data (see
GUIDATA)


fig2 = figure;


alpha = get(handles.alpha,'String');
R = get(handles.R,'String');
q = get(handles.q,'String');
zf = get(handles.zf,'String');
xb = get(handles.xb,'String');
xd = get(handles.xd,'String');


% copy axes into the new figure (this is not trivial)
new_handle = copyobj(handles.axes1,fig2);
set(new_handle, 'units', 'normalized', 'position', [0.13
0.11 0.775 0.815]);
text(0.75,0.35,['zf = ' zf]); text(0.75,0.3,['q = ' q]);
text(0.75,0.25,['xd = ' xd])
```

```
text(0.75,0.2,['xb = ' xb]); text(0.75,0.15,['R = ' R]);
text(0.75,0.1,['alpha = ' alpha])
rectangle('Position',[0.7,0.05,0.25,0.35])

% Save the graph with a unique name
hgsave(new_handle,genvarname(['mctd_' datestr(clock,
'HHMMSS')]))

% ----------------------------------------------------
--------------
function HelpMenu_Callback(hObject, eventdata, handles)
% hObject    handle to HelpMenu (see GCBO)
% eventdata  reserved - to be defined in a future version
of MATLAB
% handles    structure with handles and user data (see
GUIDATA)

% ----------------------------------------------------
--------------
function AboutMenu_Callback(hObject, eventdata, handles)
% hObject    handle to AboutMenu (see GCBO)
% eventdata  reserved - to be defined in a future version
of MATLAB
% handles    structure with handles and user data (see
GUIDATA)
% Credits and final reference (Text Box in Help/About)
help.message = {{'This GUI was created following the
McCabe and Thiele Graphical Method in: '; ...
    '';'McCabe, Smith and Harriott. Unit Operations of
Chemical Engineering, '; ...
    'McGraw-Hill, 7th Edition, 2004.'; ...
    '';'The autor wants to acknowledge the function
"McCabe-Thiele Method for an Ideal Binary Mixture" (FileID
= 4472) by Housam Binous.';...
    '';'Comments, bugs or suggestions, please write to
cgelmi@gmail.com'; ...
    '';' Claudio Gelmi, 2006'};'About this GUI'};
msgbox(help.message{1},help.message{2})
```

This program utilizes a GUI for the user to input the operation conditions necessary to compute for the actual number of plates in a distillation column using the McCabe and Thiele graphical method. The program also uses a constant relative volatility (alpha) to estimate the equilibrium curve.

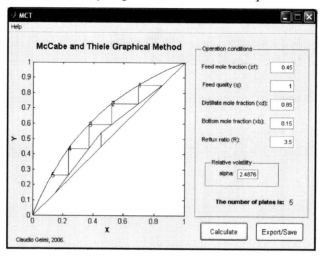

FIGURE E.5:   MCT program output.

This section does not intend to discuss in detail how to develop a MATLAB program with GUI. This section only provides an overview on the features of MATLAB concerning graphical user interface development. It is therefore advisable, for those who want to learn more about the topic, to read MATLAB 7 Creating Graphical User Interfaces documentation available at this site: *http://www.mathworks.com/access/helpdesk/help/pdf_doc/matlab/buildgui.pdf*

F

ALTERNATIVE MATRIX-BASED PROGRAMMING LANGUAGES

Considering that MATLAB can be expensive for students, there are other software programs that were developed and are available on the Internet, with features similar to MATLAB. These software choices also use matrices as the basic data type and have built-in support for complex numbers. Most importantly, they are available for free. Two of the better known software options with similar features to MATLAB are Scilab and GNU Octave.

SCILAB

Scilab, developed by the Scilab Group of INRIA, has similar syntax and features as MATLAB. Since this is a matrix-based language, it can perform matrix manipulation and mathematical operations. It also has an open programming environment where users are encouraged to develop their own functions. Finally, it comes with excellent documentation and support, and is distributed freely in both source and installer binary version for easy installation.

Home Page: *http://www.scilab.org/*

Documentation Page: *http://www.scilab.org/product/index_product.php*

Download Page: *http://www.scilab.org/download/index_download. php?page=release.html*

GNU OCTAVE

GNU Octave, like MATLAB and Scilab, is well suited for numerical computation. It also has a command line interface for performing engineering calculations and solving linear and nonlinear equations. But the best thing is, not only it is free, but modification and redistribution is also possible under the terms of the GNU General Public License (GPL) as published by the Free Software Foundation.

For those using Unix (or a similar operating system), Octave is a better choice than Scilab because it is more compatible with MATLAB. However, for Windows users, they might find Octave difficult to run. A solution is to install Octave Workshop, which is an integrated development environment for the GNU Octave programming language.

GNU Octave
Home Page: *http://www.gnu.org/software/octave*
Documentation Page: *http://www.gnu.org/software/octave/docs.html*
Download Page: *http://www.gnu.org/software/octave/download.html*

Octave Workshop
Home and Download Page: *http://www.math.mcgill.ca/loisel/octave-workshop/*

Appendix G

SATURATED STEAM TABLE

Saturated Temperature (K)	Saturated Pressure (Pa)	Specific Vol Liquid (m³/kg)	Specific Vol Vapor (m³/kg)	Enthalpy Liquid J/kg	Enthalpy Vapor J/kg	Entropy Liquid J/kg-K	Entropy Vapor J/kg-K
273.16	611.6571	0.00100	206.00451	0.611786	2500538.6045	0.0000	9154.1148
274	650.0098	0.00100	194.44199	3544.513	2502148.6706	12.9537	9131.9469
275	698.4359	0.00100	181.61636	7760.320	2504057.1322	28.3116	9105.7546
276	750.0309	0.00100	169.73335	11972.980	2505957.1655	43.6024	9079.7770
277	804.9717	0.00100	158.71719	16182.735	2507849.3393	58.8274	9054.0137
278	863.4432	0.00100	148.49866	20389.809	2509734.1839	73.9878	9028.4640
279	925.6384	0.00100	139.01454	24594.410	2511612.1937	89.0850	9003.1272
280	991.7589	0.00100	130.20700	28796.729	2513483.8287	104.1199	8978.0024
281	1062.0153	0.00100	122.02315	32996.945	2515349.5176	119.0937	8953.0886
282	1136.6269	0.00100	114.41458	37195.221	2517209.6588	134.0074	8928.3848
283	1215.8228	0.00100	107.33692	41391.710	2519064.6227	148.8619	8903.8899
284	1299.8413	0.00100	100.74952	45586.554	2520914.7535	163.6583	8879.6026
285	1388.9311	0.00100	94.61510	49779.884	2522760.3701	178.3973	8855.5218

Saturated Temperature (K)	Saturated Pressure (Pa)	Specific Vol Liquid (m³/kg)	Specific Vol Vapor (m³/kg)	Enthalpy Liquid J/kg	Enthalpy Vapor J/kg	Entropy Liquid J/kg-K	Entropy Vapor J/kg-K
286	1483.3507	0.00100	88.89945	53971.821	2524601.7684	193.0798	8831.6460
287	1583.3694	0.00100	83.57115	58162.479	2526439.2221	207.7065	8807.9739
288	1689.2674	0.00100	78.60135	62351.962	2528272.9844	222.2783	8784.5040
289	1801.3357	0.00100	73.96352	66540.369	2530103.2888	236.7958	8761.2349
290	1919.8772	0.00100	69.63326	70727.790	2531930.3504	251.2597	8738.1651
291	2045.2062	0.00100	65.58812	74914.311	2533754.3673	265.6707	8715.2929
292	2177.6491	0.00100	61.80742	79100.009	2535575.5211	280.0294	8692.6168
293	2317.5450	0.00100	58.27211	83284.960	2537393.9780	294.3365	8670.1352
294	2465.2454	0.00100	54.96465	87469.230	2539209.8900	308.5925	8647.8464
295	2621.1149	0.00100	51.86883	91652.886	2541023.3950	322.7979	8625.7488
296	2785.5317	0.00100	48.96972	95835.986	2542834.6186	336.9534	8603.8406
297	2958.8873	0.00100	46.25352	100018.588	2544643.6738	351.0594	8582.1203
298	3141.5877	0.00100	43.70749	104200.743	2546450.6624	365.1164	8560.5860
299	3334.0528	0.00100	41.31985	108382.502	2548255.6754	379.1251	8539.2360
300	3536.7176	0.00100	39.07972	112563.911	2550058.7934	393.0857	8518.0686
301	3750.0318	0.00100	36.97701	116745.013	2551860.0878	406.9988	8497.0821
302	3974.4608	0.00100	35.00240	120925.850	2553659.6204	420.8649	8476.2747
303	4210.4856	0.00100	33.14726	125106.462	2555457.4450	434.6843	8455.6446
304	4458.6031	0.00100	31.40358	129286.883	2557253.6069	448.4575	8435.1902
305	4719.3268	0.00100	29.76395	133467.151	2559048.1440	462.1850	8414.9095
306	4993.1871	0.00101	28.22149	137647.296	2560841.0869	475.8671	8394.8010
307	5280.7311	0.00101	26.76980	141827.351	2562632.4595	489.5041	8374.8628
308	5582.5238	0.00101	25.40298	146007.346	2564422.2791	503.0967	8355.0932
309	5899.1477	0.00101	24.11551	150187.308	2566210.5572	516.6449	8335.4904
310	6231.2036	0.00101	22.90228	154367.264	2567997.2994	530.1494	8316.0527
311	6579.3109	0.00101	21.75854	158547.242	2569782.5060	543.6104	8296.7784
312	6944.1076	0.00101	20.67987	162727.265	2571566.1722	557.0282	8277.6657
313	7326.2512	0.00101	19.66214	166907.357	2573348.2882	570.4033	8258.7130
314	7726.4185	0.00101	18.70154	171087.541	2575128.8401	583.7359	8239.9184

Saturated Temperature (K)	Saturated Pressure (Pa)	Specific Vol Liquid (m³/kg)	Specific Vol Vapor (m³/kg)	Enthalpy Liquid J/kg	Enthalpy Vapor J/kg	Entropy Liquid J/kg-K	Entropy Vapor J/kg-K
315	8145.3065	0.00101	17.79450	175267.839	2576907.8092	597.0265	8221.2803
316	8583.6323	0.00101	16.93768	179448.273	2578685.1731	610.2753	8202.7971
317	9042.1335	0.00101	16.12801	183628.863	2580460.9055	623.4826	8184.4670
318	9521.5690	0.00101	15.36258	187809.630	2582234.9765	636.6489	8166.2883
319	10022.7187	0.00101	14.63871	191990.592	2584007.3528	649.7744	8148.2595
320	10546.3844	0.00101	13.95387	196171.770	2585777.9978	662.8593	8130.3788
321	11093.3897	0.00101	13.30573	200353.180	2587546.8721	675.9041	8112.6446
322	11664.5807	0.00101	12.69208	204534.843	2589313.9333	688.9091	8095.0553
323	12260.8263	0.00101	12.11088	208716.774	2591079.1364	701.8745	8077.6093
324	12883.0181	0.00101	11.56022	212898.993	2592842.4337	714.8005	8060.3050
325	13532.0714	0.00101	11.03829	217081.516	2594603.7752	727.6877	8043.1408
326	14208.9251	0.00101	10.54343	221264.360	2596363.1088	740.5361	8026.1151
327	14914.5422	0.00101	10.07407	225447.541	2598120.3802	753.3461	8009.2263
328	15649.9100	0.00101	9.62873	229631.078	2599875.5330	766.1180	7992.4730
329	16416.0406	0.00101	9.20604	233814.986	2601628.5090	778.8520	7975.8536
330	17213.9712	0.00102	8.80471	237999.281	2603379.2483	791.5485	7959.3666
331	18044.7642	0.00102	8.42354	242183.979	2605127.6892	804.2077	7943.0104
332	18909.5080	0.00102	8.06138	246369.098	2606873.7687	816.8298	7926.7837
333	19809.3167	0.00102	7.71718	250554.653	2608617.4220	829.4152	7910.6848
334	20745.3310	0.00102	7.38993	254740.660	2610358.5831	841.9641	7894.7124
335	21718.7180	0.00102	7.07871	258927.135	2612097.1849	854.4768	7878.8650
336	22730.6720	0.00102	6.78262	263114.096	2613833.1587	866.9535	7863.1412
337	23782.4145	0.00102	6.50085	267301.556	2615566.4350	879.3945	7847.5395
338	24875.1945	0.00102	6.23261	271489.534	2617296.9431	891.8000	7832.0586
339	26010.2890	0.00102	5.97718	275678.045	2619024.6112	904.1703	7816.6971
340	27189.0031	0.00102	5.73386	279867.106	2620749.3668	916.5057	7801.4535
341	28412.6704	0.00102	5.50202	284056.733	2622471.1364	928.8064	7786.3266
342	29682.6532	0.00102	5.28104	288246.942	2624189.8458	941.0726	7771.3150
343	31000.3429	0.00102	5.07035	292437.751	2625905.4198	953.3046	7756.4174

Saturated Temperature (K)	Saturated Pressure (Pa)	Specific Vol Liquid (m³/kg)	Specific Vol Vapor (m³/kg)	Enthalpy Liquid J/kg	Enthalpy Vapor J/kg	Entropy Liquid J/kg-K	Entropy Vapor J/kg-K
344	32367.1603	0.00102	4.86941	296629.175	2627617.7828	965.5027	7741.6323
345	33784.5556	0.00102	4.67772	300821.233	2629326.8584	977.6670	7726.9587
346	35254.0090	0.00102	4.49479	305013.941	2631032.5696	989.7978	7712.3950
347	36777.0309	0.00103	4.32017	309207.316	2632734.8389	1001.8954	7697.9402
348	38355.1619	0.00103	4.15344	313401.376	2634433.5882	1013.9600	7683.5928
349	39989.9734	0.00103	3.99419	317596.138	2636128.7389	1025.9918	7669.3517
350	41683.0678	0.00103	3.84205	321791.620	2637820.2121	1037.9911	7655.2157
351	43436.0785	0.00103	3.69665	325987.840	2639507.9285	1049.9581	7641.1834
352	45250.6703	0.00103	3.55767	330184.815	2641191.8082	1061.8930	7627.2537
353	47128.5397	0.00103	3.42477	334382.564	2642871.7712	1073.7960	7613.4255
354	49071.4152	0.00103	3.29767	338581.106	2644547.7370	1085.6674	7599.6974
355	51081.0572	0.00103	3.17606	342780.460	2646219.6251	1097.5074	7586.0685
356	53159.2585	0.00103	3.05969	346980.643	2647887.3544	1109.3163	7572.5374
357	55307.8445	0.00103	2.94830	351181.675	2649550.8439	1121.0942	7559.1030
358	57528.6734	0.00103	2.84165	355383.576	2651210.0121	1132.8413	7545.7643
359	59823.6362	0.00103	2.73951	359586.365	2652864.7776	1144.5580	7532.5201
360	62194.6572	0.00103	2.64166	363790.061	2654515.0586	1156.2444	7519.3693
361	64643.6941	0.00103	2.54790	367994.685	2656160.7733	1167.9006	7506.3109
362	67172.7381	0.00104	2.45804	372200.257	2657801.8397	1179.5271	7493.3436
363	69783.8141	0.00104	2.37189	376406.797	2659438.1758	1191.1239	7480.4665
364	72478.9811	0.00104	2.28928	380614.326	2661069.6996	1202.6913	7467.6785
365	75260.3321	0.00104	2.21004	384822.865	2662696.3286	1214.2294	7454.9786
366	78129.9943	0.00104	2.13401	389032.435	2664317.9808	1225.7386	7442.3657
367	81090.1296	0.00104	2.06106	393243.057	2665934.5737	1237.2189	7429.8388
368	84142.9343	0.00104	1.99103	397454.753	2667546.0251	1248.6707	7417.3970
369	87290.6394	0.00104	1.92379	401667.545	2669152.2527	1260.0941	7405.0391
370	90535.5112	0.00104	1.85923	405881.456	2670753.1740	1271.4893	7392.7642
371	93879.8505	0.00104	1.79721	410096.507	2672348.7068	1282.8566	7380.5714
372	97325.9938	0.00104	1.73762	414312.721	2673938.7687	1294.1961	7368.4596
373	100876.3125	0.00104	1.68036	418530.121	2675523.2775	1305.5080	7356.4280

Saturated Temperature (K)	Saturated Pressure (Pa)	Specific Vol Liquid (m³/kg)	Specific Vol Vapor (m³/kg)	Enthalpy Liquid J/kg	Enthalpy Vapor J/kg	Entropy Liquid J/kg-K	Entropy Vapor J/kg-K
374	104533.2139	0.00104	1.62532	422748.731	2677102.1508	1316.7926	7344.4755
375	108299.1406	0.00104	1.57240	426968.573	2678675.3065	1328.0500	7332.6013
376	112176.5709	0.00105	1.52151	431189.672	2680242.6624	1339.2805	7320.8044
377	116168.0189	0.00105	1.47256	435412.051	2681804.1364	1350.4843	7309.0840
378	120276.0348	0.00105	1.42547	439635.735	2683359.6463	1361.6614	7297.4390
379	124503.2047	0.00105	1.38015	443860.749	2684909.1103	1372.8123	7285.8686
380	128852.1507	0.00105	1.33653	448087.116	2686452.4464	1383.9369	7274.3720
381	133325.5313	0.00105	1.29454	452314.863	2687989.5728	1395.0356	7262.9483
382	137926.0413	0.00105	1.25411	456544.014	2689520.4076	1406.1086	7251.5965
383	142656.4116	0.00105	1.21517	460774.595	2691044.8692	1417.1559	7240.3159
384	147519.4098	0.00105	1.17765	465006.632	2692562.8759	1428.1779	7229.1056
385	152517.8399	0.00105	1.14151	469240.152	2694074.3463	1439.1747	7217.9648
386	157654.5427	0.00105	1.10668	473475.180	2695579.1988	1450.1464	7206.8926
387	162932.3953	0.00105	1.07310	477711.744	2697077.3521	1461.0934	7195.8882
388	168354.3118	0.00106	1.04073	481949.870	2698568.7250	1472.0157	7184.9509
389	173923.2429	0.00106	1.00952	486189.586	2700053.2361	1482.9136	7174.0798
390	179642.1761	0.00106	0.97941	490430.919	2701530.8045	1493.7873	7163.2742
391	185514.1358	0.00106	0.95037	494673.897	2703001.3489	1504.6369	7152.5332
392	191542.1833	0.00106	0.92234	498918.549	2704464.7886	1515.4626	7141.8561
393	197729.4167	0.00106	0.89530	503164.903	2705921.0425	1526.2646	7131.2421
394	204078.9711	0.00106	0.86920	507412.987	2707370.0299	1537.0432	7120.6905
395	210594.0186	0.00106	0.84400	511662.831	2708811.6700	1547.7984	7110.2005
396	217277.7682	0.00106	0.81966	515914.464	2710245.8822	1558.5305	7099.7714
397	224133.4659	0.00106	0.79616	520167.916	2711672.5858	1569.2396	7089.4025
398	231164.3948	0.00106	0.77346	524423.216	2713091.7002	1579.9259	7079.0929
399	238373.8750	0.00107	0.75152	528680.395	2714503.1450	1590.5896	7068.8422
400	245765.2635	0.00107	0.73033	532939.483	2715906.8396	1601.2310	7058.6494
401	253341.9544	0.00107	0.70984	537200.512	2717302.7037	1611.8501	7048.5139
402	261107.3788	0.00107	0.69004	541463.512	2718690.6569	1622.4471	7038.4350
403	269065.0048	0.00107	0.67090	545728.514	2720070.6187	1633.0222	7028.4121

Saturated Temperature (K)	Saturated Pressure (Pa)	Specific Vol Liquid (m³/kg)	Specific Vol Vapor (m³/kg)	Enthalpy Liquid J/kg	Enthalpy Vapor J/kg	Entropy Liquid J/kg-K	Entropy Vapor J/kg-K
404	277218.3373	0.00107	0.65238	549995.551	2721442.5089	1643.5757	7018.4444
405	285570.9185	0.00107	0.63448	554264.655	2722806.2470	1654.1076	7008.5313
406	294126.3273	0.00107	0.61715	558535.857	2724161.7529	1664.6182	6998.6721
407	302888.1796	0.00107	0.60040	562809.191	2725508.9462	1675.1075	6988.8661
408	311860.1282	0.00107	0.58418	567084.689	2726847.7464	1685.5759	6979.1128
409	321045.8627	0.00108	0.56848	571362.385	2728178.0733	1696.0235	6969.4115
410	330449.1096	0.00108	0.55329	575642.312	2729499.8464	1706.4504	6959.7614
411	340073.6321	0.00108	0.53858	579924.505	2730812.9854	1716.8568	6950.1621
412	349923.2303	0.00108	0.52433	584208.997	2732117.4097	1727.2428	6940.6128
413	360001.7407	0.00108	0.51053	588495.823	2733413.0388	1737.6088	6931.1129
414	370313.0368	0.00108	0.49717	592785.019	2734699.7920	1747.9548	6921.6619
415	380861.0286	0.00108	0.48422	597076.618	2735977.5887	1758.2809	6912.2592
416	391649.6624	0.00108	0.47168	601370.657	2737246.3481	1768.5875	6902.9040
417	402682.9214	0.00108	0.45952	605667.172	2738505.9892	1778.8745	6893.5959
418	413964.8249	0.00108	0.44773	609966.199	2739756.4310	1789.1423	6884.3342
419	425499.4288	0.00109	0.43631	614267.773	2740997.5924	1799.3909	6875.1184
420	437290.8252	0.00109	0.42523	618571.933	2742229.3921	1809.6206	6865.9479
421	449343.1423	0.00109	0.41449	622878.716	2743451.7485	1819.8315	6856.8221
422	461660.5449	0.00109	0.40408	627188.158	2744664.5801	1830.0238	6847.7404
423	474247.2335	0.00109	0.39397	631500.299	2745867.8050	1840.1975	6838.7023
424	487107.4449	0.00109	0.38417	635815.175	2747061.3412	1850.3530	6829.7072
425	500245.4516	0.00109	0.37465	640132.826	2748245.1065	1860.4904	6820.7546
426	513665.5622	0.00109	0.36542	644453.291	2749419.0183	1870.6098	6811.8439
427	527372.1211	0.00109	0.35646	648776.609	2750582.9939	1880.7113	6802.9745
428	541369.5084	0.00110	0.34776	653102.819	2751736.9503	1890.7953	6794.1460
429	555662.1397	0.00110	0.33932	657431.963	2752880.8043	1900.8617	6785.3578
430	570254.4663	0.00110	0.33112	661764.079	2754014.4723	1910.9108	6776.6094
431	585150.9751	0.00110	0.32315	666099.209	2755137.8704	1920.9428	6767.9003
432	600356.1881	0.00110	0.31542	670437.395	2756250.9143	1930.9578	6759.2298
433	615874.6629	0.00110	0.30791	674778.677	2757353.5195	1940.9559	6750.5976

Saturated Temperature (K)	Saturated Pressure (Pa)	Specific Vol Liquid (m³/kg)	Specific Vol Vapor (m³/kg)	Enthalpy Liquid J/kg	Enthalpy Vapor J/kg	Entropy Liquid J/kg-K	Entropy Vapor J/kg-K
434	631710.9922	0.00110	0.30060	679123.097	2758445.6010	1950.9374	6742.0031
435	647869.8037	0.00110	0.29351	683470.699	2759527.0735	1960.9023	6733.4457
436	664355.7604	0.00111	0.28662	687821.524	2760597.8513	1970.8509	6724.9251
437	681173.5600	0.00111	0.27992	692175.615	2761657.8482	1980.7833	6716.4406
438	698327.9352	0.00111	0.27340	696533.017	2762706.9775	1990.6997	6707.9918
439	715823.6536	0.00111	0.26707	700893.773	2763745.1523	2000.6002	6699.5783
440	733665.5172	0.00111	0.26091	705257.927	2764772.2849	2010.4849	6691.1994
441	751858.3627	0.00111	0.25492	709625.525	2765788.2873	2020.3542	6682.8547
442	770407.0614	0.00111	0.24910	713996.610	2766793.0709	2030.2080	6674.5438
443	789316.5187	0.00111	0.24343	718371.230	2767786.5465	2040.0465	6666.2662
444	808591.6747	0.00112	0.23792	722749.429	2768768.6244	2049.8700	6658.0214
445	828237.5035	0.00112	0.23256	727131.255	2769739.2144	2059.6786	6649.8088
446	848259.0132	0.00112	0.22734	731516.754	2770698.2254	2069.4724	6641.6281
447	868661.2461	0.00112	0.22226	735905.974	2771645.5660	2079.2515	6633.4788
448	889449.2784	0.00112	0.21732	740298.962	2772581.1439	2089.0162	6625.3604
449	910628.2200	0.00112	0.21250	744695.767	2773504.8662	2098.7667	6617.2724
450	932203.2148	0.00112	0.20782	749096.437	2774416.6393	2108.5029	6609.2145
451	954179.4401	0.00112	0.20325	753501.022	2775316.3689	2118.2252	6601.1861
452	976562.1069	0.00113	0.19881	757909.571	2776203.9599	2127.9337	6593.1867
453	999356.4597	0.00113	0.19448	762322.134	2777079.3164	2137.6284	6585.2160
454	1022567.7761	0.00113	0.19027	766738.763	2777942.3417	2147.3097	6577.2735
455	1046201.3674	0.00113	0.18616	771159.508	2778792.9383	2156.9776	6569.3587
456	1070262.5777	0.00113	0.18216	775584.421	2779631.0078	2166.6323	6561.4713
457	1094756.7846	0.00113	0.17826	780013.554	2780456.4511	2176.2739	6553.6107
458	1119689.3985	0.00113	0.17446	784446.960	2781269.1678	2185.9026	6545.7765
459	1145065.8626	0.00114	0.17076	788884.693	2782069.0570	2195.5187	6537.9682
460	1170891.6534	0.00114	0.16715	793326.806	2782856.0165	2205.1221	6530.1856
461	1197172.2798	0.00114	0.16363	797773.353	2783629.9434	2214.7131	6522.4280
462	1223913.2835	0.00114	0.16020	802224.390	2784390.7335	2224.2918	6514.6951
463	1251120.2390	0.00114	0.15686	806679.972	2785138.2817	2233.8584	6506.9865

Saturated Temperature (K)	Saturated Pressure (Pa)	Specific Vol Liquid (m³/kg)	Specific Vol Vapor (m³/kg)	Enthalpy Liquid J/kg	Enthalpy Vapor J/kg	Entropy Liquid J/kg-K	Entropy Vapor J/kg-K
464	1278798.7530	0.00114	0.15360	811140.155	2785872.4818	2243.4130	6499.3017
465	1306954.4651	0.00114	0.15041	815604.995	2786593.2266	2252.9559	6491.6402
466	1335593.0469	0.00115	0.14731	820074.551	2787300.4076	2262.4870	6484.0018
467	1364720.2027	0.00115	0.14428	824548.880	2787993.9151	2272.0067	6476.3858
468	1394341.6687	0.00115	0.14133	829028.041	2788673.6384	2281.5151	6468.7920
469	1424463.2136	0.00115	0.13845	833512.093	2789339.4656	2291.0123	6461.2199
470	1455090.6382	0.00115	0.13564	838001.095	2789991.2832	2300.4985	6453.6691
471	1486229.7752	0.00115	0.13289	842495.109	2790628.9767	2309.9738	6446.1391
472	1517886.4896	0.00115	0.13021	846994.195	2791252.4302	2319.4384	6438.6295
473	1550066.6781	0.00116	0.12760	851498.416	2791861.5265	2328.8924	6431.1400
474	1582776.2695	0.00116	0.12504	856007.833	2792456.1470	2338.3361	6423.6701
475	1616021.2244	0.00116	0.12255	860522.511	2793036.1715	2347.7695	6416.2193
476	1649807.5354	0.00116	0.12012	865042.514	2793601.4786	2357.1928	6408.7873
477	1684141.2267	0.00116	0.11774	869567.905	2794151.9453	2366.6063	6401.3737
478	1719028.3543	0.00116	0.11542	874098.751	2794687.4471	2376.0100	6393.9779
479	1754475.0061	0.00117	0.11315	878635.118	2795207.8578	2385.4040	6386.5997
480	1790487.3013	0.00117	0.11094	883177.074	2795713.0498	2394.7887	6379.2387
481	1827071.3914	0.00117	0.10878	887724.685	2796202.8938	2404.1641	6371.8942
482	1864233.4589	0.00117	0.10666	892278.022	2796677.2588	2413.5304	6364.5661
483	1901979.7184	0.00117	0.10460	896837.153	2797136.0120	2422.8877	6357.2538
484	1940316.4160	0.00117	0.10258	901402.149	2797579.0191	2432.2362	6349.9570
485	1979249.8293	0.00118	0.10061	905973.082	2798006.1438	2441.5761	6342.6752
486	2018786.2678	0.00118	0.09868	910550.023	2798417.2480	2450.9076	6335.4080
487	2058932.0723	0.00118	0.09680	915133.046	2798812.1918	2460.2307	6328.1551
488	2099693.6155	0.00118	0.09495	919722.225	2799190.8334	2469.5458	6320.9159
489	2141077.3016	0.00118	0.09315	924317.636	2799553.0290	2478.8528	6313.6900
490	2183089.5665	0.00118	0.09139	928919.353	2799898.6328	2488.1521	6306.4772
491	2225736.8780	0.00119	0.08967	933527.455	2800227.4969	2497.4437	6299.2768
492	2269025.7353	0.00119	0.08798	938142.020	2800539.4715	2506.7279	6292.0886
493	2312962.6695	0.00119	0.08634	942763.126	2800834.4044	2516.0048	6284.9121

Saturated Temperature (K)	Saturated Pressure (Pa)	Specific Vol Liquid (m³/kg)	Specific Vol Vapor (m³/kg)	Enthalpy Liquid J/kg	Enthalpy Vapor J/kg	Entropy Liquid J/kg-K	Entropy Vapor J/kg-K
494	2357554.2437	0.00119	0.08473	947390.854	2801112.1416	2525.2746	6277.7468
495	2402807.0526	0.00119	0.08315	952025.285	2801372.5264	2534.5374	6270.5924
496	2448727.7230	0.00120	0.08161	956666.503	2801615.4003	2543.7934	6263.4485
497	2495322.9136	0.00120	0.08010	961314.590	2801840.6021	2553.0429	6256.3145
498	2542599.3151	0.00120	0.07862	965969.632	2802047.9684	2562.2859	6249.1902
499	2590563.6504	0.00120	0.07718	970631.714	2802237.3334	2571.5227	6242.0750
500	2639222.6747	0.00120	0.07577	975300.924	2802408.5288	2580.7533	6234.9686
501	2688583.1755	0.00120	0.07438	979977.351	2802561.3836	2589.9781	6227.8704
502	2738651.9725	0.00121	0.07303	984661.084	2802695.7245	2599.1972	6220.7802
503	2789435.9182	0.00121	0.07170	989352.215	2802811.3752	2608.4107	6213.6973
504	2840941.8977	0.00121	0.07040	994050.836	2802908.1571	2617.6189	6206.6215
505	2893176.8287	0.00121	0.06913	998757.042	2802985.8884	2626.8219	6199.5523
506	2946147.6620	0.00121	0.06789	1003470.929	2803044.3849	2636.0200	6192.4892
507	2999861.3813	0.00122	0.06667	1008192.592	2803083.4591	2645.2132	6185.4319
508	3054325.0037	0.00122	0.06548	1012922.132	2803102.9208	2654.4018	6178.3797
509	3109545.5796	0.00122	0.06431	1017659.647	2803102.5767	2663.5860	6171.3324
510	3165530.1929	0.00122	0.06316	1022405.240	2803082.2305	2672.7660	6164.2895
511	3222285.9613	0.00122	0.06204	1027159.014	2803041.6825	2681.9419	6157.2505
512	3279820.0363	0.00123	0.06094	1031921.075	2802980.7299	2691.1140	6150.2149
513	3338139.6039	0.00123	0.05987	1036691.528	2802899.1667	2700.2825	6143.1824
514	3397251.8842	0.00123	0.05881	1041470.484	2802796.7832	2709.4475	6136.1524
515	3457164.1318	0.00123	0.05778	1046258.051	2802673.3666	2718.6093	6129.1245
516	3517883.6364	0.00124	0.05677	1051054.344	2802528.7002	2727.7681	6122.0982
517	3579417.7226	0.00124	0.05577	1055859.476	2802362.5639	2736.9240	6115.0731
518	3641773.7506	0.00124	0.05480	1060673.564	2802174.7337	2746.0773	6108.0487
519	3704959.1161	0.00124	0.05385	1065496.725	2801964.9819	2755.2282	6101.0245
520	3768981.2507	0.00124	0.05291	1070329.081	2801733.0767	2764.3769	6094.0000
521	3833847.6224	0.00125	0.05200	1075170.755	2801478.7826	2773.5237	6086.9748
522	3899565.7357	0.00125	0.05110	1080021.872	2801201.8597	2782.6687	6079.9484
523	3966143.1320	0.00125	0.05022	1084882.558	2800902.0639	2791.8121	6072.9202

Saturated Temperature (K)	Saturated Pressure (Pa)	Specific Vol Liquid (m³/kg)	Specific Vol Vapor (m³/kg)	Enthalpy Liquid J/kg	Enthalpy Vapor J/kg	Entropy Liquid J/kg-K	Entropy Vapor J/kg-K
524	4033587.3901	0.00125	0.04935	1089752.944	2800579.1469	2800.9543	6065.8898
525	4101906.1264	0.00126	0.04850	1094633.162	2800232.8560	2810.0954	6058.8567
526	4171106.9953	0.00126	0.04767	1099523.346	2799862.9336	2819.2356	6051.8204
527	4241197.6898	0.00126	0.04686	1104423.635	2799469.1178	2828.3752	6044.7803
528	4312185.9415	0.00126	0.04606	1109334.168	2799051.1416	2837.5145	6037.7360
529	4384079.5216	0.00127	0.04528	1114255.087	2798608.7334	2846.6536	6030.6870
530	4456886.2410	0.00127	0.04451	1119186.540	2798141.6161	2855.7928	6023.6326
531	4530613.9508	0.00127	0.04375	1124128.674	2797649.5078	2864.9324	6016.5724
532	4605270.5429	0.00127	0.04301	1129081.641	2797132.1209	2874.0726	6009.5059
533	4680863.9504	0.00128	0.04228	1134045.596	2796589.1624	2883.2137	6002.4324
534	4757402.1484	0.00128	0.04157	1139020.698	2796020.3338	2892.3559	5995.3515
535	4834893.1541	0.00128	0.04087	1144007.107	2795425.3305	2901.4996	5988.2626
536	4913345.0281	0.00128	0.04018	1149004.989	2794803.8419	2910.6449	5981.1651
537	4992765.8741	0.00129	0.03951	1154014.513	2794155.5514	2919.7921	5974.0585
538	5073163.8404	0.00129	0.03885	1159035.850	2793480.1359	2928.9416	5966.9421
539	5154547.1200	0.00129	0.03820	1164069.176	2792777.2657	2938.0936	5959.8154
540	5236923.9514	0.00129	0.03756	1169114.673	2792046.6042	2947.2484	5952.6779
541	5320302.6196	0.00130	0.03693	1174172.523	2791287.8082	2956.4063	5945.5288
542	5404691.4563	0.00130	0.03632	1179242.915	2790500.5268	2965.5675	5938.3676
543	5490098.8413	0.00130	0.03571	1184326.043	2789684.4019	2974.7326	5931.1936
544	5576533.2029	0.00131	0.03512	1189422.103	2788839.0677	2983.9016	5924.0063
545	5664003.0186	0.00131	0.03454	1194531.298	2787964.1503	2993.0750	5916.8050
546	5752516.8163	0.00131	0.03396	1199653.833	2787059.2678	3002.2530	5909.5890
547	5842083.1753	0.00131	0.03340	1204789.923	2786124.0294	3011.4361	5902.3577
548	5932710.7265	0.00132	0.03285	1209939.782	2785158.0360	3020.6245	5895.1104
549	6024408.1544	0.00132	0.03231	1215103.635	2784160.8790	3029.8186	5887.8464
550	6117184.1972	0.00132	0.03177	1220281.709	2783132.1405	3039.0188	5880.5650
551	6211047.6483	0.00133	0.03125	1225474.238	2782071.3928	3048.2253	5873.2656
552	6306007.3572	0.00133	0.03073	1230681.463	2780978.1982	3057.4387	5865.9473
553	6402072.2308	0.00133	0.03023	1235903.630	2779852.1082	3066.6593	5858.6095

Saturated Temperature (K)	Saturated Pressure (Pa)	Specific Vol Liquid (m³/kg)	Specific Vol Vapor (m³/kg)	Enthalpy Liquid J/kg	Enthalpy Vapor J/kg	Entropy Liquid J/kg-K	Entropy Vapor J/kg-K
554	6499251.2345	0.00134	0.02973	1241140.992	2778692.6636	3075.8874	5851.2514
555	6597553.3931	0.00134	0.02924	1246393.808	2777499.3939	3085.1234	5843.8722
556	6696987.7926	0.00134	0.02876	1251662.346	2776271.8167	3094.3679	5836.4712
557	6797563.5811	0.00134	0.02829	1256946.880	2775009.4376	3103.6211	5829.0476
558	6899289.9701	0.00135	0.02783	1262247.691	2773711.7492	3112.8836	5821.6005
559	7002176.2363	0.00135	0.02737	1267565.070	2772378.2312	3122.1557	5814.1292
560	7106231.7228	0.00135	0.02692	1272899.314	2771008.3493	3131.4380	5806.6327
561	7211465.8403	0.00136	0.02648	1278250.730	2769601.5551	3140.7308	5799.1102
562	7317888.0693	0.00136	0.02604	1283619.633	2768157.2851	3150.0348	5791.5608
563	7425507.9613	0.00137	0.02562	1289006.349	2766674.9604	3159.3502	5783.9836
564	7534335.1406	0.00137	0.02520	1294411.213	2765153.9857	3168.6778	5776.3777
565	7644379.3062	0.00137	0.02479	1299834.569	2763593.7490	3178.0179	5768.7421
566	7755650.2331	0.00138	0.02438	1305276.774	2761993.6205	3187.3712	5761.0759
567	7868157.7749	0.00138	0.02398	1310738.193	2760352.9520	3196.7381	5753.3779
568	7981911.8655	0.00138	0.02359	1316219.206	2758671.0759	3206.1193	5745.6472
569	8096922.5208	0.00139	0.02320	1321720.203	2756947.3046	3215.5153	5737.8828
570	8213199.8413	0.00139	0.02282	1327241.587	2755180.9293	3224.9268	5730.0835
571	8330754.0142	0.00140	0.02244	1332783.775	2753371.2188	3234.3543	5722.2482
572	8449595.3154	0.00140	0.02208	1338347.197	2751517.4193	3243.7985	5714.3758
573	8569734.1124	0.00140	0.02171	1343932.299	2749618.7520	3253.2601	5706.4651
574	8691180.8665	0.00141	0.02136	1349539.542	2747674.4132	3262.7397	5698.5148
575	8813946.1352	0.00141	0.02100	1355169.401	2745683.5720	3272.2381	5690.5237
576	8938040.5754	0.00142	0.02066	1360822.370	2743645.3695	3281.7561	5682.4904
577	9063474.9459	0.00142	0.02032	1366498.962	2741558.9173	3291.2942	5674.4137
578	9190260.1104	0.00142	0.01998	1372199.705	2739423.2956	3300.8534	5666.2922
579	9318407.0404	0.00143	0.01965	1377925.149	2737237.5521	3310.4345	5658.1243
580	9447926.8188	0.00143	0.01933	1383675.864	2735000.6996	3320.0383	5649.9087
581	9578830.6430	0.00144	0.01901	1389452.443	2732711.7149	3329.6657	5641.6438
582	9711129.8281	0.00144	0.01869	1395255.501	2730369.5358	3339.3176	5633.3280
583	9844835.8111	0.00145	0.01838	1401085.676	2727973.0600	3348.9949	5624.9596

Saturated Temperature (K)	Saturated Pressure (Pa)	Specific Vol Liquid (m³/kg)	Specific Vol Vapor (m³/kg)	Enthalpy Liquid J/kg	Enthalpy Vapor J/kg	Entropy Liquid J/kg-K	Entropy Vapor J/kg-K
584	9979960.1544	0.00145	0.01807	1406943.634	2725521.1418	3358.6987	5616.5369
585	10116514.5497	0.00146	0.01777	1412830.066	2723012.5903	3368.4299	5608.0581
586	10254510.8222	0.00146	0.01747	1418745.694	2720446.1664	3378.1896	5599.5215
587	10393960.9353	0.00147	0.01718	1424691.267	2717820.5800	3387.9790	5590.9250
588	10534876.9945	0.00147	0.01689	1430667.570	2715134.4870	3397.7992	5582.2667
589	10677271.2529	0.00148	0.01661	1436675.419	2712386.4855	3407.6513	5573.5445
590	10821156.1155	0.00148	0.01633	1442715.667	2709575.1130	3417.5367	5564.7561
591	10966544.1451	0.00149	0.01605	1448789.206	2706698.8415	3427.4567	5555.8994
592	11113448.0675	0.00149	0.01578	1454896.969	2703756.0742	3437.4126	5546.9719
593	11261880.7773	0.00150	0.01551	1461039.933	2700745.1402	3447.4059	5537.9712
594	11411855.3439	0.00150	0.01524	1467219.120	2697664.2899	3457.4382	5528.8947
595	11563385.0184	0.00151	0.01498	1473435.603	2694511.6897	3467.5109	5519.7396
596	11716483.2398	0.00152	0.01473	1479690.507	2691285.4155	3477.6258	5510.5031
597	11871163.6427	0.00152	0.01447	1485985.015	2687983.4470	3487.7845	5501.1822
598	12027440.0642	0.00153	0.01422	1492320.372	2684603.6603	3497.9890	5491.7738
599	12185326.5525	0.00153	0.01397	1498697.884	2681143.8203	3508.2412	5482.2745
600	12344837.3750	0.00154	0.01373	1505118.933	2677601.5721	3518.5431	5472.6808
601	12505987.0274	0.00155	0.01349	1511584.972	2673974.4323	3528.8968	5462.9891
602	12668790.2430	0.00155	0.01325	1518097.539	2670259.7782	3539.3047	5453.1954
603	12833262.0031	0.00156	0.01301	1524658.259	2666454.8370	3549.7691	5443.2957
604	12999417.5473	0.00157	0.01278	1531268.850	2662556.6734	3560.2926	5433.2856
605	13167272.3852	0.00157	0.01255	1537931.138	2658562.1758	3570.8778	5423.1605
606	13336842.3089	0.00158	0.01233	1544647.057	2654468.0413	3581.5277	5412.9155
607	13508143.4051	0.00159	0.01210	1551418.666	2650270.7586	3592.2453	5402.5453
608	13681192.0700	0.00160	0.01188	1558248.155	2645966.5892	3603.0339	5392.0445
609	13856005.0238	0.00160	0.01166	1565137.861	2641551.5469	3613.8968	5381.4070
610	14032599.3268	0.00161	0.01145	1572090.279	2637021.3732	3624.8379	5370.6266
611	14210992.3967	0.00162	0.01123	1579108.081	2632371.5118	3635.8611	5359.6965
612	14391202.0275	0.00163	0.01102	1586194.128	2627597.0781	3646.9706	5348.6094
613	14573246.4094	0.00164	0.01081	1593351.497	2622692.8259	3658.1710	5337.3575

Saturated Temperature (K)	Saturated Pressure (Pa)	Specific Vol Liquid (m³/kg)	Specific Vol Vapor (m³/kg)	Enthalpy Liquid J/kg	Enthalpy Vapor J/kg	Entropy Liquid J/kg-K	Entropy Vapor J/kg-K
614	14757144.1510	0.00165	0.01061	1600583.497	2617653.1084	3669.4673	5325.9324
615	14942914.3027	0.00165	0.01040	1607893.700	2612471.8349	3680.8647	5314.3251
616	15130576.3831	0.00166	0.01020	1615285.966	2607142.4209	3692.3690	5302.5256
617	15320150.4075	0.00167	0.01000	1622764.481	2601657.7305	3703.9865	5290.5234
618	15511656.9186	0.00168	0.00980	1630333.795	2596010.0101	3715.7239	5278.3068
619	15705117.0210	0.00169	0.00960	1637998.865	2590190.8119	3727.5887	5265.8633
620	15900552.4195	0.00170	0.00941	1645765.114	2584190.9047	3739.5890	5253.1790
621	16097985.4607	0.00172	0.00922	1653638.487	2578000.1690	3751.7336	5240.2388
622	16297439.1801	0.00173	0.00902	1661625.531	2571607.4746	3764.0325	5227.0259
623	16498937.3549	0.00174	0.00883	1669733.478	2565000.5342	3776.4963	5213.5221
624	16702504.5625	0.00175	0.00865	1677970.353	2558165.7304	3789.1373	5199.7068
625	16908166.2478	0.00176	0.00846	1686345.096	2551087.9072	3801.9687	5185.5572
626	17115948.7991	0.00178	0.00827	1694867.724	2543750.1175	3815.0057	5171.0479
627	17325879.6352	0.00179	0.00808	1703549.510	2536133.3160	3828.2653	5156.1502
628	17537987.3059	0.00181	0.00790	1712403.222	2528215.9808	3841.7668	5140.8317
629	17752301.6086	0.00182	0.00771	1721443.408	2519973.6425	3855.5320	5125.0554
630	17968853.7256	0.00184	0.00753	1730686.770	2511378.2933	3869.5862	5108.7791
631	18187676.3853	0.00185	0.00735	1740152.628	2502397.6338	3883.9586	5091.9539
632	18408804.0568	0.00187	0.00716	1749863.537	2492994.1033	3898.6835	5074.5230
633	18632273.1836	0.00189	0.00698	1759846.082	2483123.6083	3913.8010	5056.4196
634	18858122.4708	0.00191	0.00679	1770131.968	2472733.8297	3929.3595	5037.5643
635	19086393.2432	0.00193	0.00661	1780759.491	2461761.9234	3945.4170	5017.8617
636	19317129.8969	0.00196	0.00642	1791775.610	2450131.3261	3962.0450	4997.1955
637	19550380.4849	0.00198	0.00623	1803238.930	2437747.2023	3979.3328	4975.4212
638	19786197.4885	0.00201	0.00604	1815224.121	2424489.7529	3997.3944	4952.3562
639	20024638.8677	0.00204	0.00584	1827828.736	2410204.0195	4016.3788	4927.7642
640	20265769.5346	0.00207	0.00564	1841184.199	2394683.6489	4036.4870	4901.3299
641	20509663.5117	0.00211	0.00543	1855474.527	2377643.5948	4058.0008	4872.6171
642	20756407.2571	0.00216	0.00522	1870970.606	2358670.9107	4081.3355	4840.9933
643	21006105.1540	0.00221	0.00499	1888099.110	2337127.5007	4107.1463	4805.4797

Saturated Temperature (K)	Saturated Pressure (Pa)	Specific Vol Liquid (m³/kg)	Specific Vol Vapor (m³/kg)	Enthalpy Liquid J/kg	Enthalpy Vapor J/kg	Entropy Liquid J/kg-K	Entropy Vapor J/kg-K
644	21258889.4652	0.00227	0.00474	1907600.605	2311931.5252	4136.5715	4764.4145
645	21514941.1045	0.00235	0.00446	1930972.642	2280959.7795	4171.9160	4714.5318
646	21774544.7052	0.00248	0.00411	1962246.672	2238729.7585	4219.3932	4647.3856
647	22038358.0103	0.00280	0.00348	2030864.294	2148680.2641	4324.4514	4506.5471
647.096	22064000.0000	0.00311	0.00311	2086574.196	2086574.1965	4410.4328	4410.4328

Computed based on *Revised Supplementary Release on Saturation Properties of Ordinary Water Substance by IAPWS* (September 1992).

Appendix H

IAPWS 95 C PROGRAM (BY STEPHEN L. MOSHIER)

```
/* iapws95.c
   Thermodynamic properties of water and steam
   IAPWS copyright applicable to the numerical tables and
   formulas
   reproduced herein:
      The International Association for the Properties of
Water and Steam
                        Frederica, Denmark
                         September 1996
        Release on the IAPWS Formulation 1995 for the
Thermodynamic Properties of
                Ordinary Water Substance for General and
Scientific Use
        (c) International Association for the Properties of
Water and Steam.
        Publication in whole or in part is allowed in all
countries provided that
    attribution is given to the International Association
for the Properties
        of Water and Steam.
```

```
#include <stdio.h>
#include <stdlib.h>
extern double log (double), exp (double);
extern double pow (double, double);
extern double sqrt (double);
/* Use only pow if powi not available.  */
#define powi pow
/* extern double powi (double, int); */
/* Notations:
    T : absolute temperature on the International
Temperature Scale of 1990
  rho: mass density, kg m^-3
  delta = rho / rho_c
  tau = T_c / T
*/
static double T_c = 647.096; /* Critical absolute
temperature, degrees K */
static double rho_c = 322.; /* Critical mass density, kg
m^-3 */
static double R = 0.46151805; /* Specific gas constant, kJ
kg^-1 K^-1 */
  /* Table 1 of IAPWS95.
    Numerical values of the coefficients and parameters of
      the ideal-gas part phi_0 of the dimensionless
Helmholtz free
    energy Eq(5).  */
static double gc[8] = {
  -8.32044648201,
  6.6832105268,
  3.00632,
  0.012436,
  0.97315,
  1.27950,
  0.96956,
  0.24873
```

```
};
static double ge[5] = {
  1.28728967,
  3.53734222,
  7.74073708,
  9.24437796,
 27.5075105
};
/* Table 2 of IAPWS95.
   Numerical values of the coefficients and parameters of
the
     residual part of the dimensionless Helmholtz free
energy, Eq.(6)   */
struct iapws_term_float
{
  int ci;
  int di;
  double ti;
  double ni;
};
struct iapws_term
{
  int ci;
  int di;
  int ti;
  double ni;
};
static struct iapws_term_float p[] = {
  /*  i     ci  di   ti       ni  */
  /*  0 */ /* {-1, -1, -1.0,   -1.0},*/
  /*  1 */ {-1,  1, -0.5,    0.12533547935523e-1},
  /*  2 */ {-1,  1,  0.875,  0.78957634722828e1},
  /*  3 */ {-1,  1,  1.0,   -0.87803203303561e1},
  /*  4 */ {-1,  2,  0.5,    0.31802509345418e0},
  /*  5 */ {-1,  2,  0.75,  -0.26145533859358e0},
  /*  6 */ {-1,  3,  0.375, -0.78199751687981e-2},
  /*  7 */ {-1,  4,  1.0,    0.88089493102134e-2},
};
static struct iapws_term q[] = {
  /*  i     ci  di   ti       ni  */
  /*  8 */ { 1,  1,  4,    -0.66856572307965e0},
```

```
/*   9 */ {  1,   1,   6,    0.20433810950965e0},
/*  10 */ {  1,   1,  12,   -0.66212605039687e-4},
/*  11 */ {  1,   2,   1,   -0.19232721156002e0},
/*  12 */ {  1,   2,   5,   -0.25709043003438e0},
/*  13 */ {  1,   3,   4,    0.16074868486251e0},
/*  14 */ {  1,   4,   2,   -0.40092828925807e-1},
/*  15 */ {  1,   4,  13,    0.39343422603254e-6},
/*  16 */ {  1,   5,   9,   -0.75941377088144e-5},
/*  17 */ {  1,   7,   3,    0.56250979351888e-3},
/*  18 */ {  1,   9,   4,   -0.15608652257135e-4},
/*  19 */ {  1,  10,  11,    0.11537996422951e-8},
/*  20 */ {  1,  11,   4,    0.36582165144204e-6},
/*  21 */ {  1,  13,  13,   -0.13251180074668e-11},
/*  22 */ {  1,  15,   1,   -0.62639586912454e-9},
/*  23 */ {  2,   1,   7,   -0.10793600908932e0},
/*  24 */ {  2,   2,   1,    0.17611491008752e-1},
/*  25 */ {  2,   2,   9,    0.22132295167546e0},
/*  26 */ {  2,   2,  10,   -0.40247669763528e0},
/*  27 */ {  2,   3,  10,    0.58083399985759e0},
/*  28 */ {  2,   4,   3,    0.49969146990806e-2},
/*  29 */ {  2,   4,   7,   -0.31358700712549e-1},
/*  30 */ {  2,   4,  10,   -0.74315929710341e0},
/*  31 */ {  2,   5,  10,    0.47807329915480e0},
/*  32 */ {  2,   6,   6,    0.20527940895948e-1},
/*  33 */ {  2,   6,  10,   -0.13636435110343e0},
/*  34 */ {  2,   7,  10,    0.14180634400617e-1},
/*  35 */ {  2,   9,   1,    0.83326504880713e-2},
/*  36 */ {  2,   9,   2,   -0.29052336009585e-1},
/*  37 */ {  2,   9,   3,    0.38615085574206e-1},
/*  38 */ {  2,   9,   4,   -0.20393486513704e-1},
/*  39 */ {  2,   9,   8,   -0.16554050063734e-2},
/*  40 */ {  2,  10,   6,    0.19955571979541e-2},
/*  41 */ {  2,  10,   9,    0.15870308324157e-3},
/*  42 */ {  2,  12,   8,   -0.16388568342530e-4},
/*  43 */ {  3,   3,  16,    0.43613615723811e-1},
/*  44 */ {  3,   4,  22,    0.34994005463765e-1},
/*  45 */ {  3,   4,  23,   -0.76788197844621e-1},
/*  46 */ {  3,   5,  23,    0.22446277332006e-1},
/*  47 */ {  4,  14,  10,   -0.62689710414685e-4},
/*  48 */ {  6,   3,  50,   -0.55711118565645e-9},
/*  49 */ {  6,   6,  44,   -0.19905718354408e0},
```

```
  /* 50 */ { 6,  6, 46,    0.31777497330738e0},
  /* 51 */ { 6,  6, 50,   -0.11841182425981e0},
};
struct z_coef {
  int ci;
  int di;
  int ti;
  double ni;
  int alphai;
  int betai;
  double gammai;
  int epsiloni;
};
static struct z_coef z[3] = {
  /* i     ci   di ti    ni              alphai betai
gammai epsiloni */
  /* 52 */ {-1,  3,  0,  -0.31306260323435e2, 20,   150,
1.21,   1},
  /* 53 */ {-1,  3,  1,   0.31546140237781e2, 20,   150,
1.21,   1},
  /* 54 */ {-1,  3,  4,  -0.25213154341695e4, 20,   250,
1.25,   1}
};
struct y_coef {
  double ai;
  double bi;
  double Bi;
  double ni;
  int Ci;
  int Di;
  double Ai;
  double betai;
};
static struct y_coef y[2] = {
  /* i     ai   bi  Bi       ni              Ci  Di
Ai  betai */
  /* 55 */ {3.5, 0.85, 0.2, -0.14874640856724e0, 28, 700,
0.32, 0.3},
  /* 56 */ {3.5, 0.95, 0.2,  0.31806110878444e0, 32, 800,
0.32, 0.3}
};
```

```
static double delta;   /* rho / rho_c   */
static double tau;      /* T_c / T   */
static double dm1;       /* delta - 1  */
static double d2;       /* (delta - 1)^2  */
static double t1;       /* 1 - tau */
static double t2;       /* (1 - tau)^2   */

/* Temporary variables  */
static double psi, theta, Delta, dc, de, te, ex;
static double phi_0, phi_r;
static double d_phi_d_tau_0, d_phi_d_tau_r;
static double d2_phi_d_tau2_0, d2_phi_d_tau2_r;
static double d_phi_d_delta_0, d_phi_d_delta_r;
static double d2_phi_d_delta2_0, d2_phi_d_delta2_r;
static double d2_phi_d_delta_d_tau_0, d2_phi_d_delta_d_
tau_r;
static double d_Delta_d_delta;
static double d2_Delta_d_delta2;
static double d_Deltabi_d_delta;
static double d2_Deltabi_d_delta2;
static double d_Deltabi_d_tau;
static double d2_Deltabi_d_tau2;
static double d2_Deltabi_d_delta_d_tau;
static double d_psi_d_delta;
static double d2_psi_d_delta2;
static double d2_psi_d_delta_d_tau;
static double d_psi_d_tau;
static double d2_psi_d_tau2;
/* Evaluate derived parameters used by all the routines. */
int
iapws_common (double T, double rho)
{
  if (T <= 0.0)
    return -1;
  delta = rho / rho_c;
  tau = T_c / T;
  /* (delta - 1.0) ^2 */
  dm1 = (rho - rho_c)/ rho_c;
  d2 = dm1 * dm1;
  /* 1 - tau */
  t1 = (T - T_c) / T;
```

```
  /* (tau - 1) ^ 2 */
  t2 = t1 * t1;
  return 0;
}
/* Dimensionless Helmholtz free energy, phi. */
double
iapws_helmholtz (void)
{
  double s;
  int i;
  /* Ideal gas part, phi_0 */
  s = log (delta)
  + gc[0]
  + gc[1] * tau
  + gc[2] * log (tau)
    /* should use log1p */
  + gc[3] * log (1.0 - exp (-ge[0] * tau))
  + gc[4] * log (1.0 - exp (-ge[1] * tau))
  + gc[5] * log (1.0 - exp (-ge[2] * tau))
  + gc[6] * log (1.0 - exp (-ge[3] * tau))
    /* check for exponential underflow */
  + gc[7] * log (1.0 - exp (-ge[4] * tau));
  phi_0 = s;

  /* Residual part, phi_r.  */
  s = 0.0;
  for (i = 0; i < 7; i++)
    {
      s = s + p[i].ni * powi (delta, p[i].di) * pow (tau,
p[i].ti);
    }
  for (i = 0; i < 44; i++)
    {
      dc = powi (delta, q[i].ci);
      s = s + q[i].ni * powi (delta, q[i].di) * powi (tau,
q[i].ti)
              * exp (-dc);
    }
  for (i = 0; i < 3; i++)
    {
      de = delta - z[i].epsiloni;
```

```
        te = tau - z[i].gammai;
        ex = -z[i].alphai * de * de - z[i].betai * te *
te;
        s = s + z[i].ni * powi (delta, z[i].di) * powi (tau,
z[i].ti) * exp(ex);
      }
    for (i = 0; i < 2; i++)
      {
        psi = exp (-y[i].Ci * d2 - y[i].Di * t2);
        theta = t1 + y[i].Ai * pow(d2, 0.5 / y[i].betai);
        Delta = theta * theta + y[i].Bi * pow(d2, y[i].ai);
        s = s + y[i].ni * pow(Delta, y[i].bi) * delta * psi;
      }
    phi_r = s;
    return (phi_0 + phi_r);
  }
/* Derivative of phi with respect to delta.  */
double
f_d_phi_d_delta_r (void)
{
  double s;
  int i;

  s = 0.0;
  for (i = 0; i < 7; i++)
    {
      s = s + p[i].ni * p[i].di * powi (delta, p[i].di - 1)
              * pow (tau, p[i].ti);
    }

  for (i = 0; i < 44; i++)
    {
      dc = powi (delta, q[i].ci);
      s = s + q[i].ni * exp (-dc)
              * (powi (delta, q[i].di - 1) * powi (tau,
q[i].ti)
                * (q[i].di - q[i].ci * dc));
    }

  for (i = 0; i < 3; i++)
    {
```

```
      de = delta - z[i].epsiloni;
      te = tau - z[i].gammai;
      ex = -z[i].alphai * de * de - z[i].betai * te * te;
      s = s + z[i].ni * powi (delta, z[i].di) * powi (tau,
z[i].ti) * exp(ex)
          * (z[i].di / delta  -  2.0 * z[i].alphai * de);
    }

  for (i = 0; i < 2; i++)
    {
      psi = exp (-y[i].Ci * d2 - y[i].Di * t2);
      theta = t1 + y[i].Ai * pow(d2, 0.5 / y[i].betai);
       Delta = theta * theta + y[i].Bi * pow(d2, y[i].
ai);
      d_psi_d_delta = -2.0 * y[i].Ci * dm1 * psi;
      d_Delta_d_delta = dm1 * (y[i].Ai * theta * (2.0 /
y[i].betai)
                    * pow(d2, (0.5/y[i].betai) - 1)
                  + 2.0 * y[i].Bi * y[i].ai * pow(d2,y[i].
ai - 1));
      d_Deltabi_d_delta = y[i].bi * pow(Delta, y[i].bi -
1) * d_Delta_d_delta;
       s = s + y[i].ni * (pow(Delta, y[i].bi) * (psi +
delta * d_psi_d_delta)
                  + d_Deltabi_d_delta * delta * psi);
    }
  return s;
}


/* Ideal gas part of the derivative of phi with respect
to delta.  */
double
f_d_phi_d_delta_0 (void)
{
  return (1.0 / delta);
}


/* Second derivative of phi with respect to delta.
   Residual part.  */
double
f_d2_phi_d_delta2_r (void)
```

```
{
  double s, d, dd;
  double Delta_bim1;
  int i;
  s = 0.0;
  for (i = 0; i < 7; i++)
    {
      s = s + p[i].ni * p[i].di * (p[i].di - 1) * powi
(delta, p[i].di - 2)
                * pow (tau, p[i].ti);
    }
  for (i = 0; i < 44; i++)
    {
      dc = powi (delta, q[i].ci);
      s = s + q[i].ni * exp (-dc)
              * (powi (delta, q[i].di - 2) * powi (tau,
q[i].ti)
                  * ((q[i].di - q[i].ci * dc) * (q[i].di
- 1 - q[i].ci * dc)
                      - q[i].ci * q[i].ci * dc));
    }
  for (i = 0; i < 3; i++)
    {
      de = delta - z[i].epsiloni;
      te = tau - z[i].gammai;
      ex = -z[i].alphai * de * de - z[i].betai * te * te;
      dd = powi (delta, z[i].di);
      s = s + z[i].ni * powi (tau, z[i].ti) * exp(ex)
      * (-2.0 * z[i].alphai * dd
         + 4.0 * z[i].alphai * z[i].alphai * dd * de * de
            - 4.0 * z[i].di * z[i].alphai * powi (delta,
z[i].di - 1) * de
           + z[i].di * (z[i].di - 1) * powi (delta, z[i].
di - 2));
    }
  for (i = 0; i < 2; i++)
    {
      psi = exp (-y[i].Ci * d2 - y[i].Di * t2);
      theta = t1 + y[i].Ai * pow(d2, 0.5 / y[i].betai);
        Delta = theta * theta + y[i].Bi * pow(d2, y[i].
ai);
```

```
        d_psi_d_delta = -2.0 * y[i].Ci * dm1 * psi;
        d2_psi_d_delta2 = (2.0 * y[i].Ci * d2 - 1.0) * 2.0
* y[i].Ci * psi;

        d_Delta_d_delta = dm1 * (y[i].Ai * theta * (2.0 /
y[i].betai)
                    * pow(d2, (0.5/y[i].betai) - 1)
                + 2.0 * y[i].Bi * y[i].ai * pow(d2,y[i].
ai - 1.0));

        d = (y[i].Ai / y[i].betai) * pow (d2, 0.5/y[i].
betai - 1.0);
        d2_Delta_d_delta2 = (d_Delta_d_delta / dm1)
            + d2 * (4.0 * y[i].Bi * y[i].ai * (y[i].ai - 1)
                * pow (d2, y[i].ai - 2.0)
            + 2.0 * d * d
            + y[i].Ai * theta * (4.0/y[i].betai) * (0.5/
y[i].betai - 1.0)
                * pow (d2, 0.5/y[i].betai - 2.0));

        Delta_bim1 = pow (Delta, y[i].bi - 1);
        d_Deltabi_d_delta = y[i].bi * Delta_bim1 * d_Delta_
d_delta;

        d2_Deltabi_d_delta2 = y[i].bi * (Delta_bim1 * d2_
Delta_d_delta2
                    + (y[i].bi - 1) * pow(Delta, y[i].
bi - 2)
                        * d_Delta_d_delta
* d_Delta_d_delta);

        s = s + y[i].ni * (pow(Delta, y[i].bi)
                    * (2.0 * d_psi_d_delta + delta
* d2_psi_d_delta2)
                + 2.0 * d_Deltabi_d_delta * (psi + delta
* d_psi_d_delta)
                + d2_Deltabi_d_delta2 * delta * psi);
    }
    return s;
}
```

```
/* Second derivative of phi with respect to delta.
   Ideal gas part.  */
double
f_d2_phi_d_delta2_0 (void)
{
  return (-1.0 / (delta * delta));
}


/* Derivative of phi with respect to tau.
   Residual part.  */
double
f_d_phi_d_tau_r (void)
{
  double s;
  int i;

  s = 0.0;
  for (i = 0; i < 7; i++)
    {
      s = s + p[i].ni * p[i].ti * powi (delta, p[i].di)
        * pow (tau, p[i].ti - 1);
    }

  for (i = 0; i < 44; i++)
    {
      dc = powi (delta, q[i].ci);
      s = s + q[i].ni * q[i].ti * powi (delta, q[i].di)
        * powi (tau, q[i].ti - 1) * exp (-dc);
    }
  for (i = 0; i < 3; i++)
    {
      de = delta - z[i].epsiloni;
      te = tau - z[i].gammai;
      ex = -z[i].alphai * de * de - z[i].betai * te * te;
      s = s + z[i].ni * powi (delta, z[i].di) * powi (tau,
z[i].ti) * exp(ex)
        * (z[i].ti/tau  -  2.0 * z[i].betai * te);
    }
  for (i = 0; i < 2; i++)
    {
      psi = exp (-y[i].Ci * d2 - y[i].Di * t2);
```

```
       theta = t1 + y[i].Ai * pow(d2, 0.5 / y[i].betai);
       Delta = theta * theta + y[i].Bi * pow(d2, y[i].ai);

       d_psi_d_tau = 2.0 * y[i].Di * t1 * psi; /* t1 is
(1 - tau) */

       d_Deltabi_d_tau = -2.0 * theta * y[i].bi * pow
(Delta, y[i].bi - 1.0);
       s = s + y[i].ni * delta * (d_Deltabi_d_tau * psi
                               + pow(Delta, y[i].bi) *
d_psi_d_tau);
    }
  return s;
}

/* Derivative of phi with respect to tau.
   Ideal gas part.  */
double
f_d_phi_d_tau_0 (void)
{
  double s;

  s = gc[1]
    + gc[2] / tau
    + gc[3] * ge[0] * (1.0/ (1.0 - exp (-ge[0] * tau)) - 1)
    + gc[4] * ge[1] * (1.0/ (1.0 - exp (-ge[1] * tau)) - 1)
    + gc[5] * ge[2] * (1.0/ (1.0 - exp (-ge[2] * tau)) - 1)
    + gc[6] * ge[3] * (1.0/ (1.0 - exp (-ge[3] * tau)) - 1)
    + gc[7] * ge[4] * (1.0/ (1.0 - exp (-ge[4] * tau)) - 1);
  return s;
}

/* Second derivative of phi with respect to tau.
   Residual part.  */

double
f_d2_phi_d_tau2_r (void)
{
  double s, d, dd;
  double Delta_bim1;
  int i;
```

```
   s = 0.0;
  for (i = 0; i < 7; i++)
    {
      s = s + p[i].ni * p[i].ti * (p[i].ti - 1) * powi
(delta, p[i].di)
              * pow (tau, p[i].ti - 2);
    }
  for (i = 0; i < 44; i++)
    {
      dc = powi (delta, q[i].ci);
      s = s + q[i].ni * q[i].ti * (q[i].ti - 1) * powi
(delta, q[i].di)
        * powi (tau, q[i].ti - 2) * exp (-dc);
    }
  for (i = 0; i < 3; i++)
    {
      de = delta - z[i].epsiloni;
      te = tau - z[i].gammai;
      ex = -z[i].alphai * de * de - z[i].betai * te * te;
      dd = powi (delta, z[i].di);
      d = (z[i].ti / tau) - 2.0 * z[i].betai * te;
      s = s + z[i].ni * dd * powi (tau, z[i].ti) *
exp(ex)
        * (d * d - (z[i].ti / (tau * tau)) - 2.0 * z[i].
betai);
    }
  for (i = 0; i < 2; i++)
    {
      psi = exp (-y[i].Ci * d2 - y[i].Di * t2);
      theta = t1 + y[i].Ai * pow(d2, 0.5 / y[i].betai);
      Delta = theta * theta + y[i].Bi * pow(d2, y[i].ai);
      d_psi_d_tau = 2.0 * y[i].Di * t1 * psi; /* t1 is
(1 - tau) */
      d2_psi_d_tau2 = (2.0 * y[i].Di * t2 - 1.0) * 2.0 *
y[i].Di * psi;

      Delta_bim1 = pow (Delta, y[i].bi - 1.0);
      d_Deltabi_d_tau = -2.0 * theta * y[i].bi * Delta_bim1;

      d2_Deltabi_d_tau2 = 2.0 * y[i].bi * Delta_bim1
        + 4.0 * theta * theta * y[i].bi * (y[i].bi - 1)
```

```
                            * pow(Delta, y[i].bi - 2);

              s = s + y[i].ni * delta * (d2_Deltabi_d_tau2 * psi
                                          + 2.0 * d_Deltabi_d_tau
     * d_psi_d_tau
                                               + pow (Delta, y[i].bi)
     * d2_psi_d_tau2);
         }
     return s;
  }


  /* Second derivative of phi with respect to tau.
     Ideal gas part.   */

  double
  f_d2_phi_d_tau2_0 (void)
  {
    double s, e, em1;
    int i;

    s = gc[2] / (tau * tau);

    for (i = 0; i < 5; i++)
       {
         e = exp (-ge[i] * tau);
         em1 = 1.0 - e;
         s = s + gc[3 + i] * ge[i] * ge[i] * e / (em1 * em1);
       }
    return (-s);
  }


  /* Second derivative of phi with respect to delta and tau.
     Residual part.
     The ideal gas part = 0.   */
  double
  f_d2_phi_d_delta_d_tau_r (void)
  {
    double s, d, dd;
    double Delta_bim1;
    int i;
```

```
   s = 0.0;
  for (i = 0; i < 7; i++)
    {
      s = s + p[i].ni * p[i].di * p[i].ti * powi (delta,
p[i].di - 1)
              * pow (tau, p[i].ti - 1);
    }
  for (i = 0; i < 44; i++)
    {
      dc = powi (delta, q[i].ci);
     s = s + q[i].ni * q[i].ti * powi (delta, q[i].di - 1)
        * powi (tau, q[i].ti - 1) * (q[i].di - q[i].ci *
dc)* exp (-dc);
    }
  for (i = 0; i < 3; i++)
    {
      de = delta - z[i].epsiloni;
      te = tau - z[i].gammai;
      ex = -z[i].alphai * de * de - z[i].betai * te * te;
      dd = powi (delta, z[i].di);
        s = s + z[i].ni * dd * powi (tau, z[i].ti) *
exp(ex)
          * ((z[i].di / delta) - 2.0 * z[i].alphai * de)
          * ((z[i].ti / tau) - 2.0 * z[i].betai * te);
    }
   for (i = 0; i < 2; i++)
     {
      psi = exp (-y[i].Ci * d2 - y[i].Di * t2);
      theta = t1 + y[i].Ai * pow(d2, 0.5 / y[i].betai);
      Delta = theta * theta + y[i].Bi * pow(d2, y[i].ai);
      /* Note in the following two lines, t1 is (1 - tau) */
      d_psi_d_tau = 2.0 * y[i].Di * t1 * psi;
      d_psi_d_delta = -2.0 * y[i].Ci * dm1 * psi;
       d2_psi_d_delta_d_tau = -4.0 * y[i].Ci * y[i].Di *
dm1 * t1 * psi;
       d2_psi_d_tau2 = (2.0 * y[i].Di * t2 - 1.0) * 2.0 *
y[i].Di * psi;
      Delta_bim1 = pow (Delta, y[i].bi - 1);
       d_Deltabi_d_tau = -2.0 * theta * y[i].bi * Delta_
bim1;
```

```
      d2_Deltabi_d_tau2 = 2.0 * y[i].bi * Delta_bim1
         + 4.0 * theta * theta * y[i].bi * (y[i].bi - 1)
         * pow(Delta, y[i].bi - 2.0);

      d_Delta_d_delta = dm1 * (y[i].Ai * theta * (2.0 /
  y[i].betai)
                      * pow(d2, (0.5/y[i].betai) - 1.0)
                   + 2.0 * y[i].Bi * y[i].ai * pow(d2,y[i].
  ai - 1.0));

      d_Deltabi_d_delta = y[i].bi * Delta_bim1 * d_Delta_
  d_delta;

      d2_Deltabi_d_delta_d_tau = -y[i].Ai * y[i].bi *
  (2.0/y[i].betai)
             * Delta_bim1 * dm1 * pow (d2, 0.5/y[i].betai
  - 1.0)
             - 2.0 * theta * y[i].bi * (y[i].bi - 1) * pow(Delta,
  y[i].bi - 2.0)
              * d_Delta_d_delta;

      s = s + y[i].ni * (pow(Delta, y[i].bi)* (d_psi_d_
  tau
                                   + delta *
  d2_psi_d_delta_d_tau)
                         + delta * d_Deltabi_d_delta *
  d_psi_d_tau
                      + d_Deltabi_d_tau * (psi + delta
  * d_psi_d_delta)
                      + d2_Deltabi_d_delta_d_tau *
  delta * psi);
        }
     return s;
   }


 /* Pressure in M Pa, as a function of temperature and
 density.  */
   double pressure (double T, double rho)
   {
     double s;
```

```
   iapws_common (T, rho);
   d_phi_d_delta_r = f_d_phi_d_delta_r ();
   s = (1.0 + delta * d_phi_d_delta_r) * rho * R * T;
   return (0.001 * s);
}


/* Specific isochoric heat capacity, kJ kg^-1 K^-1  */

double cv (double T, double rho)
{
   double s;

   iapws_common (T, rho);
   d2_phi_d_tau2_0 = f_d2_phi_d_tau2_0 ();
   d2_phi_d_tau2_r = f_d2_phi_d_tau2_r ();
   s = -tau * tau * (d2_phi_d_tau2_0 + d2_phi_d_tau2_r) * R;
   return s;
}


/* Speed of sound, m s^-1.  */

double sound (double T, double rho)
{
   double s;

   iapws_common (T, rho);
   d_phi_d_delta_r = f_d_phi_d_delta_r ();
   d2_phi_d_delta2_r = f_d2_phi_d_delta2_r ();
   d2_phi_d_delta_d_tau_r = f_d2_phi_d_delta_d_tau_r ();
   d2_phi_d_tau2_0 = f_d2_phi_d_tau2_0 ();
   d2_phi_d_tau2_r = f_d2_phi_d_tau2_r ();

   s = 1.0 + delta * d_phi_d_delta_r - delta * tau * d2_
phi_d_delta_d_tau_r;

   s = - (s * s) / (tau * tau * (d2_phi_d_tau2_0 + d2_phi_
d_tau2_r));
   s = s + delta * delta * d2_phi_d_delta2_r;
   s = s + 2.0 * delta * d_phi_d_delta_r;
   s = (s + 1.0) * R * T * 1000.0;
   s = sqrt (s);
```

```c
    return s;
}

/* Specific entropy, kJ kg^-1 K^-1 */
double entropy (double T, double rho)
{
  double s;

  iapws_common (T, rho);
  d_phi_d_tau_0 = f_d_phi_d_tau_0 ();
  d_phi_d_tau_r = f_d_phi_d_tau_r ();
  s = tau * (d_phi_d_tau_0 + d_phi_d_tau_r) - iapws_
helmholtz ();
  return (R * s);
}

/* Specific enthalpy, J kg^-1 */
double enthalpy (double T, double rho)
{
  double s;

  iapws_common (T, rho);
  d_phi_d_tau_0 = f_d_phi_d_tau_0 ();
  d_phi_d_tau_r = f_d_phi_d_tau_r ();
  d_phi_d_delta_r = f_d_phi_d_delta_r ();
  s = 1.0 + tau * (d_phi_d_tau_0 + d_phi_d_tau_r) + delta
* d_phi_d_delta_r;
  return (s * R * T);
}

/* Specific internal energy  */
double internal_energy (double T, double rho)
{
  double s;

  iapws_common (T, rho);
  d_phi_d_tau_0 = f_d_phi_d_tau_0 ();
  d_phi_d_tau_r = f_d_phi_d_tau_r ();
  s = tau * (d_phi_d_tau_0 + d_phi_d_tau_r) * R * T;
  return s;
}
```

```
/* Table 7 of IAPWS95.
   Thermodynamic property values  in  the  single-phase
region for selected
   values of T and rho  */
struct test
{
  double T;     /* Temperature, Kelvin */
  double rho;   /* Density, kg m^-3 */
  double p;     /* Pressure, M Pa   */
  double cv;    /* Isochoric heat capacity, kJ kg^-1 K^-1 */
  double w;     /* Speed of sound, m s^-1 */
  double s;     /* Entropy, kJ kg^-1 K^-1 */
};

#define N_TEST 11
static struct test test_vec[N_TEST] = {
 /* T         rho         p           cv        w            s */
 {300., 0.9965560e3, 0.99242e-1,      0.413018111e1,
 0.150151914e4, 0.393062642e0},
 {300.,  0.1005308e4,  0.200022514e2, 0.406798347e1,
 0.153492501e4, 0.387405401e0},
 {300.,  0.1188202e4,  0.700004704e3, 0.346135580e1,
 0.244357992e4, 0.132609616e0},
 {500.,  0.4350000e0,  0.999679423e-1, 0.150817541e1,
 0.548314253e3, 0.794488271e1},
 {500.,  0.4532000e1,  0.999938125e0, 0.166991025e1,
 0.535739001e3, 0.682502725e1},
 {500.,  0.8380250e3,  0.100003858e2, 0.322106219e1,
 0.127128441e4, 0.256690918e1},
 {500.,  0.1084564e4,  0.700000405e3, 0.307437693e1,
 0.241200877e4, 0.203237509e1},
 {647.,  0.3580000e3,  0.220384756e2, 0.618315728e1,
 0.252145078e3, 0.432092307e1},
 {900.,  0.2410000e0,  0.100062559e0, 0.175890657e1,
 0.724027147e3, 0.916653194e1},
 {900.,  0.5261500e2,  0.200000690e2, 0.193510526e1,
 0.698445674e3, 0.659070225e1},
 {900.,  0.8707690e3,  0.700000006e3, 0.266422350e1,
 0.201933608e4, 0.417223802e1},
 };
```

```
/* Compare computed values with the table above.  */
void
check_test(double ref, double y)
{
  double e;

  e = (y - ref) / ref;
  if (e < 0)
      e = -e;
  if (e > 7e-9)
      printf ("\n%.16e should be %.16e\n", y, ref);
}

/* Test program.  */

/* Compute and print tables from the IAPWS95 document.
   Calculation using the 80-bit extended-precision floating
point
   arithmetic on an Intel PC yields the same results as
double precision */
void
print_tables (void)
{
  double T, rho;
  double p, y;
  int i;

  T = 500.0;
  rho = 838.025;
  i = iapws_common (T, rho);
  if (i)
    {
      printf ("Invalid input.\n");
      exit (1);
    }

  printf ("\nComputing Table 6 of IAPWS95.\n");
  printf ("T = 500K, rho = 838.025 kg m-3\n");
  iapws_helmholtz ();
  printf ("phi_0 = %.8e,  ", phi_0);
  printf ("phi_r = %.8e\n", phi_r);
```

```
      d_phi_d_delta_0 = f_d_phi_d_delta_0 ();
      printf ("d_phi_d_delta_0 = %.8e,  ", d_phi_d_delta_0);
      d_phi_d_delta_r = f_d_phi_d_delta_r ();
      printf ("d_phi_d_delta_r = %.8e\n", d_phi_d_delta_r);

      d2_phi_d_delta2_0 = f_d2_phi_d_delta2_0 ();
      printf ("d2_phi_d_delta2_0 = %.8e,  ", d2_phi_d_delta2_0);
      d2_phi_d_delta2_r = f_d2_phi_d_delta2_r ();
      printf ("d2_phi_d_delta2_r = %.8e\n", d2_phi_d_delta2_r);

      d_phi_d_tau_0 = f_d_phi_d_tau_0 ();
      printf ("d_phi_d_tau_0 = %.8e,  ", d_phi_d_tau_0);
      d_phi_d_tau_r =  f_d_phi_d_tau_r ();
      printf ("d_phi_d_tau_r = %.8e\n", d_phi_d_tau_r);

      d2_phi_d_tau2_0 = f_d2_phi_d_tau2_0 ();
      printf ("d2_phi_d_tau2_0 = %.8e,  ", d2_phi_d_tau2_0);
      d2_phi_d_tau2_r = f_d2_phi_d_tau2_r ();
      printf ("d2_phi_d_tau2_r = %.8e\n", d2_phi_d_tau2_r);

      d2_phi_d_delta_d_tau_r = f_d2_phi_d_delta_d_tau_r ();
       printf ("d2_phi_d_delta_d_tau_r = %.8e\n", d2_phi_d_
  delta_d_tau_r);

    printf ("\nComputing Table 7 of IAPWS95.\n");
    printf (
       " T       rho                p               c_v            w
  s\n");
    for (i = 0; i < N_TEST; i++)
      {
        T = test_vec[i].T;
        printf ("%4.0f ", T);
        rho = test_vec[i].rho;
        printf ("%.6e ", rho);
        y = pressure (T, rho);
        printf ("%.8e ", y);
        if (i != 0) /* Table gives only 5 decimals */
            check_test(test_vec[i].p, y);
        y = cv (T, rho);
        printf ("%.8e ", y);
        check_test(test_vec[i].cv, y);
```

```
      y = sound (T, rho);
      printf ("%.8e ", y);
      check_test(test_vec[i].w, y);
      y = entropy (T,  rho);
      printf ("%.8e ", y);
      check_test(test_vec[i].s, y);
      printf ("\n");
   }
```

/* Table 8 and the program disagree in two computations. The

tabulated p_sigma at T=275 is 6.98451167e-04.   The program gives

agreement for the gas pressure, but for the liquid, the density

is not tabulated to a high enough precision to reproduce p_sigma.

Thus a change of 1 in the ninth decimal of the density yields

pressures differing in the third decimal:

liquid pressure at T=275, density 9.99887406e+02 = 6.98212475e-04

liquid pressure at T=275, density 9.99887407e+02 = 7.00204064e-04

Similarly for the enthalpy, the computed values for a change

of 1 in the ninth decimal bracket the tabulated value of enthalpy,

7.75972200e+00:

enthalpy at   T=275, density 9.99887406e+02 = 7.75972154e+00

enthalpy at   T=275, density 9.99887407e+02 = 7.75972355e+00

More precise density solutions for liquid pressure and enthalpy

are not identical, however, so the discrepancy is not entirely

```
      explained by rounding error in the tabulations.  */

    printf ("\nComputing Table 8.\n");
    T = 275.;
    rho = 0.550664919e-2;
    y = pressure (T, rho);
     printf ("vapor pressure at  T=%3.0f, density %.8e =
%.8e\n", T, rho, y);
     y = enthalpy (T, rho);
    printf ("enthalpy at  T=%3.0f, density %.8e = %.8e\n",
T, rho, y);
     y = entropy (T, rho);
    printf ("entropy at  T=%3.0f, density %.8e = %.8e\n",
T, rho, y);
     T = 275.;
     /* rho = 0.999887406e3; */
    rho = 0.99988740611985e3;
    p = pressure (T, rho);
     printf ("liquid pressure at T=%3.0f, density %.13e =
%.8e\n", T, rho, p);
     /* rho = 0.999887406e3; */
    rho = 0.99988740623e3;
    y = enthalpy (T, rho);
    printf ("enthalpy at  T=%3.0f, density %.10e = %.8e\n",
T, rho, y);
    rho = 0.999887406e3;
    y = entropy (T, rho);
    printf ("entropy at  T=%3.0f, density %.8e = %.8e\n",
T, rho, y);

    T = 450.;
    rho = 0.481200360e1;
    y = pressure (T, rho);
     printf ("vapor pressure at  T=%3.0f, density %.8e =
%.8e\n", T, rho, y);
     T = 450.;
    rho = 0.890341250e3;
    p = pressure (T, rho);
     printf ("liquid pressure at T=%3.0f, density %.8e =
%.8e\n", T, rho, p);
     T = 450.;
```

```
    rho = 0.481200360e1;
    y = enthalpy (T, rho);
    printf ("enthalpy at  T=%3.0f, density %.8e = %.8e\n",
T, rho, y);
    T = 450.;
    rho = 0.890341250e3;
    y = enthalpy (T, rho);
    printf ("enthalpy at  T=%3.0f, density %.8e = %.8e\n",
T, rho, y);

    T = 625.;
    rho = 0.118290280e3;
    y = pressure (T, rho);
     printf ("vapor pressure at  T=%3.0f, density %.8e =
%.8e\n", T, rho, y);
    T = 625.;
    rho = 0.567090385e3;
    p = pressure (T, rho);
     printf ("liquid pressure at T=%3.0f, density %.8e =
%.8e\n", T, rho, p);

    T = 625.;
    rho = 0.118290280e3;
    y = enthalpy (T, rho);
    printf ("enthalpy at  T=%3.0f, density %.8e = %.8e\n",
T, rho, y);
    T = 625.;
    rho = 0.567090385e3;
    y = enthalpy (T, rho);
    printf ("enthalpy at  T=%3.0f, density %.8e = %.8e\n",
T, rho, y);

    T = 647.096;
    rho = 332.0;
    p = pressure (T, rho);
     printf ("critical pressure at T=%3.0f, density %.8e =
%.8e\n", T, rho, p);

    /* Triple point pressure = 0.000611657 MPa  */
    /* Normal pressure = 0.101325 MPa */
    T = 273.16;
```

```
    rho = 891.9426578193017; /* solid */
    p = pressure (T, rho);
    printf (" solid pressure at T=%6.2f, density %.15e =
%.8e\n", T, rho, p);
    rho = 999.792520032755; /* liquid */
    p = pressure (T, rho);
    printf ("liquid pressure at T=%6.2f, density %.14e =
%.8e\n", T, rho, p);
    rho = 4.854593429e-3; /* gas */
    p = pressure (T, rho);
    printf ("   gas pressure at T=%6.2f, density %.9e =
%.8e\n", T, rho, p);
}


/* Melting curve
 *
 * IAPWS, 1993:
 * Release on the Pressure along the Melting and the
Sublimation Curves
 * of Ordinary Water Substance
 * Unrestricted publication allowed in all countries.
 * Issued by the International Association for the
Properties of
 * Water and Steam
 */


double
melting_pressure (double T)
{
  double t, p;
  if (T < 251.165)
    {
      printf ("melting_pressure: T < 251.165 K limit of
validity.\n");
      return 0.0;
    }
  /* Melting pressure of ice I */
  if ((T >= 251.165) && (T <= 273.16))
    {
      t = T / 273.16;
      p = 1.0 - 0.626000e6 * (1.0 - t*t*t)
```

```
          + 0.197135e6 * (1.0 - pow(t, 21.2)));
       return p * 0.000611657;
     }
#if 1
  if (T > 273.16)
     {
        printf ("melting_pressure: T > 273.16 K limit of
validity.\n");
        return 0.0;
     }
#else
  /* High pressure region, 209.9 M Pa and up */
  /* Melting pressure of ice III */
  if ((T >= 251.165) && (T < 256.164))
     {
       t = T / 251.165
       p = 1.0 - 0.295252 * (1.0 - pow(t, 60.);
       return p * 209.9;
     }
  /* Melting pressure of ice V */
  if ((T >= 256.164) && (T < 273.31))
     {
       t = T / 256.164;
       p = 1.0 - 1.18721 * (1.0 - pow(t, 8.));
       return p * 350.1;
     }
  /* Melting pressure of ice VI */
  if ((T >= 273.31) && (T < 355.))
     {
       t = T / 273.31;
       p = 1.0 - 1.07476 * (1.0 - pow(t, 4.6));
       return p * 632.4;
     }
  /* Melting pressure of ice VII */
  if ((T >= 355.) && (T < 715.))
     {
       t = T / 355.;
       p = 0.173683e1 * (1.0 - 1.0/t)
       - 0.544606e-1 * (1.0 - pow(t, 5.0))
       + 0.806106e-7 * (1.0 - pow(t, 22.0));
       p = 2216. * exp (p);
```

```
            return p;
        }
     if (T > 715.)
        {
          printf ("melting_pressure: Temperature exceeds 715
K.\n");
          return 0.0;
        }
    #endif
    }


   /* Saturation vapor pressure.
    *
    * Formula from IAPWS 1992:
    * "Revised  Supplementary  Release  on  Saturation
Properties
    * of Ordinary Water Substance
    * Unrestricted publication allowed in all countries.
    * Issued by the International Association for the
Properties of Water
    * and Steam"
    *
    *                           6
    *                           -           k_i
    * ln (p / p )  = tau     >  F     t
    *       s    c           -   k_i
    *                          i=1
    *
    * t = (T_c - T) / T_c
    *
    * p  = critical pressure
    *   c
    *
    * T  = critical temparature
    *   c
    *
    * T = temperature, degrees Kelvin
    *
    * tau = T_c/T, T = Kelvin temperature
    *
    */
```

```
/* Critical point pressure, M Pa  */
#define PC 22.064


double
psat (double T)
{
  double t, th, t2, t3, t4, p;

  /* Domain checks */
  if ((T >= 190.) && (T <= 273.16))
    {
      printf ("psat: T in sublimation range.\n");
      goto ice_sub;
    }
  if (T < 190.)
    {
      printf ("psat: T lower than valid function range.\n");
      return 0.0;
    }
  if (T > T_c)
    {
      printf ("psat: Temperature exceeds T_c.\n");
      return 0.0;
    }

  t = (T_c - T) / T_c;
  th = sqrt (t);
  t2 = t * t;
  t3 = t2 * t;
  t4 = t2 * t2;

  p = (1.80122502 * t3 * th - 15.9618719) * t4
    + (22.6807411 * th - 11.7866497) * t3
    + (1.84408259 * th - 7.85951783) * t;

  p = PC * exp (p * T_c / T );
  return p; /* Mega Pascals */

  /* Sublimation pressure */
 ice_sub:
```

```
   t = T / 273.16;
   th = sqrt (t);
   t3 = t * th;
   t4 = t * sqrt (th);
   p = -13.9281690 * (t3 - 1.0) / t3
     + 34.7078238 * (t4 - 1.0) / t4;
   p = 0.000611657 * exp (p);
   return p;
}


/* Iterate to find density as a function of pressure and
temperature.  */
double
density (double p0, double T, int vapor)
{
   double lr, r, hr, lp, p, hp, z;
   int iter;
   /* Boyle's law approximation for the vapor phase.  */
   if (vapor)
     {
       r = 220. * p0 / T;
       p = pressure (T, r);
       z = 0.707;
     }
   else
     {
       r = 999.79;                     /* kg / m^3 */
       p = pressure (T, r);
       z = 0.99;
     }
   /* Bracket the solution. */
   iter = 0;
   if (p > p0)
     {
       hp = p;
       hr = r;
       while (p > p0)
         {
           r = z * r;
           z = z * z;
           p = pressure (T, r);
```

```
      if (++iter > 10)
        {
          printf ("iter > 10");
          return -1;
        }
    }
  lp = p;
  lr = r;
}
else
  {
    lp = p;
    lr = r;
    z = 1.0 / z;
    while (p < p0)
      {
        r = z * r;
        z = z * z;
        p = pressure (T, r);
        if (++iter > 10)
          {
            printf ("iter > 10");
            return -1;
          }
      }
    hp = p;
    hr = r;
  }
/*  printf ("bracket %d, ", iter); */
do
  {
    /* New guess.  */
    z = (p0 - lp) / (hp - lp);
    if (z < 0.1)
      z = 0.3;
    if (z > 0.9)
      z = 0.7;
    r = lr + z * (hr - lr);
    p = pressure (T, r);
    if (p > p0)
      {
```

```
                         hp = p;
                         hr = r;
                     }
                 else
                     {
                         lp = p;
                         lr = r;
                     }
                 z = (hp - lp) / p0;
                 if (++iter > 40)
                     {
                         printf ("density: iterations > 40\n");
                         break;
                     }
            }
    while (z > 1.0e-9);
#if DEBUG
    printf ("iter = %d\n", iter);
#endif
    return (r);
}


/* Interactive test program.   */
int
main ()
{
    double T, rho, p, u, s, v, h, ps, pm;
    int vapor, input_pressure, tables;
    printf ("\nEnter 1 to compute and print tabulated IAPWS
values\n");
    printf ("or 0 to begin interactive program.   ? \n");
    scanf ("%d", &tables);
    if (tables == 1)
        {
            print_tables ();
            exit (0);
        }
    printf ("\nSteam tables according to IAPWS, 1995.\n");
    printf ("  p: Pressure in mega pascals.\n");
    printf ("  v: Specific volume, cubic meters per kilogram.\n");
    printf ("  r: Density, kilograms per cubic meter.\n");
```

```
  printf (" u: Specific internal energy, kilojoules per
kilogram.\n");
  printf (" h: Specific enthalpy, kilojoules per kilogram.\n");
  printf (" s: Specific entropy, kilojoules per kilogram
degree Kelvin.\n");
  printf (" psat: Saturation pressure, mega pascals.\
n");
  printf ("\nEnter 1 for vapor, 0 for liquid phase ? \
n");
  scanf ("%d", &vapor);
  printf ("\nEnter 1 to input the pressure, 0 to input
density ? \n");
  scanf ("%d", &input_pressure);
loop:
  printf ("Temperature (deg K) ? ");
  scanf ("%lf", &T);
  if (input_pressure == 0)
    {
      /* Input is density.  */
      printf ("Density (kilograms per cubic meters) ? ");
      scanf ("%lf", &rho);
      p = pressure (T, rho);
      printf ("p = %.9g\n", p);
    }
  else
    {
      /* Input is pressure.  */
      printf ("Pressure (MPa) ? ");
      scanf ("%lf", &p);
    }
  /* Domain checks.  */
  if (vapor && ((T < 190) || (T > 1588) || (p < 0) || (p
> 103.4)))
    {
      printf ("Limits are T <= 1588 deg K, p < 1034 bar.\n");
      goto loop;
    }
  if ((vapor == 0) && ((T < 190) || (T > 650) || (p < 0)
|| (p > 137.8)))
    {
      printf ("Limits are T <= 650 deg K, p < 1378 bar.\n");
```

```
        goto loop;
    }
  if (input_pressure == 1)
    rho = density (p, T, vapor);
  u = internal_energy (T, rho);
  h = enthalpy (T, rho);
  s = entropy (T, rho);
  printf ("r %.9g\n", rho);
  printf ("u %.9g\n", u);
  printf ("h %.9g\n", h);
  printf ("s %.9g\n", s);
  ps = psat (T);
  printf ("psat = %.9g\n", ps);
  if (vapor && (p > ps))
      printf ("Pressure of vapor exceeds saturation
pressure.\n");
  if ((vapor == 0) && (p < ps))
    printf ("Pressure of liquid is less than saturation
pressure.\n");
  if ((T >= 251.165) && (T <= 273.16))
    {
     pm = melting_pressure (T);
     printf ("melting pressure = %.9g\n", pm);
    }
  goto loop;
}
```

ANTOINE
EQUATION CONSTANTS

The Antoine equation: $\log p^{\circ} \text{ (bar)} = A - \dfrac{B}{T(\text{K}) + C}$.

Note: Choose constants that are appropriate for the given temperature range.

Antoine Equation Constants for Selected Substances

Substance	Temperature Range, K	A	B	C
Acetaldehyde	293.3 - 377.4	3.68639	822.894	-69.899
	272.9 - 307.5	5.1883	1637.083	22.317
Acetic Acid	290.26 - 391.01	4.68206	1642.54	-39.764
Acetone	259.16 - 507.60	4.42448	1312.253	-32.445
Ammonia	164.0 - 239.5	3.18757	506.713	-80.78
	239.5 - 371.4	4.86886	1113.928	-10.409
Benzene	287.70 - 354.07	4.01814	1203.835	-53.226
	333.4 - 373.4	4.72583	1660.652	-1.461
	421.56 - 554.7	4.60362	1701.073	20.806

Substance	Temperature Range, K	A	B	C
Chlorine	155.0 - 239.3	3.0213	530.591	-64.639
	239.3 - 400.2	4.28814	969.992	-12.791
Ethanol	364.8 - 513.91	4.92531	1432.526	-61.819
	292.77 - 366.63	5.24677	1598.673	-46.424
	273.0 - 351.70	5.37229	1670.409	-40.191
Formic Acid	273.6 - 307.3	2.00121	515.000	-139.408
n-Heptane	185.29 - 295.60	4.81803	1635.409	-27.338
	299.07 - 372.43	4.02832	1268.636	-56.199
Methanol	353.4 - 512.63	5.15853	1569.613	-34.846
	288.0 - 356.83	5.20409	1581.341	-33.500
	353.0 - 483.0	5.31301	1676.569	-21.728
n-Pentane	268.7 - 341.37	3.9892	1070.617	-40.454
Styrene	305.5 - 355.34	4.0593	1459.909	-59.551
	303.07 - 417.92	4.21948	1525.059	-56.379
Toluene	273.13 - 297.89	4.23679	1426.448	-45.957
	303.0 - 343.0	4.08245	1346.382	-53.508
	420.00 - 580.00	4.54436	1738.123	0.394
Water	273.0 - 303.0	5.40221	1838.675	-31.737
	304.0 - 333.0	5.20389	1733.926	-39.485
	334.0 - 363.0	5.0768	1659.793	-45.854
	344.0 - 373.0	5.08354	1663.125	-45.622
	379.0 - 573.0	3.55959	643.748	-198.043

Source: P.J. Linstrom and W.G. Mallard, Eds., **NIST Chemistry WebBook, NIST Standard Reference Database Number 69**, June 2005, National Institute of Standards and Technology, Gaithersburg MD, 20899 (*http://webbook.nist.gov*).

STUDENT'S
T-DISTRIBUTION TABLE

Significant values $t_v(\alpha)$ of t- distribution (Two Tail Areas) $\left[|t| > t_v(\alpha)\right] = \alpha$

d.f. (v)	Probability (Level of significance)					
	0.50	0.10	0.05	0.02	0.01	0.001
1	1.00	6.31	12.71	31.82	63.66	636.62
2	0.82	0.92	4.30	6.97	6.93	31.60
3	0.77	2.32	3.18	4.54	5.84	12.94
4	0.74	2.13	2.78	3.75	4.60	8.61
5	0.73	2.02	2.57	3.37	4.03	6.86
6	0.72	1.94	2.45	3.14	3.71	5.96
7	0.71	1.90	2.37	3.00	3.50	5.41
8	0.71	1.80	2.31	2.90	3.36	5.04
9	0.70	1.83	2.26	2.82	3.25	4.78
10	0.70	1.81	2.23	2.76	3.17	4.59
11	0.70	1.80	2.20	2.72	3.11	4.44
12	0.70	1.78	2.18	2.68	3.06	4.32

	0.50	**0.10**	**0.05**	**0.02**	**0.01**	**0.001**
13	0.69	1.77	2.16	2.65	3.01	4.22
14	0.69	1.76	2.15	2.62	2.98	4.14
15	0.69	1.75	2.13	2.60	2.95	4.07
16	0.69	1.75	2.12	2.58	2.92	4.02
17	0.69	1.74	2.11	2.57	2.90	3.97
18	0.69	1.73	2.10	2.55	2.88	3.92
19	0.69	1.73	2.09	2.54	2.86	3.88
20	0.69	1.73	2.09	2.53	2.85	3.85
21	0.69	1.72	2.08	2.52	2.83	3.83
22	0.69	1.72	2.07	2.51	2.82	3.79
23	0.69	1.71	2.07	2.50	2.81	3.77
24	0.69	1.71	2.06	2.49	2.80	3.75
25	0.68	1.71	2.06	2.49	2.79	3.73
26	0.68	1.71	2.06	2.48	2.78	3.71
27	0.68	1.70	2.05	2.47	2.77	3.69
28	0.68	1.70	2.05	2.47	2.76	3.67
29	0.68	1.70	2.05	2.46	2.76	3.66
30	0.68	1.70	2.04	2.46	2.75	3.65
∞	0.67	1.65	1.96	2.33	2.58	3.29

Source: Goyal, M. and Watkins, C. *Computer-Based Numerical and Statistical Techniques*, Infinity Science Press, 2007.

Appendix K REFERENCES

Black, Ken (2001). *Business Statistics: Contemporary Decision Making*, 3rd ed. Singapore: Thomson Learning Asia.

Bluman, Allan G. (2004). *Elementary Statistics: A Step by Step Approach*. 5th ed. New York: McGraw-Hill.

Chapman, Stephen J. (2004). *MATLAB Programming for Engineers*. 3rd ed. Canada: Nelson.

Chapra, Steven C. and Canale, Raymond P. (2006). *Numerical Methods for Engineers*. 5th ed. Singapore: McGraw-Hill International.

Crockett, William E. (1986). *Chemical Engineering: A Review for the P.E. Exam*. New York: John Wiley and Sons.

Dean, John A. (1999). *Lange's Handbook of Chemistry*. 15th ed. New York: McGraw-Hill Inc.

Deitel, H.M. and Deitel, P.J. (2002). *C How to Program*. 3rd ed. Singapore: Pearson Education South Asia.

Edgar, Thomas F., Himmelblau, David M. and Lasdon, Leon S. (2001). *Optimization of Chemical Processes*, 2nd ed., McGraw-Hill International.

Epperson, James F. (2002). *Introduction to Numerical Methods and Analysis*. New York: John Wiley and Sons.

Etter, Dolores M., Kuncicky, David C. and Hull, Doug (2002). *Introduction to MATLAB 6*, New Jersey: Prentice Hall.

Fausett, Laurence V. (2003). *Numerical Methods Algorithm and Applications*. Singapore: Prentice-Hall International.

Foust, A. S., Wenzel L. A., Clump, C.W., Maus, L. and Anderson L. B. (1980). *Principles of Unit Operations*. New York: John Wiley and Sons.

Gander, W. and Gautschi W. (1998). *Adaptive Quadrature – Revisited*. Zurich.

Geankoplis, Christie J. (2005). *Principles of Transport Process and Separation Processes*. 4th ed. Singapore: Pearson Education South Asia.

Hanselman, Duane and Littlefield, Bruce. (2002). *Mastering MATLAB 6*. New Jersey: Prentice Hall.

Hillier, Frederick S. and Lieberman, Gerald J. (2005). *Introduction to Operations Research*. 8th ed. Singapore: McGraw-Hill International.

Himmelblau, David M. and Riggs, James B. (2004). *Principles of Chemical Engineering*. 7th ed. Singapore: Prentice-Hall International.

IAPWS (The International Association for the Properties of Water and Steam) September 1996. Released on the IAPWS Formulation 1995 for the Thermodynamic Properties of Water and Steam, Fredericia, Denmark.

Linstrom, P.J. and Mallard, W.G., Eds. *NIST Chemistry Web Book, NIST Standard Reference Database Number 69*. June 2005. National Institute of Standards and Technology, Gaithersburg MD, 20899. http://webbook.nist.gov

Jamsa, Kris (2002). *Jamsa's C/C++/C# Programmers' Bible*. 2nd ed. Singapore: Thomas Learning Asia.

Jensen, M.E., R.D. Burman, and R.G. Allen (ed.).(1990). *Evapotranspiration and irrigation water requirements*. ASCE Manual 70.

Kent, Jeff(2005). *C++ Demystified*. Philippines: McGraw-Hill International.

Kyle, B.G. (1999). *Chemical and Process Thermodynamics*. 3rd ed. Singapore: Prentice-Hall International Edition.

Levenspiel, Octave (2001). *Chemical Reaction Engineering*. 3rd ed. New York: John Wiley and Sons.

McCabe, Warren L., Smith, Julian C. and Harriott, Peter (2006). *Unit Operations of Chemical Engineering*. 7th ed. Singapore: McGraw-Hill International.

Palm, William J. (2001). *Introduction to MATLAB 6 for Engineers*. New York: McGraw-Hill Book.

Pepito, Copernicus P. (2006). *Introduction to Turbo C Programming*. Philippines: National Book Store.

Perry, Robert H. and Green, D. (1997). *Perry's Chemical Engineers' Handbook*. 7th ed. New York: McGraw-Hill.

Peters, Max S., Timmerhaus, Klaus D. and West, Ronald E. (2004). *Plant Design and Economics for Chemical Engineers*. 7th ed. Singapore: McGraw-Hill International.

Riggs, James B. (1998). *An Introduction to Numerical Methods for Chemical Engineering*. Texas: Texas Tech University Press.

Schildt, Herbert(2000). *Programmer's Reference C/C++*. 2nd ed. New York: McGraw-Hill.

Smith J.M. and Van Ness, H.C. (2005). *Introduction to Chemical Engineering Thermodynamics*. 7th ed. New York: McGraw-Hill.

Speight, James G. (2003). *Perry's Standard Tables and Formulas for Chemical Engineers*. New York: McGraw-Hill.

Spiegel, Murray R., Schiller, John and Srinivasan, R. Alu (2001). *Probability and Statistics Schaum's Easy Outlines Series*. New York: McGraw-Hill.

Spiegel, Murray R. (2000). *Statistic Schaum's Easy Outlines Series*, New York: McGraw-Hill.

Winnick, Jack. (1997). *Chemical Engineering Thermodynamics*. New York: John Wiley and Sons.

L

OPEN WATCOM
PUBLIC LICENSE

SYBASE OPEN WATCOM PUBLIC LICENSE VERSION 1.0

USE OF THE SYBASE OPEN WATCOM SOFTWARE DESCRIBED BELOW ("SOFTWARE") IS SUBJECT TO THE TERMS AND CONDITIONS SET FORTH IN THE SYBASE OPEN WATCOM PUBLIC LICENSE SET FORTH BELOW ("LICENSE"). YOU MAY NOT USE THE SOFTWARE IN ANY MANNER UNLESS YOU ACCEPT THE TERMS AND CONDITIONS OF THE LICENSE. YOU INDICATE YOUR ACCEPTANCE BY IN ANY MANNER USING (INCLUDING WITHOUT LIMITATION BY REPRODUCING, MODIFYING OR DISTRIBUTING) THE SOFTWARE. IF YOU DO NOT ACCEPT ALL OF THE TERMS AND CONDITIONS OF THE LICENSE, DO NOT USE THE SOFTWARE IN ANY MANNER.

1.   General; Definitions. This License applies only to the following software programs: the open source versions of Sybase's Watcom C/C++ and Fortran compiler products ("Software"), which are modified versions of, with significant changes from, the last versions made commercially available by Sybase. As used in this License:

1.1   "Applicable Patent Rights" mean: (a) in the case where Sybase is the grantor of rights, (i) claims of patents that are now or hereafter acquired, owned by or assigned to Sybase and (ii) that cover subject matter contained in the Original Code, but only to the extent necessary to use, reproduce and/or distribute the Original Code without infringement; and (b) in the case where You are the grantor of rights, (i) claims of patents that are now or hereafter acquired, owned by or assigned to You and (ii) that cover subject matter in Your Modifications, taken alone or in combination with Original Code.

1.2   "Contributor" means any person or entity that creates or contributes to the creation of Modifications.

1.3   "Covered Code" means the Original Code, Modifications, the combination of Original Code and any Modifications, and/or any respective portions thereof.

1.4   "Deploy" means to use, sublicense or distribute Covered Code other than for Your internal research and development (R&D) and/or Personal Use, and includes without limitation, any and all internal use or distribution of Covered Code within Your business or organization except for R&D use and/or Personal Use, as well as direct or indirect sublicensing or distribution of Covered Code by You to any third party in any form or manner.

1.5 "Larger Work" means a work which combines Covered Code or portions thereof with code not governed by the terms of this License.

1.6 "Modifications" mean any addition to, deletion from, and/or change to, the substance and/or structure of the Original Code, any previous Modifications, the combination of Original Code and any previous Modifications, and/or any respective portions thereof. When code is released as a series of files, a Modification is: (a) any addition to or deletion from the contents of a file containing Covered Code; and/or (b) any new file or other representation of computer program statements that contains any part of Covered Code.

1.7 "Original Code" means (a) the Source Code of a program or other work as originally made available by Sybase under this License, including the Source Code of any updates or upgrades to such programs or works made available by Sybase under this License, and that has been expressly identified by Sybase as such in the header file(s) of such work; and (b) the object code compiled from such Source Code and originally made available by Sybase under this License.

1.8 "Personal Use" means use of Covered Code by an individual solely for his or her personal, private and non-commercial purposes. An individual's use of Covered Code in his or her capacity as an officer, employee, member, independent contractor or agent of a corporation, business or organization (commercial or non-commercial) does not qualify as Personal Use.

1.9 "Source Code" means the human readable form of a program or other work that is suitable for making modifications to it, including all modules it contains, plus any associated interface definition files, scripts used to control compilation and installation of an executable (object code).

1.10 "You" or "Your" means an individual or a legal entity exercising rights under this License. For legal entities, "You" or "Your" includes any entity which controls, is controlled by, or is under common control with, You, where "control" means (a) the power, direct or indirect, to cause the direction or management of such entity, whether by contract or otherwise, or (b) ownership of fifty percent (50%) or more of the outstanding shares or beneficial ownership of such entity.

2. Permitted Uses; Conditions & Restrictions Subject to the terms and conditions of this License, Sybase hereby grants You, effective on the date You accept this License and download the Original Code, a world-wide, royalty-free, non-exclusive license, to the extent of Sybase's Applicable Patent Rights and copyrights covering the Original Code, to do the following:

2.1 You may use, reproduce, display, perform, modify and distribute Original Code, with or without Modifications, solely for Your internal research and development and/or Personal Use, provided that in each instance:

   (a) You must retain and reproduce in all copies of Original Code the copyright and other proprietary notices and disclaimers of Sybase as they appear in the Original Code, and keep intact all notices in the Original Code that refer to this License; and

   (b) You must retain and reproduce a copy of this License with every copy of Source Code of Covered Code and documentation You distribute, and You may not offer or impose any terms on such Source Code that alter or restrict this License or the recipients' rights hereunder, except as permitted under Section 6.

   (c) Whenever reasonably feasible you should include the copy of this License in a click-wrap format, which requires affirmative acceptance by clicking on an "I accept" button or similar mechanism. If a click-wrap format is not included, you must include a statement that any use (including without limitation reproduction, modification or distribution) of the Software, and any other affirmative act that you define, constitutes acceptance of the License, and instructing the user not to use the Covered Code in any manner if the user does not accept all of the terms and conditions of the License.

2.2 You may use, reproduce, display, perform, modify and Deploy Covered Code, provided that in each instance:

   (a) You must satisfy all the conditions of Section 2.1 with respect to the Source Code of the Covered Code;

   (b) You must duplicate, to the extent it does not already exist, the notice in Exhibit A in each file of the Source Code of all Your Modifications, and cause the modified files to carry prominent notices stating that You changed the files and the date of any change;

   (c) You must make Source Code of all Your Deployed Modifications publicly available under the terms of this License, including the license grants set forth in Section 3 below, for as long as you Deploy the Covered Code or twelve (12) months from the date of initial Deployment, whichever is longer. You should preferably distribute the Source Code of Your Deployed Modifications electronically (e.g. download from a web site);

(d) if You Deploy Covered Code in object code, executable form only, You must include a prominent notice, in the code itself as well as in related documentation, stating that Source Code of the Covered Code is available under the terms of this License with information on how and where to obtain such Source Code; and

(e) the object code form of the Covered Code may be distributed under Your own license agreement, provided that such license agreement contains terms no less protective of Sybase and each Contributor than the terms of this License, and stating that any provisions which differ from this License are offered by You alone and not by any other party.

2.3   You expressly acknowledge and agree that although Sybase and each Contributor grants the licenses to their respective portions of the Covered Code set forth herein, no assurances are provided by Sybase or any Contributor that the Covered Code does not infringe the patent or other intellectual property rights of any other entity. Sybase and each Contributor disclaim any liability to You for claims brought by any other entity based on infringement of intellectual property rights or otherwise. As a condition to exercising the rights and licenses granted hereunder, You hereby assume sole responsibility to secure any other intellectual property rights needed, if any. For example, if a third party patent license is required to allow You to distribute the Covered Code, it is Your responsibility to acquire that license before distributing the Covered Code.

3.   Your Grants. In consideration of, and as a condition to, the licenses granted to You under this License, You hereby grant to Sybase and all third parties a non-exclusive, royalty-free license, under Your Applicable Patent Rights and other intellectual property rights (other than patent) owned or controlled by You, to use, reproduce, display, perform, modify, distribute and Deploy Your Modifications of the same scope and extent as Sybase's licenses under Sections 2.1 and 2.2.

4.   Larger Works. You may create a Larger Work by combining Covered Code with other code not governed by the terms of this License and distribute the Larger Work as a single product. In each such instance, You must make sure the requirements of this License are fulfilled for the Covered Code or any portion thereof.

5.   Limitations on Patent License. Except as expressly stated in Section 2, no other patent rights, express or implied, are granted by Sybase herein. Modifications and/or Larger Works may require additional patent licenses from Sybase which Sybase may grant in its sole discretion.

6.   Additional Terms. You may choose to offer, and to charge a fee for, warranty, support, indemnity or liability obligations and/or other rights consistent with this License ("Additional Terms") to one or more recipients of Covered Code. However, You may do so only on Your own behalf and as Your sole responsibility, and not on behalf of Sybase or any Contributor. You must obtain the recipient's agreement that any such Additional Terms are offered by You alone, and You hereby agree to indemnify, defend and hold Sybase and every Contributor harmless for any liability incurred by or claims asserted against Sybase or such Contributor by reason of any such Additional Terms.

7.   Versions of the License. Sybase may publish revised and/or new versions of this License from time to time. Each version will be given a distinguishing version number. Once Original Code has been published under a particular version of this License, You may continue to use it under the terms of that version. You may also choose to use such Original Code under the terms of any subsequent version of this License published by Sybase. No one other than Sybase has the right to modify the terms applicable to Covered Code created under this License.

8.   NO WARRANTY OR SUPPORT. The Covered Code may contain in whole or in part pre-release, untested, or not fully tested works. The Covered Code may contain errors that could cause failures or loss of data, and may be incomplete or contain inaccuracies. You expressly acknowledge and agree that use of the Covered Code, or any portion thereof, is at Your sole and entire risk. THE COVERED CODE IS PROVIDED "AS IS" AND WITHOUT WARRANTY, UPGRADES OR SUPPORT OF ANY KIND AND SYBASE AND SYBASE'S LICENSOR(S) (COLLECTIVELY REFERRED TO AS "SYBASE" FOR THE PURPOSES OF SECTIONS 8 AND 9) AND ALL CONTRIBUTORS EXPRESSLY DISCLAIM ALL WARRANTIES AND/OR CONDITIONS, EXPRESS OR IMPLIED, INCLUDING, BUT NOT LIMITED TO, THE IMPLIED WARRANTIES AND/OR CONDITIONS OF MERCHANTABILITY, OF SATISFACTORY QUALITY, OF FITNESS FOR A PARTICULAR PURPOSE, OF ACCURACY, OF QUIET ENJOYMENT, AND NONINFRINGEMENT OF THIRD PARTY RIGHTS. SYBASE AND EACH CONTRIBUTOR DOES NOT WARRANT AGAINST INTERFERENCE WITH YOUR ENJOYMENT OF THE COVERED CODE, THAT THE FUNCTIONS CONTAINED IN THE COVERED CODE WILL MEET YOUR REQUIREMENTS, THAT THE OPERATION OF THE

COVERED CODE WILL BE UNINTERRUPTED OR ERROR-FREE, OR THAT DEFECTS IN THE COVERED CODE WILL BE CORRECTED. NO ORAL OR WRITTEN INFORMATION OR ADVICE GIVEN BY SYBASE, A SYBASE AUTHORIZED REPRESENTATIVE OR ANY CONTRIBUTOR SHALL CREATE A WARRANTY. You acknowledge that the Covered Code is not intended for use in the operation of nuclear facilities, aircraft navigation, communication systems, or air traffic control machines in which case the failure of the Covered Code could lead to death, personal injury, or severe physical or environmental damage.

9. LIMITATION OF LIABILITY. TO THE EXTENT NOT PROHIBITED BY LAW, IN NO EVENT SHALL SYBASE OR ANY CONTRIBUTOR BE LIABLE FOR ANY DIRECT, INCIDENTAL, SPECIAL, INDIRECT, CONSEQUENTIAL OR OTHER DAMAGES OF ANY KIND ARISING OUT OF OR RELATING TO THIS LICENSE OR YOUR USE OR INABILITY TO USE THE COVERED CODE, OR ANY PORTION THEREOF, WHETHER UNDER A THEORY OF CONTRACT, WARRANTY, TORT (INCLUDING NEGLIGENCE), PRODUCTS LIABILITY OR OTHERWISE, EVEN IF SYBASE OR SUCH CONTRIBUTOR HAS BEEN ADVISED OF THE POSSIBILITY OF SUCH DAMAGES, AND NOTWITHSTANDING THE FAILURE OF ESSENTIAL PURPOSE OF ANY REMEDY. SOME JURISDICTIONS DO NOT ALLOW THE LIMITATION OF LIABILITY OF INCIDENTAL OR CONSEQUENTIAL OR OTHER DAMAGES OF ANY KIND, SO THIS LIMITATION MAY NOT APPLY TO YOU. In no event shall Sybase's or any Contributor's total liability to You for all damages (other than as may be required by applicable law) under this License exceed the amount of five hundred dollars ($500.00).

10. Trademarks. This License does not grant any rights to use the trademarks or trade names "Sybase" or any other trademarks or trade names belonging to Sybase (collectively "Sybase Marks") or to any trademark or trade name belonging to any Contributor("Contributor Marks"). No Sybase Marks or Contributor Marks may be used to endorse or promote products derived from the Original Code or Covered Code other than with the prior written consent of Sybase or the Contributor, as applicable.

11. Ownership. Subject to the licenses granted under this License, each Contributor retains all rights, title and interest in and to any Modifications made by such Contributor. Sybase retains all rights, title and interest in and to the Original Code and any Modifications made by or on behalf of Sybase ("Sybase Modifications"), and such Sybase Modifications will not be automatically subject to this License. Sybase may, at its sole discretion, choose to license such Sybase Modifications under this License, or on different terms from those contained in this License or may choose not to license them at all.

12. Termination.

12.1 Termination. This License and the rights granted hereunder will terminate:

(a) automatically without notice if You fail to comply with any term(s) of this License and fail to cure such breach within 30 days of becoming aware of such breach;

(b) immediately in the event of the circumstances described in Section 13.5(b); or

(c) automatically without notice if You, at any time during the term of this License, commence an action for patent infringement (including as a cross claim or counterclaim) against Sybase or any Contributor.

12.2 Effect of Termination. Upon termination, You agree to immediately stop any further use, reproduction, modification, sublicensing and distribution of the Covered Code and to destroy all copies of the Covered Code that are in your possession or control. All sublicenses to the Covered Code that have been properly granted prior to termination shall survive any termination of this License. Provisions which, by their nature, should remain in effect beyond the termination of this License shall survive, including but not limited to Sections 3, 5, 8, 9, 10, 11, 12.2 and 13. No party will be liable to any other for compensation, indemnity or damages of any sort solely as a result of terminating this License in accordance with its terms, and termination of this License will be without prejudice to any other right or remedy of any party.

13. Miscellaneous.

13.1 Government End Users. The Covered Code is a "commercial item" as defined in FAR 2.101. Government software and technical data rights in the Covered Code include only those rights customarily provided to the public as defined in this License. This customary commercial license in technical data and software is provided in accordance with FAR 12.211 (Technical Data) and 12.212 (Computer Software) and, for Department of Defense purchases, DFAR 252.227-7015 (Technical Data -- Commercial Items) and 227.7202-3 (Rights in Commercial Computer Software or Computer Software Documentation). Accordingly, all U.S. Government End Users acquire Covered Code with only those rights set forth herein.

13.2 Relationship of Parties. This License will not be construed as creating an agency, partnership, joint venture or any other form of legal association between or among you, Sybase or any Contributor, and You will not represent to the contrary, whether expressly, by implication, appearance or otherwise.

13.3 Independent Development. Nothing in this License will impair Sybase's or any Contributor's right to acquire, license, develop, have others develop for it, market and/or distribute technology or products that perform the same or similar functions as, or otherwise compete with, Modifications, Larger Works, technology or products that You may develop, produce, market or distribute.

13.4 Waiver; Construction. Failure by Sybase or any Contributor to enforce any provision of this License will not be deemed a waiver of future enforcement of that or any other provision. Any law or regulation which provides that the language of a contract shall be construed against the drafter will not apply to this License.

13.5 Severability. (a) If for any reason a court of competent jurisdiction finds any provision of this License, or portion thereof, to be unenforceable, that provision of the License will be enforced to the maximum extent permissible so as to effect the economic benefits and intent of the parties, and the remainder of this License will continue in full force and effect. (b) Notwithstanding the foregoing, if applicable law prohibits or restricts You from fully and/or specifically complying with Sections 2 and/or 3 or prevents the enforceability of either of those Sections, this License will immediately terminate and You must immediately discontinue any use of the Covered Code and destroy all copies of it that are in your possession or control.

13.6 Dispute Resolution. Any litigation or other dispute resolution between You and Sybase relating to this License shall take place in the Northern District of California, and You and Sybase hereby consent to the personal jurisdiction of, and venue in, the state and federal courts within that District with respect to this License. The application of the United Nations Convention on Contracts for the International Sale of Goods is expressly excluded.

13.7 Entire Agreement; Governing Law. This License constitutes the entire agreement between the parties with respect to the subject matter hereof. This License shall be governed by the laws of the United States and the State of California, except that body of California law concerning conflicts of law.

Where You are located in the province of Quebec, Canada, the following clause applies: The parties hereby confirm that they have requested that this License and all related documents be drafted in English. Les parties ont exigé que le présent contrat et tous les documents connexes soient rédigés en anglais.

EXHIBIT A.

"Portions Copyright (c) 1983-2002 Sybase, Inc. All Rights Reserved.

This file contains Original Code and/or Modifications of Original Code as defined in and that are subject to the Sybase Open Watcom Public License version 1.0 (the 'License'). You may not use this file except in compliance with the License. **BY USING THIS FILE YOU AGREE TO ALL TERMS AND CONDITIONS OF THE LICENSE.** A copy of the License is provided with the Original Code and Modifications, and is also available at *http://www.sybase.com/developer/opensource*.

The Original Code and all software distributed under the License are distributed on an 'AS IS' basis, WITHOUT WARRANTY OF ANY KIND, EITHER EXPRESS OR IMPLIED, AND SYBASE AND ALL CONTRIBUTORS HEREBY DISCLAIM ALL SUCH WARRANTIES, INCLUDING WITHOUT LIMITATION, ANY WARRANTIES OF MERCHANTABILITY, FITNESS FOR A PARTICULAR PURPOSE, QUIET ENJOYMENT OR NON-INFRINGEMENT.

Please see the License for the specific language governing rights and limitations under the License."

Retrieved from "*http://www.openwatcom.org/index.php/Open_Watcom_Public_License*"

Appendix M ABOUT THE CD-ROM

System/Software Requirements
- IBM PC Compatible
- Microsoft Windows 98 or higher
- CD-ROM disk drive
- 16 MB RAM
- Hard disk with enough space

The CD-ROM contains:
- All of the C and C++ Program Listings, including header files in the book.
- All executable files of compiled C and C++ programs.
- All the M-files, MAT-files, MEX-files, and MATLAB Engines.
- Open Watcom C/C++ Compiler ver. 1.6 – Installer for Windows.
- Open Watcom Reference Manuals.
- Figures in the Book.

All of the C and C++ Program Listings, including header files in the book.

```
\Program Listings\Chapter 1\PL1_1.c        Program Listing 1.1
\Program Listings\Chapter 1\PL1_2.c        Program Listing 1.2
\Program Listings\Chapter 1\PL1_3.c        Program Listing 1.3
```

```
\Program Listings\Chapter 1\PL1_4.c        Program Listing 1.4
\Program Listings\Chapter 1\PL1_5.c        Program Listing 1.5
\Program Listings\Chapter 1\PL1_6.c        Program Listing 1.6
\Program Listings\Chapter 1\PL1_7.c        Program Listing 1.7
\Program Listings\Chapter 1\PL1_8.c        Program Listing 1.8
\Program Listings\Chapter 1\PL1_9.c        Program Listing 1.9
\Program Listings\Chapter 1\PL1_10.c       Program Listing 1.10
\Program Listings\Chapter 1\multiply.h     Multiply Header File

\Program Listings\Chapter 2\PL2_1.c        Program Listing 2.1
\Program Listings\Chapter 2\PL2_2.c        Program Listing 2.2
\Program Listings\Chapter 2\PL2_3.c        Program Listing 2.3
\Program Listings\Chapter 2\PL2_4.c        Program Listing 2.4
\Program Listings\Chapter 2\PL2_5.c        Program Listing 2.5
\Program Listings\Chapter 2\PL2_6.c        Program Listing 2.6
\Program Listings\Chapter 2\PL2_7.c        Program Listing 2.7
\Program Listings\Chapter 2\PL2_8.c        Program Listing 2.8

\Program Listings\Chapter 3\PL3_1.c        Program Listing 3.1
\Program Listings\Chapter 3\PL3_2.c        Program Listing 3.2
\Program Listings\Chapter 3\PL3_3.c        Program Listing 3.3
\Program Listings\Chapter 3\PL3_4.c        Program Listing 3.4
\Program Listings\Chapter 3\PL3_5.c        Program Listing 3.5
\Program Listings\Chapter 3\PL3_6.c        Program Listing 3.6
\Program Listings\Chapter 3\benz_tol.h     Benzene-Toluene  Header
                                           File
\Program Listings\Chapter 3\humidity.h     Humidity Header File
\Program Listings\Chapter 3\pipe.h         Pipe Header File
\Program Listings\Chapter 3\steam.h        Steam Header File

\Program Listings\Chapter 4\PL4_1.c        Program Listing 4.1
\Program Listings\Chapter 4\PL4_2.c        Program Listing 4.2
\Program Listings\Chapter 4\PL4_3.c        Program Listing 4.3
\Program Listings\Chapter 4\PL4_4.c        Program Listing 4.4
\Program Listings\Chapter 4\PL4_5.c        Program Listing 4.5
\Program Listings\Chapter 4\benz_tol.h     Benzene-Toluene  Header
                                           File
\Program Listings\Chapter 4\pipe.h         Pipe Header File
\Program Listings\Chapter 4\steam.h        Steam Header File

\Program Listings\Chapter 5\PL5_1.cpp      Program Listing 5.1
```

\Program Listings\Chapter 5\PL5_2.cpp	Program Listing 5.2
\Program Listings\Chapter 5\PL5_3.cpp	Program Listing 5.3
\Program Listings\Chapter 5\PL5_4.cpp	Program Listing 5.4
\Program Listings\Chapter 5\PL5_5.cpp	Program Listing 5.5
\Program Listings\Chapter 5\PL5_6.cpp	Program Listing 5.6
\Program Listings\Chapter 5\PL5_7.cpp	Program Listing 5.7
\Program Listings\Chapter 5\PL5_8.cpp	Program Listing 5.8
\Program Listings\Chapter 5\PL5_9.cpp	Program Listing 5.9
\Program Listings\Chapter 5\MULTIPLY.hpp	Program Listing 5.10
\Program Listings\Chapter 5\MULTIPLY.cpp	Program Listing 5.11
\Program Listings\Chapter 5\PL5_12.cpp	Program Listing 5.12
\Program Listings\Chapter 5\WAT_ACET.hpp	Program Listing 5.13
\Program Listings\Chapter 5\WAT_ACET.cpp	Program Listing 5.14
\Program Listings\Chapter 5\BENZ_TOL.hpp	Program Listing 5.15
\Program Listings\Chapter 5\BENZ_TOL.cpp	Program Listing 5.16
\Program Listings\Chapter 5\METH_WAT.hpp	Program Listing 5.17
\Program Listings\Chapter 5\METH_WAT.cpp	Program Listing 5.18
\Program Listings\Chapter 5\DISTILL.hpp	Program Listing 5.19
\Program Listings\Chapter 5\DISTILL.cpp	Program Listing 5.20
\Program Listings\Chapter 5\PL5_21.cpp	Program Listing 5.21

Other C Program included on the CD-ROM

APPENDIX H - IAPWS 95 C Program (by Stephen L. Moshier)
\Program Listings\Others\C Sample Programs\steam.zip

All executable files of compiled C and C++ programs

\Program Listings\Chapter 1\PL1_1.exe	Program Listing 1.1
\Program Listings\Chapter 1\PL1_2.exe	Program Listing 1.2
\Program Listings\Chapter 1\PL1_3.exe	Program Listing 1.3
\Program Listings\Chapter 1\PL1_4.exe	Program Listing 1.4
\Program Listings\Chapter 1\PL1_5.exe	Program Listing 1.5
\Program Listings\Chapter 1\PL1_6.exe	Program Listing 1.6
\Program Listings\Chapter 1\PL1_7.exe	Program Listing 1.7
\Program Listings\Chapter 1\PL1_8.exe	Program Listing 1.8
\Program Listings\Chapter 1\PL1_9.exe	Program Listing 1.9
\Program Listings\Chapter 1\PL1_10.exe	Program Listing 1.10
\Program Listings\Chapter 2\PL2_1.exe	Program Listing 2.1
\Program Listings\Chapter 2\PL2_2.exe	Program Listing 2.2

```
\Program Listings\Chapter 2\PL2_3.exe        Program Listing 2.3
\Program Listings\Chapter 2\PL2_4.exe        Program Listing 2.4
\Program Listings\Chapter 2\PL2_5.exe        Program Listing 2.5
\Program Listings\Chapter 2\PL2_6.exe        Program Listing 2.6
\Program Listings\Chapter 2\PL2_7.exe        Program Listing 2.7
\Program Listings\Chapter 2\PL2_8.exe        Program Listing 2.8

\Program Listings\Chapter 3\PL3_1.exe        Program Listing 3.1
\Program Listings\Chapter 3\PL3_2.exe        Program Listing 3.2
\Program Listings\Chapter 3\PL3_3.exe        Program Listing 3.3
\Program Listings\Chapter 3\PL3_4.exe        Program Listing 3.4
\Program Listings\Chapter 3\PL3_5.exe        Program Listing 3.5
\Program Listings\Chapter 3\PL3_6.exe        Program Listing 3.6

\Program Listings\Chapter 4\PL4_1.exe        Program Listing 4.1
\Program Listings\Chapter 4\PL4_2.exe        Program Listing 4.2
\Program Listings\Chapter 4\PL4_3.exe        Program Listing 4.3
\Program Listings\Chapter 4\PL4_4.exe        Program Listing 4.4
\Program Listings\Chapter 4\PL4_5.exe        Program Listing 4.5

\Program Listings\Chapter 5\PL5_1.exe        Program Listing 5.1
\Program Listings\Chapter 5\PL5_2.exe        Program Listing 5.2
\Program Listings\Chapter 5\PL5_3.exe        Program Listing 5.3
\Program Listings\Chapter 5\PL5_4.exe        Program Listing 5.4
\Program Listings\Chapter 5\PL5_5.exe        Program Listing 5.5
\Program Listings\Chapter 5\PL5_6.exe        Program Listing 5.6
\Program Listings\Chapter 5\PL5_7.exe        Program Listing 5.7
\Program Listings\Chapter 5\PL5_8.exe        Program Listing 5.8
\Program Listings\Chapter 5\PL5_9.exe        Program Listing 5.9
\Program Listings\Chapter 5\PL5_12.exe       Program Listing 5.12
\Program Listings\Chapter 5\PL5_21.exe       Program Listing 5.21
```

All the M-files, MAT-files, MEX-files, and MATLAB Engines

```
\Program Listings\Chapter 6\flue.mat         Flue Temp Distribution
\Program Listings\Chapter 6\PL6_1.m          Program Listing 6.1
\Program Listings\Chapter 6\PL6_2.m          Program Listing 6.2
\Program Listings\Chapter 6\PL6_3.m          Program Listing 6.3
\Program Listings\Chapter 6\PL6_4.m          Program Listing 6.4
\Program Listings\Chapter 6\PL6_5.m          Program Listing 6.5
\Program Listings\Chapter 6\PL6_6.m          Program Listing 6.6
\Program Listings\Chapter 6\PL6_7.m          Program Listing 6.7
```

```
\Program Listings\Chapter 6\PL6_8.m        Program Listing 6.8
\Program Listings\Chapter 6\PL6_9.m        Program Listing 6.9
\Program Listings\Chapter 6\PL6_10.m       Program Listing 6.10
\Program Listings\Chapter 6\PL6_11.m       Program Listing 6.11
\Program Listings\Chapter 6\PL6_12.m       Program Listing 6.12
\Program Listings\Chapter 6\PL6_13.m       Program Listing 6.13
\Program Listings\Chapter 6\PL6_14.m       Program Listing 6.14
\Program Listings\Chapter 6\PL6_15.m       Program Listing 6.15
\Program Listings\Chapter 6\multiply.m     Multiply M-File
                                           Function

\Program Listings\Chapter 7\PL7_1.m        Program Listing 7.1
\Program Listings\Chapter 7\PL7_2.m        Program Listing 7.2
\Program Listings\Chapter 7\coefdet.m      Program Listing 7.3
\Program Listings\Chapter 7\stderr.m       Program Listing 7.4
\Program Listings\Chapter 7\PL7_5.m        Program Listing 7.5
\Program Listings\Chapter 7\PL7_6.m        Program Listing 7.6
\Program Listings\Chapter 7\odeapp.m       Program Listing 7.7
\Program Listings\Chapter 7\PL7_8.m        Program Listing 7.8

\Program Listings\Chapter 8\sat_p2t.m      Program Listing 8.1
\Program Listings\Chapter 8\PL8_2.m        Program Listing 8.2
\Program Listings\Chapter 8\PL8_2a.m       Program Listing 8.2a
\Program Listings\Chapter 8\PL8_2b.m       Program Listing 8.2b
\Program Listings\Chapter 8\Enthal_MgSO4.m Program Listing 8.3
\Program Listings\Chapter 8\phase.m        Program Listing 8.4
\Program Listings\Chapter 8\PL8_5.m        Program Listing 8.5
\Program Listings\Chapter 8\PL8_6.m        Program Listing 8.6
\Program Listings\Chapter 8\PL8_7.m        Program Listing 8.7
\Program Listings\Chapter 8\PL8_8.m        Program Listing 8.8
\Program Listings\Chapter 8\reactor.m      Program Listing 8.9
\Program Listings\Chapter 8\PL8_10.m       Program Listing 8.10
\Program Listings\Chapter 8\series.m       Program Listing 8.11
\Program Listings\Chapter 8\PL8_12.m       Program Listing 8.12
\Program Listings\Chapter 8\PL8_13.m       Program Listing 8.13

\Program Listings\Chapter 9\CtoF.c         Program Listing 9.2
\Program Listings\Chapter 9\dewpoint.c     Program Listing 9.3
\Program Listings\Chapter 9\muller.c       Program Listing 9.4
\Program Listings\Chapter 9\simplex.c      Program Listing 9.5
\Program Listings\Chapter 9\enthalpy_l.c   Program Listing 9.6
```

```
\Program Listings\Chapter 9\enthalpy_v.c  Program Listing 9.7
\Program Listings\Chapter 9\PL9_8.m       Program Listing 9.8
\Program Listings\Chapter 9\PL9_9.c       Program Listing 9.9
\Program Listings\Chapter 9\PL9_9.exe     Executable File of PL
                                          9.10
\Program Listings\Chapter 9\PL9_10.c      Program Listing 9.10
\Program Listings\Chapter 9\PL9_10.exe    Executable File of PL
                                          9.10
```

Other MATLAB Programs included on the CD-ROM

```
APPENDIX E - Distillation Program (MCT M-file by Claudio
Gelmi)
\Program Listings\Others\MATLAB Sample Programs\MATLAB_
Distill.zip
```

IAPWS 97 MATLAB Program Package (by Francois Brissette)

```
\Program Listings\Others\MATLAB Sample Programs\MATLAB_
WaterProp.zip
```

Open Watcom C/C++ Compiler ver. 1.6 – Installer for Windows

```
\OpenWatcom C_C++\open-watcom-c-win32-1.6.exe
```

Open Watcom Reference Manuals

```
\OpenWatcom C_C++\c_readme.pdf   Getting Started
\OpenWatcom C_C++\cguide.pdf     User's Guide
\OpenWatcom C_C++\clib.pdf       Watcom C Library Reference
\OpenWatcom C_C++\clr.pdf        Language Reference
\OpenWatcom C_C++\cpplib.pdf     Class Library Reference
\OpenWatcom C_C++\guitool.pdf    Graphical Tools User's
                                 Guide
\OpenWatcom C_C++\pguide.pdf     Programmer's Guide
\OpenWatcom C_C++\tools.pdf      Tools User's Guide
```

Figures in the Book

```
\Figures\Chapter 1\FIG1_1.tif    Figure 1.1
\Figures\Chapter 1\FIG1_2.tif    Figure 1.2
\Figures\Chapter 1\FIG1_3.tif    Figure 1.3
\Figures\Chapter 1\FIG1_4.tif    Figure 1.4
\Figures\Chapter 1\FIG1_5.tif    Figure 1.5
```

```
\Figures\Chapter 1\FIG1_6.tif      Figure 1.6
\Figures\Chapter 1\FIG1_7.tif      Figure 1.7
\Figures\Chapter 1\FIG1_8.tif      Figure 1.8
\Figures\Chapter 1\FIG1_9.tif      Figure 1.9
\Figures\Chapter 1\FIG1_10.tif     Figure 1.10

\Figures\Chapter 2\FIG2_1.tif      Figure 2.1
\Figures\Chapter 2\FIG2_2.tif      Figure 2.2
\Figures\Chapter 2\FIG2_3.tif      Figure 2.3
\Figures\Chapter 2\FIG2_4.tif      Figure 2.4
\Figures\Chapter 2\FIG2_5.tif      Figure 2.5
\Figures\Chapter 2\FIG2_6.tif      Figure 2.6
\Figures\Chapter 2\FIG2_7.tif      Figure 2.7
\Figures\Chapter 2\FIG2_8.tif      Figure 2.8
\Figures\Chapter 2\FIG2_9.tif      Figure 2.9
\Figures\Chapter 2\FIG2_10.tif     Figure 2.10
\Figures\Chapter 2\FIG2_11.tif     Figure 2.11
\Figures\Chapter 2\FIG2_12.tif     Figure 2.12

\Figures\Chapter 3\FIG3_1.tif      Figure 3.1
\Figures\Chapter 3\FIG3_2.tif      Figure 3.2
\Figures\Chapter 3\FIG3_3.tif      Figure 3.3
\Figures\Chapter 3\FIG3_4.tif      Figure 3.4
\Figures\Chapter 3\FIG3_5.tif      Figure 3.5
\Figures\Chapter 3\FIG3_6.tif      Figure 3.6
\Figures\Chapter 3\FIG3_7.tif      Figure 3.7
\Figures\Chapter 3\FIG3_8.tif      Figure 3.8
\Figures\Chapter 3\FIG3_9.tif      Figure 3.9
\Figures\Chapter 3\FIG3E_1.tif     Figure 3E.1 (Lab Exercise)

\Figures\Chapter 4\FIG4_1.tif      Figure 4.1
\Figures\Chapter 4\FIG4_2.tif      Figure 4.2
\Figures\Chapter 4\FIG4_3.tif      Figure 4.3
\Figures\Chapter 4\FIG4_4.tif      Figure 4.4
\Figures\Chapter 4\FIG4_5.tif      Figure 4.5
\Figures\Chapter 4\FIG4_6.tif      Figure 4.6
\Figures\Chapter 4\FIG4_7.tif      Figure 4.7
\Figures\Chapter 4\FIG4_8.tif      Figure 4.8
\Figures\Chapter 4\FIG4_9a.tif     Figure 4.9a
\Figures\Chapter 4\FIG4_9b.tif     Figure 4.9b
\Figures\Chapter 4\FIG4_9c.tif     Figure 4.9c
```

```
\Figures\Chapter 4\FIG4_10.tif    Figure 4.10
\Figures\Chapter 4\FIG4_11.tif    Figure 4.11
\Figures\Chapter 4\FIG4_12.tif    Figure 4.12
\Figures\Chapter 4\FIG4_13.tif    Figure 4.13
\Figures\Chapter 4\FIG4_14.tif    Figure 4.14
\Figures\Chapter 4\FIG4_15.tif    Figure 4.15
\Figures\Chapter 4\FIG4E_1.tif    Figure 4E.1 (Lab Exercise)
\Figures\Chapter 4\FIG4E_2.tif    Figure 4E.2 (Lab Exercise)

\Figures\Chapter 5\FIG5_1.tif     Figure 5.1
\Figures\Chapter 5\FIG5_2.tif     Figure 5.2
\Figures\Chapter 5\FIG5_3.tif     Figure 5.3
\Figures\Chapter 5\FIG5_4.tif     Figure 5.4
\Figures\Chapter 5\FIG5_5.tif     Figure 5.5
\Figures\Chapter 5\FIG5_6.tif     Figure 5.6
\Figures\Chapter 5\FIG5_7.tif     Figure 5.7
\Figures\Chapter 5\FIG5_8.tif     Figure 5.8
\Figures\Chapter 5\FIG5_9.tif     Figure 5.9
\Figures\Chapter 5\FIG5_10.tif    Figure 5.10
\Figures\Chapter 5\FIG5_11.tif    Figure 5.11

\Figures\Chapter 6\FIG6_1.tif     Figure 6.1
\Figures\Chapter 6\FIG6_2.tif     Figure 6.2
\Figures\Chapter 6\FIG6_3.tif     Figure 6.3
\Figures\Chapter 6\FIG6_4.tif     Figure 6.4
\Figures\Chapter 6\FIG6_5.tif     Figure 6.5
\Figures\Chapter 6\FIG6_6.tif     Figure 6.6
\Figures\Chapter 6\FIG6_7.tif     Figure 6.7
\Figures\Chapter 6\FIG6_8.tif     Figure 6.8
\Figures\Chapter 6\FIG6_9.tif     Figure 6.9
\Figures\Chapter 6\FIG6_10.tif    Figure 6.10
\Figures\Chapter 6\FIG6_11.tif    Figure 6.11
\Figures\Chapter 6\FIG6_12.tif    Figure 6.12
\Figures\Chapter 6\FIG6_13.tif    Figure 6.13
\Figures\Chapter 6\FIG6_14.tif    Figure 6.14
\Figures\Chapter 6\FIG6_15.tif    Figure 6.15
\Figures\Chapter 6\FIG6_16.tif    Figure 6.16
\Figures\Chapter 6\FIG6_17.tif    Figure 6.17
\Figures\Chapter 6\FIG6_18.tif    Figure 6.18
\Figures\Chapter 6\FIG6_19.tif    Figure 6.19
\Figures\Chapter 6\FIG6_20.tif    Figure 6.20
```

```
\Figures\Chapter 6\FIG6_21.tif      Figure 6.21
\Figures\Chapter 6\FIG6_22.tif      Figure 6.22
\Figures\Chapter 6\FIG6_23.tif      Figure 6.23
\Figures\Chapter 6\FIG6_24.tif      Figure 6.24


\Figures\Chapter 7\FIG7_1.tif       Figure 7.1
\Figures\Chapter 7\FIG7_2.tif       Figure 7.2
\Figures\Chapter 7\FIG7_3.tif       Figure 7.3
\Figures\Chapter 7\FIG7_4.tif       Figure 7.4
\Figures\Chapter 7\FIG7_5.tif       Figure 7.5
\Figures\Chapter 7\FIG7_6.tif       Figure 7.6
\Figures\Chapter 7\FIG7_7.tif       Figure 7.7
\Figures\Chapter 7\FIG7_8.tif       Figure 7.8
\Figures\Chapter 7\FIG7_9.tif       Figure 7.9
\Figures\Chapter 7\FIG7_10.tif      Figure 7.10
\Figures\Chapter 7\FIG7_11.tif      Figure 7.11
\Figures\Chapter 7\FIG7_12.tif      Figure 7.12
\Figures\Chapter 7\FIG7_13.tif      Figure 7.13
\Figures\Chapter 7\FIG7_14.tif      Figure 7.14


\Figures\Chapter 8\FIG8_1a.tif      Figure 8.1a
\Figures\Chapter 8\FIG8_1b.tif      Figure 8.1b
\Figures\Chapter 8\FIG8_2.tif       Figure 8.2
\Figures\Chapter 8\FIG8_3.tif       Figure 8.3
\Figures\Chapter 8\FIG8_4a.tif      Figure 8.4a
\Figures\Chapter 8\FIG8_4b.tif      Figure 8.4b
\Figures\Chapter 8\FIG8_5.tif       Figure 8.5
\Figures\Chapter 8\FIG8_6.tif       Figure 8.6
\Figures\Chapter 8\FIG8_7.tif       Figure 8.7
\Figures\Chapter 8\FIG8_8.tif       Figure 8.8
\Figures\Chapter 8\FIG8_9a.tif      Figure 8.9
\Figures\Chapter 8\FIG8_9b.tif      Figure 8.9
\Figures\Chapter 8\FIG8_10a.tif     Figure 8.10a
\Figures\Chapter 8\FIG8_10b.tif     Figure 8.10b
\Figures\Chapter 8\FIG8_11.tif      Figure 8.11


\Figures\Chapter 9\FIG9_1.tif       Figure 9.1
\Figures\Chapter 9\FIG9_2.tif       Figure 9.2
\Figures\Chapter 9\FIG9_3.tif       Figure 9.3
\Figures\Chapter 9\FIG9_4.tif       Figure 9.4
\Figures\Chapter 9\FIG9_5.tif       Figure 9.5
```

```
\Figures\Chapter 9\FIG9_6.tif      Figure 9.6
\Figures\Chapter 9\FIG9_7.tif      Figure 9.7
\Figures\Chapter 9\FIG9_8.tif      Figure 9.8
\Figures\Chapter 9\FIG9_9.tif      Figure 9.9
\Figures\Chapter 9\FIG9_10.tif     Figure 9.10
\Figures\Chapter 9\FIG9_11.tif     Figure 9.11
\Figures\Chapter 9\FIG9_13a.tif    Figure 9.13a
\Figures\Chapter 9\FIG9_13b.tif    Figure 9.13b
\Figures\Chapter 9\FIG9_14a.tif    Figure 9.14a
\Figures\Chapter 9\FIG9_14b.tif    Figure 9.14b
```

INDEX